Philosophy of Engineering and Technology

Volume 31

The Philosophy of Engineering and Technology book series provides the multifaceted and rapidly growing discipline of philosophy of technology with a central overarching and integrative platform. Specifically it publishes edited volumes and monographs in: the phenomenology, anthropology and socio-politics of technology and engineering the emergent fields of the ontology and epistemology of artifacts, design, knowledge bases, and instrumentation engineering ethics and the ethics of specific technologies ranging from nuclear technologies to the converging nano-, bio-, information and cognitive technologies written from philosophical and practitioners perspectives and authored by philosophers and practitioners. The series also welcomes proposals that bring these fields together or advance philosophy of engineering and technology in other integrative ways. Proposals should include: A short synopsis of the work or the introduction chapter. The proposed Table of Contents The CV of the lead author(s). If available: one sample chapter. We aim to make a first decision within 1 month of submission. In case of a positive first decision the work will be provisionally contracted: the final decision about publication will depend upon the result of the anonymous peer review of the complete manuscript. We aim to have the completework peer-reviewed within 3 months of submission. The series discourages the submission of manuscripts that contain reprints of previous published material and/or manuscripts that are below 150 pages/75,000 words. For inquiries and submission of proposals authors can contact the editor-in-chief Pieter Vermaas via: p.e.vermaas@tudelft.nl, or contact one of the associate editors.

More information about this series at http://www.springer.com/series/8657

Albrecht Fritzsche • Sascha Julian Oks
Editors

The Future of Engineering

Philosophical Foundations, Ethical Problems and Application Cases

Springer

Editors
Albrecht Fritzsche
Friedrich-Alexander-Universität
Erlangen-Nürnberg
Nürnberg, Germany

Sascha Julian Oks
Friedrich-Alexander-Universität
Erlangen-Nürnberg
Nürnberg, Germany

ISSN 1879-7202 ISSN 1879-7210 (electronic)
Philosophy of Engineering and Technology
ISBN 978-3-030-08159-1 ISBN 978-3-319-91029-1 (eBook)
https://doi.org/10.1007/978-3-319-91029-1

Printed on acid-free paper

This Springer imprint is published by the registered company Springer International Publishing AG part of Springer Nature.
The registered company address is: Gewerbestrasse 11, 6330 Cham, Switzerland

Contents

Contributors

José Aravena-Reyes Faculty of Engineering, Civil Construction Department, Federal University of Juiz de Fora, Juiz de Fora, MG, Brazil

Erik W. Aslaksen Gumbooya Pty Ltd, Allambie Heights, NSW, Australia

Cecile Badenhorst Faculty of Education, Memorial University of Newfoundland, St. John's, NL, Canada

Roberto Bartholo Management & Innovation Area, Production Engineering Program, Universidade Federal do Rio de Janeiro, Rio de Janeiro, Brazil

Tuna Baskoy Department of Politics and Public Administration, Ryerson University, Toronto, Ontario, Canada

Terry Bristol Institute for Science, Engineering and Public Policy, Portland State University, Portland, OR, USA

Sarah Fell Institute for Manufacturing, University of Cambridge, Cambridge, UK

Nico Formanek Höchstleistungsrechenzentrum, Stuttgart, Germany

Maarten Franssen Delft University of Technology, Delft, The Netherlands

Albrecht Fritzsche Institute of Information Systems, Friedrich-Alexander-Universität Erlangen-Nürnberg, Nürnberg, Germany

Nolen Gertz Department of Philosophy, University of Twente, Enschede, The Netherlands

Kathrin Goldammer Reiner Lemoine Institut, Berlin, Germany

Institut für Technikfolgenabschätzung und Systemanalyse, Karlsruher Institut für Technologie, Karlsruhe, Germany

Bruno Gransche Institute of Advanced Studies FoKoS, University of Siegen, Siegen, Germany

Rafaela Hillerbrand Institut für Technikfolgenabschätzung und Systemanalyse, Karlsruher Institut für Technologie, Karlsruhe, Germany

Hidekazu Kanemitsu Kanazawa Institute of Technology, Nonoichi, Ishikawa, Japan

Peter Kroes Delft University of Technology, Delft, The Netherlands

Bocong Li School of Humanities and Social Science, University of Chinese Academy of Sciences, Beijing, People's Republic of China

Glen Miller Department of Philosophy, Texas A&M University, College Station, TX, USA

Cecilia Moloney Faculty of Engineering and Applied Science, Memorial University of Newfoundland, St. John's, NL, Canada

Andy Neely Institute for Manufacturing, University of Cambridge, Cambridge, UK

Jay Odenbaugh Department of Philosophy, Lewis and Clark College, Portland, OR, USA

Sascha Julian Oks Institute of Information Systems, Friedrich-Alexander-Universität Erlangen-Nürnberg, Nürnberg, Germany

Zachary Pirtle Engineering Management and Systems Engineering, George Washington University, Washington, DC, USA

Domício Proença Jr Management & Innovation Area, Production Engineering Program, Universidade Federal do Rio de Janeiro, Rio de Janeiro, Brazil

Kristen Psaty Privacy and Public Policy, Facebook, Menlo Park, CA, USA

Janna Rosales Faculty of Engineering and Applied Science, Memorial University of Newfoundland, St. John's, NL, Canada

Viola Schiaffonati Dipartimento di Elettronica, Informazione e Bioingegneria, Politecnico di Milano, Milano, Italy

Édison Renato Silva Management & Innovation Area, Production Engineering Program, Universidade Federal do Rio de Janeiro, Rio de Janeiro, Brazil

Zoe Szajnfarber Engineering Management and Systems Engineering, George Washington University, Washington, DC, USA

Nan Wang School of Humanities and Social Science, University of Chinese Academy of Sciences, Beijing, People's Republic of China

Mark Thomas Young Department of Philosophy, University of Bergen, Bergen, Norway

Sjoerd Zwart Delft University of Technology, Delft, The Netherlands

Eindhoven University of Technology, Eindhoven, The Netherlands

Chapter 1
Translations of Technology and the Future of Engineering

Albrecht Fritzsche and Sascha Julian Oks

Abstract Philosophers of technology, engineers, and other experts involved with the same subject matter look at technology in different ways. This paper explores what happens if conflicts and misunderstandings between them cannot be resolved. The exchange between the different expert groups on philosophical questions concerning technology is described as a continuous practice of coping with diversity. This practice can be described as translation, because it connects otherwise unrelated expressions of meaning. It lays the foundation for any further productive treatment of technology in society and future possibilities for the development of engineering. The chapters of this book are used as an illustration of the many different faces and levels of translation in the field.

Keywords Empirical turn · Post-normal engineering · Diversity · Translation

1.1 Bringing Philosophers and Engineers Together

Over the past decades, interest in technology among philosophers has strongly increased, with many understanding that they "need to come into closer contact with the real world of technology, or at least how that world is manifested in technological discourse" (Mitcham 1994, p. 135). Since the late twentieth century, the philosophy of technology has undergone an empirical turn (Kroes and Meijers 2000) and, "with an unprecedented seriousness and determination [...] started to engage with the practice of technology and engineering" (Franssen et al. 2016, p. 4). At the same time, engineers have developed increased sensitivity for the larger philosophical and ethical questions which permeate their daily work (Robison 2017). Associations of engineers have published numerous codes of ethics, as well as other documents in which they claim a leading role in decision making processes

A. Fritzsche (✉) · S. J. Oks
Institute of Information Systems, Friedrich-Alexander-Universität Erlangen-Nürnberg, Nürnberg, Germany
e-mail: albrecht.fritzsche@fau.de

A. Fritzsche, S. J. Oks (eds.), *The Future of Engineering*, Philosophy of Engineering and Technology 31, https://doi.org/10.1007/978-3-319-91029-1_1

about technical development and take responsibility for its consequences (Harris Jr. et al. 2013). Engineering departments at universities and other research facilities nowadays employ philosophers to teach students and add another layer of reflection to their research activities. In addition, the increasing exchange between engineering, industrial design, management, computer science and other disciplines has lately resulted in various new approaches to philosophical issues in industrial contexts which add further breadth to the discussion (e.g. Halpin and Monnin 2013; Crocker 2012; Guliciuc and Guliciuc 2010).

All this gives philosophers of technology and engineers many opportunities to meet and talk. It often remains unclear, however, what actually happens on such occasions. Do they engage in a common dialogue? Do they just provide source data to one another from which they then continue with their own lines of research? Or is there something else going on? It would be a severe oversimplification to assume that bringing people together to discuss technology leads to a common understanding of the topic and joint activities to solve certain problems. The situation seems to be particularly complicated when philosophers and engineers are involved, since they are both used to making very comprehensive claims about the scope of their research, but from very different points of view. In the philosophy of science, authors like Kuhn (1962) and Feyerabend (1962) talk about incommensurability: a fundamental mismatch of taxonomies, methods, interests and agendas which inhibit the collaboration between different generations or communities of researchers. Philosophers and engineers are likely to be affected by a similar phenomenon in their treatments of technology, no matter how much time they spend with each other. Even if one person is qualified in both fields, he or she may only be able to wear either the hat of the philosopher or the hat of the engineer at one time, but not both, because there is just no way to make them fit. When philosophers and engineers engage in a common dialogue, they may therefore eventually find out that, although they claim to address the same things, the added value they provide to each other's work remains marginal. Instead of starting one discussion which covers philosophical and engineering issues at the same time, philosophers and engineers may in fact merely be able to work side by side, but on their own, unable to make appropriate use of the input they receive from one another.

Most researchers in the field will probably respond that such concerns sound a little dramatic, considering the vast amount of publications which have resulted from joint efforts of philosophers and engineers during the last years. Nevertheless, it seems worthwhile to take a short look at the interplay between philosophers and engineers without assuming that it leads to any kind of integration. On the following pages, our analysis explores this negative view: that the boundary between philosophy and engineering is not crossed, and that occasions which bring experts from both fields together rather create a heterotopic setting, filled with experiences of contrast and ambiguity (Foucault 1971). We do not really consider this as a problem, but rather as an interesting opportunity for a wider treatment of technology, breaking with many lines of thought which have dominated the industrial era. The digital age forces us to look beyond the horizon of mass production, standardized operation and repetitivity and to search for better ways to give account of ongoing change and floating points of reference in technical design and application. Treating the views of philosophers, engineers, and other experts concerned with technology as a true, irreducible manifold can be a first step into this direction and open up new horizons to discuss the future of engineering.

1.2 Translation in the Context of Technology

A conceptual framework for assessing the implications of whether philosophy and engineering are able to integrate together is the notion of translation. Translation has already become a highly popular term in post-colonial studies and surrounding fields of research. We believe that it also offers a lot of potential for a deeper philosophical treatment of technology, because it allows us to avoid inappropriate assumptions of consistence and coherence. In line with the views of translation proposed by Benjamin (1968), Derrida and Venuti (2001), and others, we leave the idea behind that translation is a simple mapping procedure of words and phrases in different languages to reproduce meaning. Translation is more than a search for the right words. It is an attempt to give insight into another way of seeing the world which would otherwise remain inaccessible to the audience. Benjamin (1968) states that translators are concerned with works of art which cannot be reduced to any limited set of descriptive information, as they are inherently connected to the cultural conditions of their production and the personal situation of the author who felt the urge to express him- or herself with it in the particular span of time when it happened. No translation can claim to capture the full meaning of the original. But it can create awareness of its existence and insight into its significance among people who would otherwise not notice it because it is not accessible in their own culture. Through translation, the translator shares his or her own impression of the original by letting the audience go through a similar experience as he or she did in perceiving it, or at least the specific aspects of it which the translator chose to focus on.

This view of translation stands in a long tradition of scholars who favour metaphrasing over paraphrasing. At the same time, however, it takes an important step beyond this tradition as it emphasizes the futility of the attempt to be fully authentic. Translation must be seen as a work of art in its own right. Like the original, it creates an experience which is unique and new to its audience. The original and its translation stand next to each other as artistic expressions in different cultural settings which have a common point of reference, relating what would otherwise remain unrelated and allowing the members of the different cultures to recognize themselves in opposition to the other. Modern approaches to translation following this line of thought therefore like to speak about cultural translation (Roessner and Italiano 2012), setting it apart from conventional translation by recoding a message. Translation can in this sense hardly be discussed in terms of right and wrong, accurate or inaccurate, but rather in terms of connection and separation or appreciation and ignorance of the original and its whole context of creation.

When it comes to technology, a translational approach turns the attention from the usual considerations of design, function and application towards the motivations and choices which go along with the construction and usage of technical devices. It makes us ask why anyone cares to engage with technology, how this happens, and what it means for the relation of this person to others with a similar engagement. Such questions only play a minor role when technology is embedded in an industrial context and separated from aspects of human life which are not related to professional work. Studies of technology which take this context as given help us to understand the behaviour of people who have already agreed to participate and contribute to a common endeavour. We might say that these people have "accepted the rules of the game" or "learned to speak the same language", following some kind of legislative texts or dictionaries which define a common vocabulary. There can be no doubt that this is necessary wherever people are supposed to work together, which we describe as industry. At the same time, however, it obscures our view to all the other people who are for one reason or another not part of the workforce. They appear at best as disturbances and pathologies in the study, because their behaviour is inappropriate in the given context.

Our current view of engineering has emerged alongside the development of organized professional interaction in the course of the industrial revolution. The last years, however, have brought a different side of engineering to our attention which does not fit this history. Michelfelder (2017) describes it as post-normal engineering. It is characterized by the growing importance of value judgments, uncertainty of technical development, continuous negotiations of acceptable risks, participatory design, and open innovation. Post-normal engineering seems to indicate a turning point in industry at which it becomes impossible to ignore what has so far been excluded. Organizational structures of professional work have penetrated human life so deeply that there are no residual spaces left where the disturbances and pathologies could be hidden, forcing us to take a closer look at everything which does not fit to the way how things are designed to be. This is the point where the concept of translation becomes relevant, to explain how variety can continue to exist without integration, how logical conclusions are affected by different values, how a constant need for negotiation can be necessary, and how users can be considered as creative forces in technology alongside designers and builders, even if they may not appear as authors in the same light.

Franssen and Koller (2016) have outlined how the philosophy of language and the study of speech acts in particular can inspire new approaches in the philosophy of technology. The study of translation seems to be a step in the same direction, but with an important difference regarding the range of the statements which are made. We do not presuppose any kind of universality regarding language. Quite in the contrary, we expect that every statement belongs to a given context, and that this context cannot be explicated in its entirety, which confronts us with an unresolvable variety of views. This turns our research interest to the question of how it is possible that the different efforts taken by different people who have something to do with the same technical devices lead to a common achievement when the devices are applied and technology leaves a mark, finds an expression in the world. In other words: we ask how translation in technology can actually proceed.

Technology involves human intention and material structure at the same time. According to Kroes (2010), they are connected through the notion of the function of a technical artefact. During the design of an artefact, engineers make decisions about the material representation of the artefact's function. Kroes (2012) calls this process translation, while noting that the reasoning behind it needs to be better understood. After all that has been said before, it seems that this requires the expansion of the scope of the investigation. So far, research does not seem to have recognized how much translation is involved in every engagement with technology before, during and after the design process. Engineering is not an abstract, combinatorial exercise. Its outcome is a much richer expression of the engineer's view of the world, the perceived need for change in human life, the risks and uncertainties which are involved, the possibilities of development, its benefits, and the control which the designer has over the further usage of the artefact. Considering this, technical devices have to be considered as true works of art, constantly re-interpreted based on different views of the world, which find their expression in the further treatment of the artefact. Borrowing an argument from methodical culturalism (Janich 2006), we would say that engineering is too often studied in the wrong order, starting with the formal process and then adding further aspects about its embeddedness in the world, which leads to claims about intentionality and technical function preceding their materialization in a fixed structure, which might not be helpful in the long run. Again, this might be explained by the fact that technology is often studied in contexts where external influences on the conceptualizations of function have been frozen down in a rigid industrial structure. The digital transformation, however, can be expected to let such structures disappear. Systemic interactions between different agents and ongoing updates of software during the product lifecycle turn manufacturing into a constant, never-ending process, blurring the boundaries of technical artefacts and the distinction of the contributions made by different people. This raises the demand for alternate, more dynamic approaches in the study of technology.

Fritzsche (2017) has suggested the metaphor of a dancefloor to gain a better understanding of translation in technology. Dancing is a cultural practice in which people come together to share an experience, which seems to be the very opposite of an industrial activity which can be broken down to a determinate sequence of jobs which are repetitively performed to produce an output. Every dance is different. It is an ephemeral, momentary work or art which allows dancers to express their experience of the music they hear and the setting in which they have come together. Dancing can be described as a translation on someone else's work of art into one's own (see also Klein 2012). Steps and gestures interpret rhythm, melody and harmonies, and there is no inherent, physically binding criterion which would determine the interpretation. The forms of expression used by musicians and dancers are fundamentally different. In some way, the movement of the dancers is supposed to relate to the music they hear, but how exactly this happens is up to them. And while there are sequences of steps which define a specific dance style over time in a given culture, there is always enough space left for the dancers to make their own, personal choices, and the opportunity to deviate from the rules, ignore certain options, set the emphasis differently, and contribute to the further development of the style. One might think of the music as the design of the dance, which affords a certain dancing behaviour, and the rules of the dance are its organizational structure. When dancing is moved from the dancefloor to the stage, it can be said to be industrialized, reminding us of the parallels in the development of bourgeois culture and industry during the last centuries. Output, coordination, and repeatability become more important than individual expression, the references between music and movement are frozen down to one specific relation, determined by an expert, a choreographer who, like an engineer, takes charge of the entire scenario and makes it work. Back on the dancefloor, everyone takes responsibility, contributes to innovation, and looks for an alignment with the other people who are involved. This describes the kinds of processes which need further attention in the philosophy of technology.

1.3 Foundations of Futures

Pitt and Shew (2017) describe the ongoing research in the philosophy of technology as an exploration of different spaces for the future. The translational perspective turns the attention to spaces which are not given and explored, but emerging from the engagement of different people with a common subject matter. Translation as a creative act creates conditions under which development becomes possible and progress can be made. In the context of Actor-Network-Theory, Callon (1986) has described this as a process of negotiation and delimitation. For us, it is important to emphasize its results are not durable. They have a temporary quality, depicting an agreement which has been reached in the moment and needs to be revised again and again. The process of negotiation and delimitation is constantly re-initiated, turning it into a practice which is less interesting because of its output and more because of the fact that people commit to it. Again, the metaphor of a dance comes to mind, which never stops attracting people, no matter how much they have achieved with it before. This kind of practice is the main focus of our investigation.

Bhabha (2004) has coined the notion of the "third space" in-between cultures to describe where the staging of cultural difference proceeds. This third space is considered as an essential prerequisite for development and insight, as it allows people to act out their conflicts, to set themselves apart from each other and, by this, also to define the common ground between them. Wherever philosophers and engineers come together to discuss technology, they meet in such a space, due to the different training, experience, interests, and problems which affect their treatments of technology. Even if they all leave afterwards more confused than informed, the fact that they take time to address technology together and experience the full range of questions to be asked and answers to be given adds something important to research in both areas. It makes us aware that the phrase technology is a hollow shell if it is not connected to experience and that it requires a continuous effort to make technology fit in human life.

The forum for Philosophy, Engineering and Technology (fPET) provides a perfect opportunity to study a third space. Every second year, fPET assembles philosophers, engineers, and other experts to discuss philosophical aspects of technology design, management, and innovation. The participants are invited to share their experiences and opinions in an open exchange across disciplines and explore the possibilities to create connections between their work and others. The fPET meetings do not only give insight into the latest state of research in various fields; they also show how well the lines of thought pursued by the different experts resonate with each other, where they take steps into the same direction and where they go different ways in their treatment of technology.

The chapters of this book are inspired by the fPET conference hosted in 2016 by Friedrich-Alexander University in Nuremberg, Germany, which assembled researchers from all continents and major industrial regions. In addition to philosophers and engineers, the organizers also invited management scholars, designers, and artists to give account of the increasing importance of other professions to the development of technology. Furthermore, fPET 2016 involved decision makers from industry, who presented practical problems of engineering to explore solutions together with academic scholars. The participants of the meeting therefore had many opportunities to reach out across the various boundaries of their professional domains to exchange with others, discover common interests, and acknowledge differences. We believe that this kind of interaction is highly important for research to keep in touch with the actual practice of engineering and to set the foundations for any further engagement in the subject matter. What we will observe as engineering in the future depends on the way how we relate our current treatments of the subject to each other.

1.4 The Contributions to this Book

In selecting the contributions to this book, we have tried to recapture the atmosphere of fPET 2016, the liveliness of the discourse, the contrasts between the participants, and the surprises resulting from the juxtaposition of their perspectives. The list of authors includes philosophers, engineers, and managers, experts from highly developed countries and others which are still going through a period of rapid change. All authors have impressed us with the quality of their argument and the originality of their perspective, which will hopefully give readers an idea of the many different facets and layers of translation which deserve attention in our field.

In the next chapter, Zachary Pirtle, Jay Odenbaugh, and Zoe Szajnfarber explore the benefits of pluralism on a practical level. Their contribution, called ""The One, the Few or the Many?": Using Independence As a Strategy in Engineering Development and Modeling", deals with problem solving strategies in engineering which take different analytical approaches in account. In particular, they compare the RAND

Corporation's parallel path strategy with Richard Levins and William Wimsatt's approach to model independence for using multiple models to assess the same system. They draw important conclusions about the desirability of independence in solution approaches with respect to the uncertainty of the given problem situation.

Sjoerd Zwart, Maarten Franssen, and Peter Kroes also take a look at engineering practice in their contribution called "Practical Inference—A Formal Analysis", focusing on the question to what extent technical norms can be said to have a truth-value, and under what conditions practical inferences are deductively valid. The authors rely on dynamic logic (PDL), in particular the deontic version of PDL by John Jules Meyer. Bringing the argumentation to a close, Zwart, Franssen, and Kroes address issues of the reliability instead of truth-value, and the defeasibility of practical inferences as they occur in engineering practice.

Mark Thomas Young turns the attention to the role of tacit knowledge in engineering design. In his contribution called "Intuition and Ineffability: Tacit Knowledge and Engineering Design", he outlines the nature and role of intuition by examining the limitations of attempts to formalize the practice of engineering design. Young identifies correspondences between characteristics of intuition in engineering design and phenomenological aspects of Michael Polyani and Harry Collins' notion of tacit knowledge. Third and final, the solution of an "ineffability problem" is proposed by a phenomenological understanding of tacit knowledge.

Terry Bristol picks up on the topic of knowledge with his outline of "The Engineering Knowledge Research Program". This program is part of a larger effort to articulate a philosophy of engineering and an engineering worldview. Bristol explains the conflicting priorities of engineering when it is understood primarily as a field of science on the one hand and the other as an accumulation of methods for design and problem solving. Bristol postulates a paradigm shift from the scientific worldview to the engineering worldview and matches it with other paradigm shifts in traditional disciplines.

Viola Schiaffonati continues with her work on "Philosophy of Engineering and the Quest for a Novel Notion of Experimentation". She investigates different practices of computer engineering and in particular autonomous robotics with a focus on experimentation. Explorative experiments are introduced as a suitable form for the investigation of novel ideas or techniques without the typical constraints of rigorous experimental methodologies. Schiaffonati reviews extant literature to develop a wider framework of experimentation which allows a more appropriate treatment of the epistemological issues involved in this topic.

Knowledge generation also plays an important role in Nico Formanek's chapter on "Demarcating Simulation". His interest is directed at the outstanding characteristics of computer simulations which give them a special status in research and design. Formanek shows how arguments from philosophical studies of mathematics can be applied to computer simulation. His main concern is the extent to which human justificatory capabilities still remain central in this context.

Wang Nan and Li Bocong continue with their chapter called "Three Stages of Technical Artifacts' Life Cycle: Based on a Four Factors Theory". They direct the attention towards the lifespan of technical artefacts. Like living beings, technical artefacts have a lifetime with a beginning and an end, albeit the kind of mortality is radically different. Wang and Li divide the life cycle of technical artefacts in three stages: creation, vitality, and ending. Reinterpreting Aristotle's notion of four causes, the authors draft a theory of four factors affecting technical artefacts in their lifespan.

José Aravena-Reyes also takes a look back at ancient Greek philosophy in his contribution called "*Métis*: Reconfiguring the Philosophy of Engineering". He revisits the basic concepts used by Plato and Aristotle in their treatments of technology and argues that métis has so far received too little attention. He conducts an etymological analysis to gain a better understanding of métis as a cunning and inventive mode of thinking. Connecting the concept of métis to modern works on technology, Aravena-Reyes seeks to open a new perspective on the role of engineering which is also applicable in ethnic contexts in developing countries, such as the Yanomami culture.

The chapter by Tuna Baskoy deals with "Thorstein B. Veblen's Philosophy of Technology and Modern Capitalism". While Veblen is well known for his social and economic studies, his work as a philosopher of technology receives very little attention. Baskoy argues that Veblen has a lot to say about technology which could enlighten the current discourse in social sciences. She reviews Veblen's writings from a contemporary point of view, highlighting in particular the embeddedness of technology in a wider social context and the dynamic relationship between politics, economy, and engineering.

The wider social context of engineering also plays an important role in Rafaela Hillerbrand and Kathrin Goldammer's contribution. Their chapter is called "Energy Technologies and Human Well-being. Using Sustainable Design for the Energy Transition". The authors investigate the relationship between sustainability and individual well-being and show that current ideas about what constitutes sustainability need revision. They develop a value-sensitive design approach for energy systems based on the concept of central human capabilities.

Erik W. Aslaksen looks in the chapter "Technology, Society, and Survival" at the role of engineering as a medium for human interaction. He discusses recent developments in information technology and their potential to influence societal and political processes, both positively and negatively. His particular interest is directed at scenarios where small groups in society controlling the availability of information. Aslaksen explores mechanisms which can prevent such scenarios from becoming reality, threatening democracy, and to continuity of social evolution.

Andy Neely, Sarah Fell, and Albrecht Fritzsche discuss the strategy behind engineering research and teaching at the University of Cambridge. Their chapter has the title "Manufacturing with a big M – The Grand Challenges of Engineering in Digital Societies from the Perspective of the Institute for Manufacturing at Cambridge University". According to the authors, engineering is in danger of losing its autonomy as a discipline due to increasing interdependencies with other professions. In order to ensure that engineers can continue doing the works as before, new content has to be added to engineering curricula, taken from design as well as business and management studies.

Cecilia Moloney, Cecile Badenhorst, and Janna Rosales discuss engineering education from a different perspective in their chapter called "Fostering Subjectivity in Engineering Education: Philosophical Framework and Pedagogical Strategies". They advocate a paradigm shift in engineering education to foster the subjectivity in these studies such that engineers gain a better understanding of the connections between their own life values and motivations and their career choice and development. They present insights from a pilot offering of a co-curricular course for engineering graduate students and discuss their implications for future engineering education.

With the chapter on "Managing the State of the Art of Engineering: Learning from Medicine", Édison Renato Silva, Roberto Bartholo, and Domício Proença Jr. take another step beyond the boundaries of engineering as a discipline. They discuss the state of the art in medical science and present possible learning opportunities for the future of engineering. To this end, both disciplines are continuously compared and collated for potential fields of transferability. This is illustrated by an example: the Patient-Intervention-Comparison-Outcome (PICO) protocol.

Glen Miller leads the discourse of this volume towards the thematic focus on ethics in the field of engineering. In his chapter on "What Ethics Owes Engineering", he argues that ethics and engineering can influence each other in both directions. Engineering is not only affected by ethical considerations; the opposite is also true. Miller gives three examples how engineering affects ethics, regarding the separation of intentions and ends, the time needed to satisfy basic needs, and the understanding of desire and its social implications.

Hidekazu Kanemitsu reports interesting developments from his home country in his chapter on "New Trends in Engineering Ethics – A Japanese Perspective". Even though the subject of engineering ethics is widespread in the Japanese education system, current curricula "lack the normative sources to evaluate the moral design of problems, and sometimes fails to motivate students to learn engineering or/and engineering ethics by emphasizing on the negative aspects of engineering". In the further course of the chapter, Kanemitsu discusses aspirational ethics in addition to the preventive ethics as a potential solution.

Nolen Gertz takes us on another kind of journey through different schools of thought on technical design in his text on "Nietzsche, Postphenomenology, and Nihilism-Technology Relations". Questioning the idea of leisure as liberation, he reviews the positions of Aristotle, Marx, Nietzsche, and Ihde on the relation between human individuals, technology and the world. He identifies various parallels, which are then used as a basis to describe a new set of human-technology relations. These so-called "Nihilism-Technology Relations" give a new access point to study technology in human life which focuses more on responsibility than leisure as liberation.

Bruno Gransche continues with his chapter called "Assisting Ourselves to Death – A Philosophical Reflection on Lifting a Finger with Advanced Assistive Systems", which is concerned with advanced assistive systems and their role as intermediaries between humans and the world. Drawing on the works of Hannah Arendt, Gransche investigates the changing relationship between humans and their labour they perform and raises strong concerns against the idea of a comprehensive service, which destroys the possibility for further self-development.

Kristen Psaty concludes the book with her contribution on "Engineering Privacy on the Scaffolds: An Existentialist Examination of Privacy by Design". Psaty looks at the possibilities of engineers to provide privacy in digital environments by design. Based on the works of Jean-Paul Sartre, she argues that engineers need to consider the users of their systems as the others, in a relationship which allows the feeling of shame. This, however, is not easy to achieve in the context of modern information and communication technologies.

1.5 Our Gratitude

We are indebted to a lot of people who helped us make this book possible. This includes everyone involved in the organization of fPET 2016 at the Institute of Philosophy and the Institute of Information Systems of Friedrich-Alexander University Erlangen-Nuremberg, and all our supporters at the Association of German Engineers (VDI), Fraunhofer IIS, acatech, and Technology Market Strategies (TMS). The team at Springer has been very helpful and cooperative in every step of the publication process, and our special thanks go, of course, out to Pieter Vermaas, the general editor of the POET book series for his advice and encouragement. Most of all, however, we would like to thank everyone who attended fPET 2016 and contributed to our common effort, in particular the authors who have written the texts for this book. Keep on dancing!

References

Benjamin, W. (1968). The task of the translator. In H. Arendt (Ed.), *Illuminations* (pp. 69–82). New York: Schocken Books.

Bhabha, H. K. (2004). *The location of culture*. London: Routledge.

Callon, M. (1986). Some elements of a sociology of translation. In J. Law (Ed.), *Power, action and belief: A new sociology of knowledge* (pp. 196–233). London: Routledge/Kegan Paul.

Crocker, G. (2012). *A managerial philosophy of technology: Technology and humanity in symbiosis*. London: Palgrave Macmillan.

Derrida, J., & Venuti, L. (2001). What is a "relevant" translation? *Critical Inquiry, 27*(2), 174–200.

Feyerabend, P. (1962). Explanation, reduction and empiricism. In H. Feigl & G. Maxwell (Eds.), *Scientific explanation, space, and time* (Minnesota studies in the philosophy of science, Vol. III, pp. 28–97). Minneapolis: University of Minneapolis Press.

Foucault, M. (1971). *The order of things*. New York: Vintage Books.

Franssen, M., & Koller, S. (2016). Philosophy of technology as a serious branch of philosophy: The empirical turn as a starting point. In M. Franssen, P. Vermaas, P. E. Kroes, & A. W. M. Meijers (Eds.), *Philosophy of technology after the empirical turn* (pp. 31–61). Amsterdam: Springer.

Franssen, M., Vermaas, P. E., Kroes, P., & Meijers, A. W. (2016). Editorial introduction: Putting the empirical turn into perspective. In M. Franssen, P. Vermaas, P. E. Kroes, & A. W. M. Meijers (Eds.), *Philosophy of technology after the empirical turn* (pp. 1–10). Amsterdam: Springer.

Fritzsche, A. (2017). Open innovation and the core of the engineer's domain. In D. P. Michelfelder, B. Newberry, & Q. Zhu (Eds.), *Philosophy and engineering: Exploring boundaries, expanding connections* (pp. 255–266). Amsterdam: Springer.
Guliciuc, V., & Guliciuc, E. (Eds.). (2010). *Philosophy of engineering and artifact in the digital age*. Newcastle upon Tyne: Cambridge Scholars Publishing.
Halpin, H., & Monnin, A. (Eds.). (2013). *Philosophical engineering: Toward a philosophy of the web*. Chichester: Wiley.
Harris, C. E., Jr., Pritchard, M. S., Rabins, M. J., James, R., & Englehardt, E. (2013). *Engineering ethics: Concepts and cases*. Boston: Cengage Learning.
Janich, P. (2006). *Was ist Information? Kritik einer Legende*. Frankfurt: Suhrkamp.
Klein, G. (2012). Toward a theory of cultural translation in dance. In S. Manning & L. Ruprecht (Eds.), *New German dance studies* (pp. 247–259). Chicago: University of Illinois Press.
Kroes, P. (2010). Engineering and the dual nature of technical artefacts. *Cambridge Journal of Economics, 34*(1), 51–62.
Kroes, P. (2012). *Technical artefacts: Creations of mind and matter: A philosophy of engineering design*. Berlin: Springer.
Kroes, P., & Meijers, A. (Eds.). (2000). *The empirical turn in the philosophy of technology*. Amsterdam: JAI-Elsevier.
Kuhn, T. (1962). *The structure of scientific revolutions*. Chicago: University of Chicago Press.
Michelfelder, D. P. (2017). Risk, disequilibrium, and virtue. *Technology in Society, 52*, 32–38.
Mitcham, C. (1994). *Thinking through technology: The path between engineering and philosophy*. Chicago: University Of Chicago Press.
Pitt, J. C., & Shew, A. (Eds.). (2017). *Spaces for the future: A companion to philosophy of technology*. New York: Routledge.
Robison, W. L. (2017). *Ethics within engineering*. London/New York: Bloomsbury.
Roessner, M., & Italiano, F. (2012). Translation: An introduction. In F. Italiano & M. Roessner (Eds.), *Translation. Narration, media and the staging of differences* (pp. 9–18). Bielefeld: Transcript.

Chapter 2
"The One, the Few or the Many?": Using Independence As a Strategy in Engineering Development and Modeling

Zachary Pirtle, Jay Odenbaugh, and Zoe Szajnfarber

> *"No one will deny that a problem cannot be fully formulated until it is well on its way to solution. The real difficulty, the nub of a problem lies somewhere amongst the subproblems...The nature of the problem can only be found by examining it through proposed solutions and it seems likely that its examination through one, and only one, proposal gives a very biased view. It seems probable that at least two radically different solutions need to be attempted in order to get, through comparison of subproblems, a clear picture of the 'real nature' of the problem" (Marples 1961, "The Decisions of Engineering Design. p. 64, source found due to Lenfle 2011).*

Abstract There are choices about the number of ways to approach and understand a problem. Sometimes finding the one right analytical approach is sufficient. Other times, such as with the Manhattan Project, the use of many approaches is desirable. Increasing independence among multiple analytical approaches, i.e. using a pluralistic approach, can be a good strategy to get knowledge to make decisions and understand a system. We considered two frameworks that have attempted to provide advice to engineering and scientific practitioners on when and how to use multiple analytical approaches. The RAND Corporation's parallel path strategy, as described by R.R. Nelson, is a way of using independent engineering efforts to explore what parts of the design space are feasible, as well as what the cost and schedule would be for different designs. Richard Levins and William Wimsatt's focus on model independence provides motivation and insights for using multiple models to assess the same system. While these approaches may appear different, both rely on using

Z. Pirtle (✉) · Z. Szajnfarber
Engineering Management and Systems Engineering, George Washington University, Washington, DC, USA

J. Odenbaugh
Department of Philosophy, Lewis and Clark College, Portland, OR, USA

A. Fritzsche, S. J. Oks (eds.), *The Future of Engineering*, Philosophy of Engineering and Technology 31, https://doi.org/10.1007/978-3-319-91029-1_2

a group of analytical approaches where the individual members are independent – or different from—one another. Comparing these two approaches provides suggestions about how to utilize independence to address uncertainties in design and model-systems. We argue that the deliberate creation of independence among engineering developments and models should be tied to key uncertainties in the model or system. With relatively low uncertainty, choosing one approach may be acceptable. Both suggest that there can be (but are not always) benefits from using multiple approaches, which can increase accuracy and reduce cost. Using a few independent approaches – as opposed to many – may be more desirable when there are only a few bounded uncertainties about the system.

Keywords Pluralism · Model independence · Parallel paths · Richard Levins · Richard Nelson

2.1 Introduction

There is often a tension in the amount of analysis or engineering development one would like to do in an ideal world versus what is cost effective and practicable.[1] One aspect of this relates to whether to use multiple analytical approaches to assess a problem. Intuitively, we know that exploring multiple analytical or engineering pathways can be useful, as is highlighted by the Manhattan Project's work on parallel paths to enrich uranium using different types of diffusion (thermal and magnetic) (Lenfle 2011). Modelers also want access to new and different ways of assessing a problem, including comparing results across multiple models (Levins 1966; Wimsatt 2007; Lloyd 2015). However, in the name of efficiency, many engineers seek to or are pressured to focus their efforts, eliminating duplication in engineering developments or modeling (Lenfle and Loch 2010; Hounshell 2000). In both areas, a similar claim is made: parallel engineering efforts are criticized for wasting resources on redundant efforts; and modelers are often asked to provide just one model of a given system, to avoid creating redundant models. The struggle between the pursuit of an ideal versus a practical analysis approach occurs throughout engineering and science. To better understand this tension, the topics of multiple engineering paths and models are worth focusing on because analysts have already established frameworks for when multiple analytical approaches are desirable.

Despite their being of very different natures, the question on whether to pursue multiple models or engineering approaches is akin to asking how many different analytical approaches one wants to use. 'Analytical approach' is here being used as a catch-all term to describe single or multiple units of engineering developments or models. Models are commonly seen as analytical (Giere 1999), whereas engineering developments are seen as more physical instantiations of analytical and conceptual designs (Baldwin and Clark 2000). While engineering developments are physical,

[1] The views expressed here are the authors' and do not necessarily represent the views of NASA or the United States Government.

progress on a development gives analytical insight into what the final design will be and how long and how much it will cost to build. The engineering parallel path literature discussed later will provide a further description on how one can abstract an engineering development as an analysis approach, thus allowing it to be compared on a somewhat apples-to-apples basis with models as analysis approaches.

When is 'one' analytical approach enough, be it in terms of engineering developments or models? If you need more than one analytical approaches, there is an additional choice on whether to pursue a few (2–4) or many (>5) different paths. How should engineers decide on the number of analytical approaches, and how should you choose how those approaches are different from – or independent of – one another? Engineers often have some intuitive sense of when to use independent approaches but there is little formal guidance and research on when and how to do it. We offer some theoretical considerations and a framework that can help in choosing when one is enough, versus the desirability of pursuing a few or many separate efforts. We focus our guidance on sufficiently complex systems and engineering efforts wherein it can be analytically difficult to establish the best path forward. Some relatively simple systems or designs can be assessed without the need to rely on multiple analytical means.

We have developed advice on whether to go multiple by assessing two frameworks coming from separate literature streams, including the RAND Corporation's work on parallel paths for engineering development and Richard Levins/William Wimsatt's advocacy of independence in modeling. Both give advice on dealing with complex systems. In studying the work of RAND, Sylvain Lenfle's prescription for dealing with complex developments is to forgo a McNamara-like 'rational' single development approach, to instead pursue an old 1950s approach, the *parallel path strategy* (Alchian and Kessel 1954; Klein and Meckling 1958; Hitch and McKean 1960; Hounshell 2000; Lenfle and Loch 2010; Lenfle 2011). This approach is similar to Richard Levins' 1960s *strategy of analysis using independent models*, where multiple analysis approaches assess the same phenomenon, with shared conclusions across models being treated as more likely true. Both attempt to offer answers on how many analytical approaches are needed and, for some conditions, conclude that 2–4 developments or models should be used depending on the nature of complexity of the design or system being explored.[2]

We discuss each framework in order to provide a deeper perspective on:

What are the rationales for using multiple analytical approaches, be it in engineering developments or models? How do these means serve as a 'force multiplier,' where the whole is greater than the sum of its parts?

[2] Levins evaluated models based on the realism, generality, and precision and claimed there was a tradeoff between these three features, as discussed later. This led to pursuing three types of models for his primary analysis case in the 'Strategy', but he did not have a general rule calling for three different models. However, he was concerned with the manageability of models, and the use of 2–4 models may be at the upper end of what's manageable. He recognized there could be other dimensions for evaluating models. For Nelson, different assumptions could yield different results on the desired number of paths. Also, we note that while Levins and others use the term 'robustness' to refer to this process of agreement across multiple lines of evidence, we focus on the term independence because it is less commonly used and is less likely to be misconstrued (Lempert et al. 2006).

To what extent is independence a key factor underlying the motivation for using multiple? How do different types of independence affect the results?
What general guidance is there for deciding whether to implement "One, Few, or Many" analytical approaches in both modeling and engineering systems development?

To anticipate our argument, both frameworks give perspective on having 'one, few, or many' based upon their different reasons for having independence. This is a feature of both frameworks that is underdeveloped, and to which we build upon in our interpretation here. Choosing one, few, or many, should depend upon the nature of the uncertainties that a development team is facing. If one uses multiple, then the choice of which approaches to use should be tied to the types of independence used in the problem at hand. If you're dealing with a relatively simple system, one may be sufficient. Critically, there can also be a situation where various parts of a system are poorly understood, which leads to many approaches being needed. Alternatively, schedule pressure resulting from some type of Manhattan project-like national emergency can lead to a practical rationale to use multiple approaches to explore the design space. We build on the two approaches here by emphasizing that the benefits of going 'multiple' only accrue if the approaches are independent from one another in relevant ways. The analysis of what the key uncertainties are and what types of independence are needed should be led by a subject matter expert with a deep knowledge of the material. Cost and the intellectual ability to manage and understand multiple analytical results across different paths needs to be strongly considered as opportunity costs, but the judicious increase of independence among analytical approaches may be a strategy worth implementing more often than is done today. With these concepts in mind, we will now discuss the two central frameworks of parallel path developments and model independence.

2.2 Framework 1: The Parallel Path Strategy for Engineering Development

David Hounshell (2000) provides the best history of the parallel path strategy, which we draw from here.[3] This was a strategy meant to help guide the military in deciding how many different engineering developments to fund to achieve a given

[3] As has been explored by Lenfle and Loch, the 1950s RAND trandition of espousing parallel paths was in part pushed aside for cultural reasons for much of the last 50 years. Lenfle convincingly argues that this RAND literature stopped being cited in the 1960s due to U.S. Secretary of Defense Robert McNamara's desire for a more stream-lined management approach (Lenfle and Loch 2010). This McNamara-led approach led to the creation of the 'rational' Stage Gate model of innovation, which assumes that innovation proceeds in a step-by-step fashion and that engineering managers should solely pursue development of a single effort at a time in order to keep costs small. While this stage gate linear model has been much discredited by historians and scholars of innovation, no competing theory of innovation has successfully replaced it (Godin 2006; Szajnfarber 2011).

function. The most famous articulation of the strategy was in the work of Nelson (1961, 1959), who was an economist with graduate training in engineering from MIT. He described the problem that motivated the strategy in his 1962 "Uncertainty, Learning, and the Economics of Parallel Research and Development Efforts:"

"Assume that the Air Force is interested in developing an advanced fighter aircraft and that a certain performance must be achieved if the plane is to have a capability significantly greater than planes currently in force. There are several competing designs. All of them have at least some promise, though considering cost, expected performance, and expected development time, some proposals are more promising than others. However, the estimate of the relevant parameters [such as technical performance and development cost and schedule] are known to be subject to considerable error. What should the Air Force do?" (1962, p. 352).

The Air Force may have received multiple proposals for building the airplane: the different designs may each seem plausible, but perhaps the overall dynamics of the final design for some of the proposal will be more complicated than expected. It can be difficult to know in the beginning which design should be chosen. In other words, the problem the strategy tries to alleviate is: "Yet despite the unreliability of estimates, choices must be made. Given the riskiness of R[esearch] and D[evelopment], what is a good way to make choices?" (p. 352).

Nelson frames an answer based on two approaches. He says that one option is to pick the best initial option, despite the lack of maturity in early estimates. This is a prelude of what Lenfle later calls a 'rational' approach, where a manager or engineer tries to pick the right design at the beginning. However, Nelson feels the uncertainty is too great for this to be a reasonable choice. This leads him to explore the parallel paths strategy as an alternative. His suggested strategy is to pursue multiple alternative paths to accomplish a 'certain performance,' which implies accomplishing a shared, desired function. Nelson goes on to discuss how early estimates about a development program will have high uncertainty, with initial cost, schedule and performance estimates being doubtful. The uncertainty about what designs could work is the key motivator here. One can also be uncertain as to whether any of the designs can meet a given functional goal; perhaps a goal is not achievable. Given that "the early stages of development are usually the cheapest stages," and that uncertainty about a development/design's potential decreases as development continues, funding parallel paths can be a way to cut down the overall cost of development for high complexity projects.

Nelson admits limitations to his argument and in generalizing it to all of engineering.[4] Recent studies look favorably on Nelson's analysis (Scherer 2011), with

[4]The two main caveats are about certainty of the design and the role that schedule pressure plays in motivating his argument. First, the argument is not a certain one, and it is always possible that another approach could be successful. For example, he said "It is true that the atomic-bomb project, the method that actually produced chain-fissionable material for the first bombs was considered relatively unpromising early in the program...But it might well be argued that had all our money and effort been allocated to this latter method, it might have produced material not only just sooner, but sooner than the former method" (p. 362). However, Nelson's argument and calculations suggest that the use of parallel approaches will be right more often than not. Second, Nelson

Lenfle and others deeming the parallel path strategy to likely be useful for complex, schedule focused developments. Based on his caveated assumptions, Nelson predicted that 2–4 projects would be the ideal number of projects for the given complexity of systems. He says the key distinguishing assumptions for his analysis are:

1. "the cost of running a project during the period of competition; [where low prototyping costs make running multiple paths more feasible]
2. the expected improvement in estimates during the period of competition; [with greater insight post-prototyping enabling choice of a lower life cycle cost effort]
3. the difference among the cost and performance estimates of the competing projects, [with greater uncertainty increasing the value of going multiple] and
4. the design similarities and differences of the competing projects." [where the more similar the projects are, the less value there is in using multiple; this ties to our discussion of independence below] (p. 361, emphasis added).

Nelson argued that the parallel paths approach is best used to fund projects initially before down-selecting based on a few prototype development efforts. There can be diminishing returns when one funds many paths, as more of them are likely to explore the same design and cost uncertainties. While Nelson does not cite this, there are present day examples of bringing multiple efforts to a fully operational state, such as the use of multiple providers for space launch vehicles.

Despite these caveats, Nelson feels his conclusions still show that "we should be wary of damning the wastefulness of independent and competitive efforts" (p. 363). We want to more precisely interpret what types of independence Nelson had in mind here. Nelson says he wants to make sure that there's enough decentralization and diversity of approaches so that there is a de facto amount of what we will call independence involved amongst the collective paths. Implicitly he seems to view the projects as needing to have different odds of success, which might come from them working based upon different types of physical principles or being managed by different types of organizations. His approach also implies some ability to combine and share insights across approaches: Lenfle 2011 builds on this in his discussion of how the Manhattan Project was able to combine the different uranium enrichment efforts (thermal diffusion, magnetic diffusion, and creation of plutonium) to greater overall benefit, merging and changing the paths to get the best outcome. In this sense, there can be a synergy across the multiple paths that leads to a 'force multiplier', wherein the end result is superior collectively than it would have been with just the individual pats.

In sum, Nelson advocated 2–4 different engineering developments be brought to a prototype phase given his assumptions based on very complex engineering developments, which would reduce costs and accelerate time to completion. The parallel path strategy was originally conceived as a way of getting knowledge about different possible designs, how much they'd cost and how long they'd take to accomplish.

---

indicated that time concerns may be the biggest reason to embrace a parallel paths approach (p. 361): if one does not care about time, then you can wait for more research to help provide clarity on what development option should be pursued.

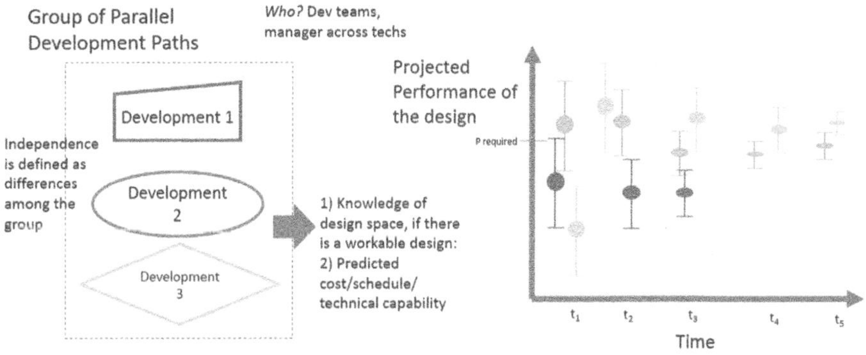

Fig. 2.1 Schematic of the parallel path framework. Downselecting at different times can yield different results based on the characteristics of the project and the remaining uncertainty in a design. Knowledge of the cost and schedule for the development will also grow over time

We deepen the interpretation of the parallel path strategy by emphasizing that the different development paths (later referred to as analytical approaches) must be independent if there is to be a sufficiently rich exploration of the possible design space such that the benefits of a parallel strategy will accrue. The overall schematic shown in Fig. 2.1 describes the strategy. It shows how the expected performance of a given project may change as it evolves over its life cycle, and the best device may be different at the end than what was there at the beginning. The ultimate functional performance as well as its cost and schedule become better known as time goes on.

2.3 Framework 2: Model Independence Among Ensembles in Levins and Wimsatt's Robustness Work

There has been a growing amount of literature specifically exploring the virtue of using independent models, much of it originating with the work of Richard Levins and William Wimsatt (Lloyd 2015; Weisberg 2006). This 'strategy' to use multiple, independent models developed out of a concern to get more accurate knowledge about complex systems. Population biologist Richard Levins presented this in his 1966 article "The Strategy of Model Building in Population Biology." In contrast to efforts to create a single, comprehensive model of a system, Levins presents three different models that he uses to analyze an ecological system. The 'Strategy' he employs is explicitly about trying to create multiple, independent models of an ecological system that have differing levels of realism and vary in their level of generality and precision. Each of Levins' models are based in evolutionary theory and ecology and can be used to examine similar, specific claims, such as how species evolve in uncertain environments. Levins has caveats about the inherent limits of each model, but notes that each model is designed to use different assumptions to analyze the same system.

His goal is to get conclusions that are supported by multiple, independent models: insights so agreed upon are more likely to be accurate. Levins states his goal as: "[I]f these models, despite their different assumptions, lead to similar results we have what we can call a robust theorem which is relatively free of the details of the model. Hence our truth is the intersection of independent lies." (Levins 1966, p. 423). Again, while there are still ongoing academic debates about the virtue of exploring a system using independent models, the rough consensus of the literature is that there is a positive epistemic benefit from engaging in and pursuing this type of independence. The literature discussing Levins' paper has been growing, with there now being over 1350 citations listed on Google Scholar by both ecologists and scholars studying the epistemology and methodology of modeling (Orzack 2012; Levins 1993; Orzack and Sober 1993, Odenbaugh 2006, Parker 2011). Beginning in the 1970s, Wimsatt (1981) saw robustness as applying to any type of scientific reasoning or activity, not just including model building and analysis.

We here build on Levins work by deepening our interpretation of how different the models should be from one another: they must be meaningfully independent from one another in order for there to be an epistemic confidence gained from their agreeing on certain claims. Pirtle et al. (2018) describes two alternative modeling strategies, one where there is a group multiple, relatively simple models, in contrast to a single, exquisite model. Both could look at the same phenomena and see things at different levels of resolution. Because independent but smaller models have different vantage points, they might be better able to represent and see features of the system under study. In this sense, the combined set of models can be a 'force multiplier,' where the combination of multiple results improves the work of the whole. This general strategy is captured below in Fig. 2.2, which shows how the collective performance of an ensemble of independent models might be stronger than an individual one.

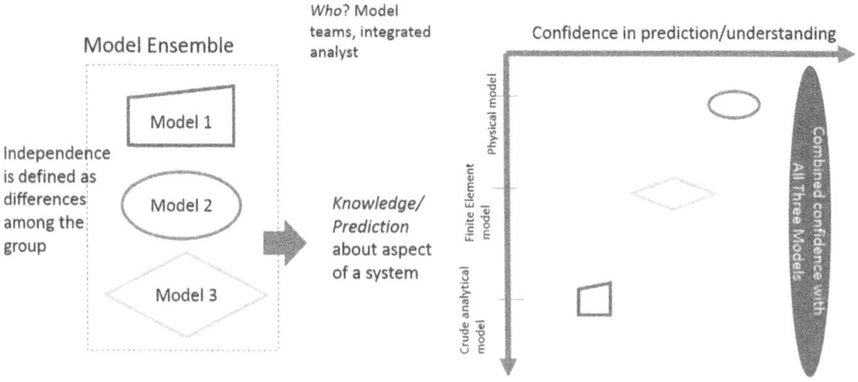

Fig. 2.2 Schematic of the Model Independence Framework. The implication is that the collected confidence of a prediction can be greater if it is derived from multiple, independent sources

2.4 Analysis

Table 2.1 summarizes the key issues within each approach, and prepares for the subsequent analysis contrasting the two strategies. We'll now describe the rationale for going multiple in both approaches, but it is first important to re-introduce the concept of independence. We highlighted that this is a critical part of both

Table 2.1 Summary of frameworks and their relation to independence

Parallel path (PP) approach	Model independence (MI) approach
Goals: Primary: Increased confidence of choosing a design can achieve a given function. Secondary: knowing the cost, schedule and ultimate capability of a project	*Goal:* Increased confidence in knowing about an empirical system
Key Proponent: 1950s/60s RAND Corporation (inc. Richard Nelson)	*Key Proponent:* 1960s Population Biologist Richard Levins; William Wimsatt 1980s on
Famous Examples: Atlas and Titan rockets, Nuclear Submarines, Manhattan Project, NASA Commercial Crew	*Famous Examples*: Ecology, models of group selection, climate models
Key rationales for going multiple:	*Key rationales for going multiple:*
§ 1) Delays assessment until better information is provided (developments always acquire information as they get farther along the development pathway). Potentially lowers overall cost	§ 1) Multiple views are less likely to miss something – using 'unrelated chunks of physics' (Hacking 1981) are likely to see different phenomena
§ 2) If there is (some) independence, some technologies or projects might work better or worse for a variety of reasons	§ 2) higher confidence in shared results
§ 3) if there is cross pollination alongside independent projects, we might learn more about what works and what is possible as a result of communication, thus improving all designs	§ 3) can be less costly, depending on the situation
Common number of paths chosen: When there is a significant uncertainty and schedule and other needs important, finds 2-3 paths to be desirable	*Common number of paths chosen:* Levins used 3 models, with different levels of abstraction. Number of models is limited by the mind's ability to understand
How independence underlies the rationales	*How independence underlies the rationales*
§ For 1), it is not applicable	§ For 1 and 2) to 'see differently' makes you more likely to see new features. Independence helps prevent being fooled by some error or false datapoint
§ For 2), independence can mean that the projects face different road blocks, or need different breakthroughs, some of which may be easier to achieve	§ For 3) Agreement can be a value multiplier (as long as relationship is known and has the right kind of difference, as with interferometry (Hacking 1981))
§ For 3), exploring different parts of the tradespace can enable more learning	

frameworks that should be more deeply developed such that the overall strategies produce a greater benefit. If all the parts of a group of models or a technology portfolio were exactly identical, that is, if they lacked independence, then it'd not be a very usefully diverse group. We characterize independence as concerning the differences among individual analytical approaches (be it models or specific technology projects). There can be many types of independence as we'll elucidate shortly. In both frameworks, the value of going multiple requires that the members of the group must be independent from one another in important ways.

2.4.1 The Rationale for Exploring Multiple Paths

For the parallel path (PP) framework, funding multiple prototype developments means that the decision on the ultimate development path is delayed until more information is developed from prototyping activities that occur over time. At the moment when decisions get made, more information is available to pick out the best design. Insofar as it relies on getting more information about a design before proceeding, this approach does not require having independence, or differences among the parallel paths. But the ability to pick a better option does rely on there being key differences among the paths, where some will ultimately be better in some important respect. The second key rationale for PP is that the different paths chosen may each have different strengths or weaknesses, which are not known in advance. In this case, independence is key, as the differences among the different paths may be the key items needed for the portfolio to function. The third rationale for doing PP is that lessons can be shared among the different parts of the path, in a way such that the overall success may be greater than the sum of the parts. In this case, the paths being different from one another is again important, as it is the learning that occurs across the combination of the different contexts which affects the extent to which the paths are as a whole stronger. If the engineering developments are identical (and the educational backgrounds and skills of those working on it are the same) then there is likelihood that using multiple approaches will be beneficial. If the paths are independent and there is an opportunity to share insights/lessons across them, then the PP strategy offers a chance to reduce overall costs due to the collective benefits of the second and third rationales, as well as by waiting for more information before making a decision.

The model independence (MI) framework has a different rationale. As Levins and others argued, different models may be more likely to capture some important part of the system that others do not. This is a positive feature that was described in Fig. 2.3, which captures the notion that different vantage points may be needed to see a feature of a complex system (Hacking 1981). Beyond seeing some unique feature, the second rationale for pursuing independent models is that they are less likely to believe an illusory phenomenon is actually representative of the world. Model predictions may always be based on some false artifact, wherein the model may have some false assumption that partially generates an inaccurate claim. Using

General Categories of Independence:	**Modeling** *Dimensions of Indepndence Among the Models/Modelers:*	**Parallel Path Developments** *Dimensions of independence among technical designs/project team characteristics:*
How: Causal Mechanisms	•Basic Physics. Core causal mechanisms that underly the model.	•Physical principles underlying design; how initial designs and prototypes will reflect a final realized design.
Who: Historical/Sociological Angles	•Epistemic culture behind their work, goals, amount of collaboration	•Epistemic culture behind their work, goals, amount of collaboration
What: Aspects of Representation	•Parameters •Scope and idealizations •Numerical coding	•Form, including different materials, sizes •Type and quantity of interdependencies between subsystems
Context: Knowledge	•Data •Team member experience, tacit knowledge of the analyst	•Team member experience, Tacit knowledge of technicians and manufacturers •Different infrastructure and tools •Data on design operation and function
Desiderata (Why):	*Model Goals* •Generality •Realism (comprehensiveness, idealizations) •Precision •Mangeability	*Development Goals:* •'Faster, Better, Cheaper' – pick one •One-off build, vs multiple construction •Exquisite, cutting edge vs standard technology Different 'Ilities", such as resilience, etc.

Fig. 2.3 Types of independence in modeling and parallel path development

multiple independent models means that they may not all share the same incorrect feature that contributed to the problem. Collectively, using multiple models may allow you to understand a phenomenon more easily, potentially allowing for costs to be reduced in the process of doing analysis. By relying on multiple models, an analyst may be more likely to rely on non-model-based sources of analysis, such as multiple data sources and ways of scoping the problem which will make them more likely to accurately understand the system.

2.4.2 A Common Call for Pluralism

Looking across the two approaches, let us first look at the core similarity. Due to challenges of accurately seeing/predicting in a complex situation, both espouse a type of *pluralism* in advocating multiple ways of approaching the same problem. We'll refer to the parallel paths approach here as PP, and Levins' independence in modeling approach as model independence, or MI. The starkest description of the differing rationales here is:

PP provides more confidence that, out of a set of developments or technology projects, one of them will perform suitably in meeting some desired shared function. A secondary goal is to generate knowledge about the associated cost, schedule and eventual technical performance of each development.

MI provides more confidence in a claim because it comes from independent models.

In the sense espoused by the original authors, the concepts work differently. MI relies more on *multiple determination* from different lenses, where each arrives at the same conclusions, whereas PP focuses on extrapolating/choosing from given paths, which involves choosing the best or most viable path out of a statistically random draw. The secondary goal of PP, to gain insight about the cost and technical capability of designs before choosing a final design, can have an interactive benefit among the paths in a way that is more similar to the MI approach. If multiple PP developments indicate that a system can be developed in a short timeframe and for a given budget, then it is more likely true that that is true.

2.4.3 Key Difference in the Two Strategies

There is one tentative difference in how the strategies work that may not be as great as it appears. In the original framing, MI is also more about representation in a way that PP is not. The model is representing an actual system, and is trying to represent features of it and then to predict aspects of the system's behavior. However, it is known that design involves representation itself (Poznic 2016) and each parallel path is a design effort trying to represent what may become a fully developed system. Thus, there is a key sense in which the PP approach works by contrasting comparative design representations, although this is a sense not directly found in Nelson's writing. Part of the uncertainty in design is representational: struggles are often due to an engineer's inability to resolve in our minds what a final realized design will be. The independent parts of the path are all attempting to represent what the final design and development process will look like, where the depth and quality of their representation improves over time. The interactive benefits of PP – where multiple groups can share insights from working on different designs – works partly as a way of thinking through what the final design will be despite each effort's partial way of understanding it. Sharing insights across the path is akin to debating what aspects of each representation are more likely accurate, and to explore what the design space should be. In this way, both the PP and MP approaches can be seen as advocating for multiple approaches on the basis of seeking independent representations, which collectively help understand the system under study, which can include the possible design space. Determining the desired number of analytical approaches to pursue can be a matter of deciding what aspects of the system are in need of being represented (or developed) in different ways.

2.4.4 Types of Independence that Distinguish Amongst Analytical Approaches

Even though we have shown that Nelson and Levins both rely on independence, the nature of independence was not fully articulated in either work. Comparing what types of independence matter to each framework gives insight into what types of independence can matter generally and how independence can be used proactively. For the context of engineering developments and modeling, Fig. 2.3 shows our synthesis of different dimensions of independence in modeling and for parallel paths, which we will explicate shortly. In the left column, we create higher level categories that help to connect the types of independence in modeling to parallel paths. We use these categories to make these lower-level dimensions of independence in the modeling and development contexts to be more comparable across each approach.

For models and analysis methods, assessing how models are independent from one another is not trivial and involves both conceptual and technical analysis. Pirtle et al. 2010 assessed climate modeling, which is an area where multiple models are used to assess the same system, and laid out ways of explaining how models are different from one another.[5] Based on Pirtle et al. (2018), we reinterpret the sources of independence mentioned there as:

Basic Physics: Core causal mechanisms that underly the model can differ. For example, Levins (1966) used one model that included genetic inheritance as part of its assumptions, while two others excluded it. Including the wrong causal variables can lead to erroneous results, and it can be difficult to identify the right causal mechanisms to model initially. Using models with different assumptions of the underlying physics helps hedge against error here.

Historical/sociological differences, including the epistemic culture behind their work, goals, amount of collaboration (Cetina 2008). The goals of a modeling effort can also significantly affect the outputs of the analysis (McNie et al. 2016).

Parameters: Specific parameters/instantiations in a model instantiate the causal assumptions of the overall model. Models can have different values for parameters that describe the same causal feature.

Scope and idealizations: Models can have different boundaries (physical and temporal). Predicted outputs of the model can diverge significantly based upon changing scope. Models can also have simplifying assumptions, or idealizations, that are introduced to make the situation more tractable (Pirtle et al. 2018).

Numerical coding: The coding approach to physical equations (such as finite element modeling) can affect how the physics of the model manifest. Different truncation schemes can also affect model accuracy.

Team member experience: includes tacit knowledge of the coder and analyst (Nightingale 2009).

Knowledge and Data: data coming from different sources may be subject to different or similar errors and may capture different, relevant causal factors.

These are what we call the physical characteristics of the models, which can in turn be different from one another. As is elaborated on in Odenbaugh (2003) and

[5] Since that time, the climate science community has written a significant amount of literature on that topic, with the lead scholar of independence analysis, Knutti (2018), offering up much more detailed approaches to align different climate models in deliberately independent ways. There are also recent examples in the engineering literature on the role of interdepednence across parts of a portfolio of technology projects (Wicht and Szajnfarber 2014).

Pirtle et al. (2018), in the 'why' section of the above table, we also list out different epistemic goals for the models, including efforts to make models maximally realistic, general or precise. These are the objectives a modeler may have in mind in trying to shape and build the model.

In the right column, we synthesize an initial description of how technologies or paths in a parallel path strategy maybe independent from one another (with more details in Pirtle, in preparation). A technology portfolio manager can get more synergy across the projects by having projects that are more independent in the following ways. These dimensions different levels relate to areas of knowledge, partly inspired by Vincenti's categories of knowledge (Vincenti 1990; Houkes 2009):

Physical (causal) principles underlying a design. The Manhattan Project's use of both thermal diffusion and magnetic diffusion to enrich uranium (Lenfle 2011) is an example of different causal mechanisms. Kroes (1998) partly references the need to reflect on causal mechanisms of designs by exploring the often indirect connection between form and function, noting that Vincenti (1990) referred to the operational principle of a device as mediating that relationship.

Disciplinary approach and historical context. Epistemic culture (Cetina 2008) reflects the knowledge that individual team members have and how their disciplinary and historical background can shape their work.

Different form and structure, including materials, sizes, geometries: These are contingent, physical instantiations of the design but that do not reflect different causal mechanisms. Treating form as different from function is common in engineering. Two projects can meaningfully differ if they use different materials or geometries, which shape how the device works, and can contribute to challenges in making the design work properly.[6]

Type and quantity of interdependencies between systems: a system can be broken down into different parts, each of which can relate to the other parts of a system in different ways. Baldwin and Clark (2000) study the modularity of different systems, arguing that modular or severable subsystems can shape the evolution of an artifact.

Team member experience: separate from disciplinary approaches, the knowledge gained by technicians and engineers about how to design, build and operate an artifact is important (Nightingale 2009). Swapping out technicians or engineers who have been working on a project can lead to different results.

[6]This feature emerges in a case study we have performed, wherein two groups tried to build the same sensor but had different pathways for what piece of superconducting metal would be used in the instrument. Sometimes the exact shape of the superconducting bilayer would differ based upon which technician was handling the software. The resolution of the sensor could differ significantly based upon the pathway taken, but there was not sufficient understanding of the phenomena to enable predicting the outcome of the physical instantiation independent of letting the development teams create the sensors. For more details on these dimensions, see Pirtle (in preparation).

Different infrastructure and tools: the ability to build a design depends on the nature of the associated technological infrastructure (Pirtle, in preparation). The use of different teams that have different supporting infrastructures may shape project decision making.

Data on design operation and function: different teams may have access to different amounts or kinds of data for relevant systems. Having different projects use different datasets can affect the results of the design effort.

Moreover, just as there are different epistemic goals of models in the 'why' category, there can also be general goals for engineering artifacts as well. Here is a not comprehensive set of examples, but helps illustrate macro-level goals that can be important differences among projects. There is a saying that it is impossible for an engineer to simultaneously make a project 'faster, better and cheaper,' but that one must pick two (McCurdy 2001). Different engineering developments can be focused on different goals, such as with some projects being designed to be faster from the beginning and other projects being designed to be cheaper or more reliable. There are also significant differences in whether an artifact is a one-off build, or something that must be able to be frequently replicated, which can greatly affect the long-term production, operations and maintenance costs of the system. If an artifact is designed to operate in uncertain and changing environments, there may be significant differences how the design space is explored.

2.4.4.1 Guidance on a 'one, a few, or many?'

Both frameworks give perspective on having 'one, a few, or many' based upon their different reasons for having independence. Deciding on the number of analytical approaches, be they models or engineering developments, should depend upon the nature and magnitude of the uncertainties that a modeling or development team is facing. We identify a common theme across Levins and Nelson that suggests additional paths should create independence surrounding those uncertainties. (Loch et al. 2011 and Lenfle 2011 make similar points about uncertainty being a motivator for using parallel paths, but do not consider independence among the paths as a key design consideration that should be cultivated). To that end, we will sketch how to decide on "one, few, or many" analytical approaches based upon tying it to different types of uncertainty. First, in the engineering context: suppose there is uncertainty about how to produce or model a key component of an engineered system, such as the process for trying to enrich uranium for a nuclear bomb. The greatest uncertainty there was about the difficulty of knowing the right physical principles that would work to enrich uranium. Given that uncertainty, then getting a parallel path approach that has paths that use fundamentally different types of causal mechanisms to achieve the desired function is likely best. The Manhattan Project's use of thermal and magnetic diffusion and the creation of a new element plutonium is a good example of this. In modeling, a similar point could apply to systems where you do not know what the dominant causal factors are, and thus may seek multiple models that rely on different physical principles. If the greatest uncertainty is about a parameter, then models that treat and assess that parameter in different ways can be essential. If there seems to be a deep uncertainty about the system overall, the use of models of fundamentally different aspects of representation (such as a physical model alongside a numerical model, see Pirtle 2010) may be appropriate.

Levins also recommended using models at different levels of the system, wherein the scope of the models is sufficiently different that they resolve key parts of the system in different ways. For parallel paths, there can be reason to seek independence in terms of how a design and its internal parts relate and connect across subsystems and components. For example, one could be worried that while all the components of a system may be understandable in their physics, the aggregate and emergent behavior of the integrated design may be uncertain.[7] In such situations, having different development paths use different types of modularity and interconnections between parts of the system could help ensure that an eventual design can come forward that will be successful, without negative emergent results.

Further research on case studies where multiple paths are used can be helpful in making a more generalizable set of which categories of independence are most important. However, the different categories discussed here can still be reflected on by engineers and managers to assess how multiple analytical approaches are independent from one another in each dimension. At minimum, assessments of independence will need to occur at multiple levels: engineering developments can have parallel paths at levels ranging from components, subsystems, interdependences and full design concepts, and models can differ in terms of the level of scale applied to different parts of the model and differences across the dimensions discussed above. The desire for 'multiple' may be greater as uncertainty is higher. Other practical concerns may influence the desired number of approaches used, such as when there is some type of Manhattan Project-like national emergency. Costs and the intellectual challenge to process multiple results needs to be strongly considered, but increasing independence among analytical approaches may be a more important analytical strategy than is commonly pursued today. And, if one is already funding multiple approaches, one needs to ensure that they are sufficiently independent such that the benefit of using multiple approaches is fully realized. The value of deliberately encouraging independence will always come at an associated cost, and the relevant decision maker may be best positioned to develop the expertise in order to judge its overall value.

[7] From some of our other research (Szajnfarber 2014; Pirtle, in preparation), we have studied a case where two groups tried to build the same functional device at the same time, but went about designing their system differently. For one development team, it turned out that the way of setting up the subsystems of the system, aligning electronic readout systems with the X ray detectors, was a significant barrier to progress. The other team set up the interfaces between its subsystems differently and was able to succeed, although both teams were able to exchange helpful knowledge and information across both paths.

2.5 Conclusion

Increasing independence within groups is a way of using different analytical approaches to better attain desired outcomes, in part by gaining new types of knowledge.[8] The RAND Corporation's parallel path strategy, as described by R.R. Nelson, is a way of using independent engineering efforts to attempt to explore what parts of the design space are feasible, as well as what costs and schedules are possible under different designs. We considered a similar strategy which has tried to provide practical guidance to scientific practice: Richard Levins and William Wimsatt's advocacy of increased independence in modeling. The analytical approaches they focus on are models, and they focus on encouraging individual models to be independent – or different from—one another. Both point to challenges or inability to precisely rationalize and control individual engineering and modeling efforts, which reflects a type of intellectual humility about understanding the world. We have synthesized their views to suggest the pursuit of independence should be part of how we try to mitigate key uncertainties. Nelson's analytical analysis shows that there can be (but are not always) sufficiently complex cases wherein the costs are reduced based on using multiple pathways, and one may assume that using multiple models can at times have a similar benefit. Further, in both cases, there can be knowledge-sharing across independent teams or modeling groups/approaches, which means that features of how one implements and pursues multiple paths can differ.

Multiple analytical approaches can often be helpful, in both engineering and models, but the exact numbers depend on the uncertainties involved. Nelson's highly contingent assumptions suggested 2–4 developments can be ideal. The concerns for manageability that Levins lays out suggests that 3–5 overall could be an upper limit for some situations. For both, the type of independence needed ideally should align with the types of uncertainty being seen overall. In situations without significant uncertainty, one is fine. In situations where things are so complex that that there's little clue on how to go forward, many approaches may be needed. A thoughtful and deliberate approach to embed independence among groups of models or technologies can be important, and both of these frameworks can be used to guide thinking on 'one, few or many'.

[8]To what extent is knowledge really the desired outcome across these two strategies? One might say that it the resemblance is superficial: that model independence is about truth, not about optimizing designs; and the parallel paths strategy is about increasing practical odds of success of yielding an outcome from a portfolio. In this sense, it wouldn't be clear that that there is much commonality across the two strategies, and that it may not make sense to refer to them both as employing diverse analytical approaches. There is some merit to this, as one could employ these strategies while holding different intentions. However, as described above, the articulated rationale behind the parallel path strategy is about how it helps decision makers due to the knowledge that is accrued. The parallel path approach wants to know what designs are possible and at what cost and schedule. Engineers can face major uncertainties about what exactly their final design will be, and in this sense both approaches are focused on the creation of knowledge.

References

Alchian, A. A., & Kessel, R. (1954). *A proper role of systems analysis*. RAND Corporation. D-2057.
Baldwin, C. Y., & Clark, K. B. (2000). *Design rules: The power of modularity* (Vol. 1). Cambridge, MA: MIT press.
Cetina, K. K. (2009). *Epistemic cultures: How the sciences make knowledge*. Cambridge, MA: Harvard University Press.
Giere, R. N. (1999). *Science without laws*. Chicago: University of Chicago Press.
Godin, B. (2006). The linear model of innovation the historical construction of an analytical framework. *Science, Technology & Human Values, 31*(6), 639–667.
Hacking, I. (1981). Do we see through a microscope? *Pacific Philosophical Quarterly, 62*(4), 305.
Hitch, C. J., & McKean, R. N. (1960). *The economics of defense in the nuclear age*. Cambridge, MA: Harvard University Press.
Hounshell, D. A. (2000). The medium is the message, or how context matters: The RAND Corporation builds an economics of innovation, 1946–1962. In *Systems, experts and computers. The systems approach in management and engineering in World War II and after* (pp. 255–310). Cambridge, MA: MIT Press.
Houkes, W. (2009). The nature of technological knowledge. In A. Meijers (Ed.), *Handbook of philosophy of technology and engineering sciences* (pp. 309–350). Dordrecht: Springer.
Klein, B., & Meckling, W. (1958). Application of operations research to development decisions. *Operations Research, 6*(3), 352–363.
Kroes, P. (1998). Technological explanations. *Techné: Research in Philosophy and Technology., 3*(3), 124–134.
Knutti, R., (2018). Climate model confirmation: From philosophy to predicting climate in the real world. In *Climate modelling* (pp. 325–359). Palgrave Macmillan
Lempert, R. J., et al. (2006). A general, analytic method for generating robust strategies and narrative scenarios. *Management Science, 52*, 514–528.
Lenfle, S. (2011). The strategy of parallel approaches in projects with unforeseeable uncertainty: The Manhattan case in retrospect. *International Journal of Project Management, 29*(4), 359–373.
Lenfle, S., & Loch, C. (2010). Lost roots: How project management came to emphasize control over flexibility and novelty. *California Management Review, 53*(1), 32–55.
Levins, R. (1966). The strategy of model building in population biology. *American Scientist, 54*(4), 421–431.
Levins, R. (1993). A response to Orzack and Sober: Formal analysis and the fluidity of science. *The Quarterly Review of Biology, 68*, 547–555.
Lloyd, E. A. (2015). Model robustness as a confirmatory virtue: The case of climate science. *Studies in History and Philosophy of Science Part A, 49*, 58–68.
Loch, C. H., et al. (2011). *Managing the unknown: A new approach to managing high uncertainty and risk in projects*. Hoboken: Wiley.
Marples, D. L. (1961). The decisions of engineering design. *IRE Transactions on Engineering Management, EM-8*(2), 55–71.
Mccurdy, H. E. (2001). *Faster, better, cheaper: Low-cost innovation in the US space program*. Baltimore: JHU Press.
McNie, E. C., et al. (2016). Improving the public value of science: A typology to inform discussion, design and implementation of research. *Research Policy., 45*(4), 884–895.
Nelson, R. R. (1959). The simple economics of basic scientific research. *The Journal of Political Economy, 67*, 297–306.
Nelson, R. R. (1961). Uncertainty, learning, and the economics of parallel research and development efforts. *The Review of Economics and Statistics, 43*, 351–364.
Nightingale, P. (2009). Tacit knowledge and engineering design. In Meijers, A. (Ed.), *Handbook of philosophy of technology and engineering sciences* (pp. 351–374). North Holland.

Odenbaugh, J. (2003). Complex systems, trade-offs, and theoretical population biology: Richard Levin's "strategy of model building in population biology" revisited. *Philosophy of Science, 70*(5), 1496–1507.

Odenbaugh, J. (2006). The strategy of "the strategy of model building in population biology". *Biology and Philosophy, 21*(5), 607–621.

Orzack, S. H. (2012). The philosophy of modelling or does the philosophy of biology have any use? *Philosophical Transactions of the Royal Society of London B: Biological Sciences, 367*, 170–180.

Orzack, S. H., & Sober, E. (1993). A critical assessment of Levins's the strategy of model building in population biology (1966). *The Quarterly Review of Biology, 68*(4), 533–546.

Parker, W. S. (2011). When climate models agree: The significance of robust model predictions. *Philosophy of Science, 78*(4), 579–600.

Pirtle, Z. (2010). How the models of engineering tell the truth. In I. van de Poel & D. Goldberg (Eds.), *Philosophy and engineering: An emerging agenda*. Dordrecht: Springer.

Pirtle, Z., et al. (2010). What does it mean when climate models agree? A case for assessing independence among general circulation models. *Environmental Science & Policy, 13*(5), 351–361.

Pirtle, Z., et al. (2018). Engineering model independence: A pluralist strategy to encourage independence among models. *Techné: Research in Philosophy and Technology., 22*, 2.

Pirtle, Z. (in preparation). *Engineering independence: An analysis framework for systems engineering in complex development programs*. Dissertation, Department of Engineering Management and Systems Engineering, The George Washington University

Poznic, M. (2016). Modeling organs with organs on chips: Scientific representation and engineering design as modeling relations. *Philosophy & Technology, 29*(4), 1–15.

Scherer, F. (2011). *Parallel R&D paths revisited*. Cambridge, MA: Harvard University, John F. Kennedy School of Government.

Szajnfarber, Z. (2011). *Innovation pathways in technology intensive government organizations: Insights from NASA*. Dissertation from the Massachusetts Institute of Technology.

Szajnfarber, Z. (2014). Space science innovation: How mission sequencing interacts with technology policy. *Space Policy, 30*(2), 83–90.

Vincenti, W. G. (1990). *What engineers know and how they know it: Analytical studies from aeronautical engineering*. Baltimore: John Hopkins University.

Weisberg, M. (2006). Forty years of 'the strategy': Levins on model building and idealization. *Biology and Philosophy, 21*(5), 623–645.

Wicht, A., & Szajnfarber, Z. (2014). Portfolios of promise: A review of R&D investment techniques and how they apply to technology development in space agencies. *Space Policy, 30*(2), 62–74.

Wimsatt, W. C. (1981). Robustness, reliability and overdetermination. In M. B. Brewer & B. E. Collins (Eds.), *Scientific inquiry and the social sciences* (pp. 124–163). San Francisco: Jossey-Bass publishers.

Wimsatt, W. C. (2007). *Re-engineering philosophy for limited beings: Piecewise approximations to reality*. Cambridge, MA: Harvard University Press.

Chapter 3
Practical Inference—A Formal Analysis

Sjoerd Zwart, Maarten Franssen, and Peter Kroes

Abstract Most engineering reasoning in practice is about how to achieve some predetermined end. Despite its paramount importance, this form of reasoning has hardly been investigated in the literature.[a] The aim of this paper is therefore to explore the question to what extent technical norms can be said to have a truth-value, and under what conditions practical inferences are deductively valid. We take technical norms to be sentences of the form 'If you want A, and you are in a situation B, then you ought to do X'. Von Wright's standard example of making a hut habitable is our paradigm for practical inferences, where an obligation to act is deduced from an intention to realize an end, and an empirical constraint on how this end can be achieved. Our instrument of analysis is dynamic logic (PDL), since actions are aimed at changing the world. PDL already suffices to provide truth-conditions for technical norms. To accommodate the obligation in practical inferences we draw on John Jules Meyer's deontic version of PDL. By paraphrasing 'person *P* wants' with 'person *P* imposes an obligation on herself,' we can give a plausible definition of the validity of practical inferences. In the discussion section, we address the issues of the reliability instead of truth-value of technical norms, and of the defeasibility of practical inferences as they occur in engineering practice.

Keywords Engineering means-end knowledge · From intentions and constraints to obligations to act · Practical inferences and logical validity · Technical norms and truth conditions · Deontic dynamic logic semantics

[a]The work of Yanjing Wang (e.g. 2016) forms the welcome exception to this claim.

S. Zwart (✉)
Delft University of Technology, Delft, The Netherlands

Eindhoven University of Technology, Eindhoven, The Netherlands
e-mail: S.D.Zwart@tudelft.nl

M. Franssen · P. Kroes
Delft University of Technology, Delft, The Netherlands

A. Fritzsche, S. J. Oks (eds.), *The Future of Engineering*, Philosophy of Engineering and Technology 31, https://doi.org/10.1007/978-3-319-91029-1_3

3.1 Introduction

Reasoning about means and ends is part and parcel of engineering practice. The aim of this paper is to explore to what extent this kind of reasoning may be deductively valid. Means-ends inferences are a form of what is called in logic 'practical inference' and the question whether this form of reasoning is deductively valid is still an unresolved issue. The locus classicus for a definition and analysis of practical inference is Von Wright's 'Practical Inference' (1963a) and that is where we start our discussion. The (by now standard) example of a practical inference that he uses is the following:

(PI) A wants to make the hut habitable
Unless the hut is heated it will not become habitable
- -
A must heat the hut

Von Wright characterizes the structure of this practical inference in the following way. The first premise states an end of *action*, that is, "we want to attain the end as a result or consequence of something which we do" (1963a, p. 160). The second premise states that a causal relationship of a particular kind exists between this end and some action, namely that this action is necessary to realize the end. This makes the action a (necessary) means to the end stated in the first premise. The conclusion, finally, expresses what Von Wright calls a practical necessity, which he describes as a "necessity of doing something under which an agent is, if he is to attain some end *of his own*" (1972, p. 43).[1]

Is this kind of means-end inference logically valid? Intuitively, the answer to this question is not straightforward. The first premise describes a goal or an end of A, the second an empirical necessity and the conclusion a means. How can a practical argument, relating such different kind of elements be conclusive in terms of truth values? What makes practical inferences logically interesting is that a causal necessity between physical facts or events is "transferred" to a practical necessity for an agent to perform an action. In other words, a necessity in the physical world is lifted to a necessity in the intentional domain. This is achieved by combining a premise that describes a state of affairs in the intentional domain – "One wants to make the hut habitable" – with a premise that describes a state of affairs in the physical domain – "Unless the hut is heated, it will not become habitable". The conclusion states a necessity in the intentional domain.[2]

[1] In 'Practical inference' Von Wright writes that he deliberately uses the word 'must' instead of 'ought' because it is "somehow stronger" (1963a, p. 161). His intuitions on this point seem not to have been very clear, however, since elsewhere he downplays this distinction between 'must' and 'ought to' (e.g. 1963b, p. 101).

[2] It could be objected that 'habitable' is a concept that involves notions from the intentional domain, but this is not what Von Wright seems to have had in mind, for he writes that the relation between

Von Wright maintains that practical inferences are logically conclusive, albeit that they are "logically valid pieces of argumentation in their own right" (1963c, p. 167). So, practical inferences have to be added to the traditional forms of logically valid reasoning and cannot be reduced to them. He explains their peculiar form of logical validity by invoking the notion of practical necessitation, that is, "necessitation of the will to action through want and understanding" (idem, p. 170). In a practical inference, a want, an understanding and a decision to act are united in such a way that "the practical necessitation of the will to action must at the same time be a logical necessitation" (idem, p. 171). Thus, Von Wright appears to solve the issue of the logical validity of a practical inference by fiat, by introducing a new kind of logical validity. This may be considered an ad hoc solution. Can we do better? We argue that we can. But before going into the details of how this may be done, we have a brief look at Niiniluoto's work on technical norms; he is one of the few people who have taken up Von Wright's work on practical inferences.

In his (1993) paper 'The aim and structure of applied research', Niiniluoto identifies *rules of actions* as forming the results of design science. A problem for this position, however, is how results of this kind can classify as knowledge, given that rules are generally taken to be normative statements. To answer this question, Niiniluoto identifies such rules with *technical norms* as introduced by Von Wright in *Norm and action* (1963b). The example of a technical norm given by Von Wright, as a match to his stock example of a practical inference, is

(TN) If you want to make the hut habitable, you must heat it.

Von Wright's notion of technical norms forms an interesting parallel to his conception of practical inference, since it is precisely a statement of this sort that would, when combined with the statement 'You want to make the hut habitable', give, by the deductively valid argument of modus ponens, the conclusion 'You must heat the hut'. But the truth of the conclusion follows only if the premises are true, and for that to be the case, they must be of the right kind to have a truth-value. Von Wright was not convinced that technical norms have a truth-value, and perhaps for that reason rejected this approach to settle the validity of a practical inference. In *Norm and action* Von Wright pronounced "the relation of the technical norm to truth and falsehood" to be "not clear to me" (p. 103). Niiniluoto did not follow Von Wright in this. According to Niiniluoto technical norms classify as 'some kind of rule of action' *and* have a truth-value, so that they can be regarded as knowledge, the sort of knowledge produced by design science. He does not, however, explain on what grounds he accepts what Von Wright could not accept. Niiniluoto simply claims,

temperature and habitability of a hut "is a causal fact about the living conditions of men" (1963a, p. 160). If we interpret "living conditions" as "physical living conditions", the second premise articulates a necessity in the physical world. Thus "habitable" should be read here as a physical property rather than a means to get the hut inhabited.

without further clarification, that a technical norm of the form 'If you want to make the hut habitable, you must heat it' is true if and only if a statement expressing an empirical necessity of the corresponding form 'heating the hut is necessary for making the hut habitable' is true (p. 12). If this is accepted, then an argument that has as premises 'You want *A*', 'You are in situation *B*' and 'If you want *A* and you are in situation *B* then you ought to do X' and as conclusion 'You ought to do X' is deductively valid simply for being an instance of modus ponens. Indeed Niiniluoto also classifies statements like the conclusion of this argument as descriptive and either true or false (idem, p. 12).[3]

In the following we chose to address the question whether a practical inference can be characterized as a deductively valid form of reasoning, and under which assumptions it is valid, rather than to be satisfied with declaring it to be a valid kind of syllogism *sui generis*. We interpret the 'want', 'understanding' and 'decision to act' statements such that the syllogism is deductively valid, and we discuss how faithful these interpretations are to the statements as formulated by Von Wright. The basic idea, upon which the validity of the practical inferences is based, runs parallel to validity of standard syllogisms, which is based on intuitive set theory. For instance, if all elements of a set have some property, then if you select one element it must have this property, too. Practical inference appears to follow the same pattern. If all paths from our state of the world to a state in which φ is the case include some world change event α, then any individual path realizing φ will contain α; thus, in those circumstances, an obligation in our state of the world to realize φ implies an obligation to do α. The validity of this reasoning aspect of practical inference rests on the same ground as some validity claims in first-order logic.

We proceed as follows. In sect. 3.1 we briefly discuss the complexity of the conceptual framework that is required for representing the various ingredients of practical inference, and for formally interpreting the notion of practical necessity, and the character of a technical norm. In sect. 3.2 we clarify the character of statements that refer to actions, such as technical norms and practical necessities. For this, we put Segerberg's (1980) advice to practice and take recourse to dynamic logic. In the following section we address the normative or intentional aspects of practical inferences and technical norms. We show that in a modified form practical inferences are logically valid. For this we use dynamic deontic logic. In the concluding section we discuss our results and the limitations of our approach regarding practical inference and means-end reasoning in engineering practices.

[3] Niiniluoto additionally considers other forms of technical norms, which correspond to empirical connections between means and ends that are weaker than necessity, e.g. connections where a means is sufficient but not necessary for an end and connections where a means is only probabilistically sufficient.

3.2 Conceptual Choices

In the process of modelling practical inferences, at least three different 'depths' can be chosen for the level of analysis. The simplest formalization does not include actions as a logical category and takes the world to exist in just a single state. It does, however, include a distinction between causal or empirical *necessity*, of the sort that is expressed in natural laws, and practical necessity or *obligation*. We illustrate the various options for formal analysis by different adaptations of Von Wright's stock example (PI). In the simplest version, which we call the *Static-State version*, all explicit reference to action falls out, and so does reference to wanting, since wanting amounts to an anticipation of a change of the state of the world, whereas there is no room for such change in the analysis. Therefore we get:

(PI^S) The hut should be habitable
Unless the hut is heated is will not be habitable
- -
The hut should be heated

In this version practical inference can be formalized using standard deontic logic. The formal representation of the *Static State* version of practical inference comes out as:

(PI^S) $O(Hab(h))$
$\Box\,(Hab(h) \rightarrow Heat(h))$

$O(Heat(h))$

Here $\Box$ is the modal operator signifying empirical necessity and O the modal operator signifying practical necessity or obligation. Further 'Hab' and 'Heat' refer to 'being habitable' and 'being heated', respectively, and '*h*' refers to the hut in question. Note that, because at this level of analysis action is not a distinct logical category 'being heated' is ambiguous between 'being heated by someone who is actively busy with heating the hut' and 'being equipped with a functioning heating installation'.

(PI^S) can easily be made valid by adding to the logic a bridging axiom that links the two modal operators. The following axiom, which is widely accepted in deontic logic and called the *principle of deontic logic*, will do:

(B_O): For any A, B: $\Box(A \rightarrow B) \rightarrow (OA \rightarrow OB)$

Thus, by paraphrasing practical inference in classical modal-deontic logic in this way, causal or empirical necessity is transferred to practical necessity or obligation. Von Wright introduces the notion of *anankastic statements* to express empirical necessities, such as expressed by the second premise. Anankastic statements do not include intentional actions *as intentional*.[4] They may include actions but then, only the physical manifestation of the action matters and not the intention that accompanies the action.

This Static-State approximation of practical inference implemented in classical deontic logic we consider too crude, because it does not include changing states of the world according to ones will or plan. In (PI^S) the inference fails to represent a *practical* necessity, which Von Wright describes as a "necessity of doing something under which an agent is, if he is to attain some end of his own" (1972, p. 43). To come closer to the intended meaning of 'practical inference', minimally it must be possible for the state of the world to change, to accommodate the idea of achieving a desired state. One version of Von Wright's practical inference which allows for changing states of the world is:

(PI^D) The hut should become habitable
Unless the hut is heated it will not become habitable
- -
The hut should be heated

This *Dynamic-State version* is an improvement with respect to the Static-State one as it allows us to take account of changes in the actual state of the world (PC_D, first premise), but it is not expressible in standard modal and deontic logic. For this version we need a form of *dynamic logic*,[5] which provides means for an exact paraphrase of anankastic statements.

The most complicated version of practical inference, which does include actions of individuals, we call *Action versions*. Von Wright's paradigmatic inference (PI) presented in the Introduction corresponds to the Action version. A complete formalization requires that individuals may change states of the world by performing goal-directed actions. Beyond this version lie still completer versions of practical reasoning that contain goals, agenda's, planning, deliberations of advantages and disadvantages, knowledge, belief and so forth. In the next section we will sketch a formalization of an Action version of practical inference. In the final Discussion section, we will address the issue to what extent this version can take care of the aspect of wanting or of the goal-directedness of action.

The distinction that we have made here between Static-State, Dynamic-State and Action versions of practical inference also applies to technical norms. Just as for

[4]Von Wright: "A statement to the effect that something is (or is not) a necessary condition of something else I shall call an anankastic statement." (1963b, p. 10); "Laws of nature and other anankastic propositions are, on the whole, not concerned with action; but these we have decided not to call 'norms'." (idem, p. 13).

[5]Von Wright characterized his work in the 1960s as "a turn in logic away from a traditional interest in what is, the static, to that which comes to be, the dynamic" (1999, p. 30).

practical inference, the version of a technical norm presented in the Introduction is an Action version of the notion of a technical norm as introduced by Von Wright. The simplest or Static-State version of a technical norm is:

(TN^S) If the hut *has to be* habitable it *should* be heated

The corresponding expression in the language of deontic logic is: $O(Hab(h)) \rightarrow O(Heat(h))$. In this logic, the truth of the anankastic statement $\Box(Hab(h) \rightarrow Heat(h))$ in combination with the bridge axiom (B_O) is sufficient for the truth of (TN^S).

The intermediate *Dynamic-State version* of Von Wright's technical norm example is:

(TN^C) If the hut should *become* habitable, then it *should be heated*

(TN^C) strongly suggests that at the moment the hut in question is not habitable and some action should be performed to change that situation.

There is a significant difference between Von Wright's formulation of a technical norm and Niiniluoto's formulation, which is (our emphasis):

(TN_N) If you want *A and you believe you are* in situation *B* then you ought to do X.

Niiniluoto adds a clause with an epistemic operator to Von Wright's technical norm, which is understandable but complicates matters considerably. What if your beliefs are wrong and you are actually in situation *C*? Should you then not do X? It might be that in *C*, X is also a necessary action for *A*. And what description of *B* should suffice to do X? If Oedipus wanted to become king of Thebe and if it was necessary for that to marry the widowed queen of Thebe, should Oedipus marry Jocasta or should he marry his mother? With an epistemic operator included in technical norms, a complete belief-desire-intention framework, such as KARO,[6] is required for an analysis, but such frameworks are too complex for our purposes. As our goal is to formulate a semantics of technical norms as part of as objective as possible engineering rules and know-how, we leave the epistemic side of things out from our account. This is in line with the treatment of Von Wright, who takes the *truth* of an anankastic statement to be what places the practical necessity on the actor, irrespective of whether the actor knows that statement to be true. Our *Action* version of the technical norm reads therefore:

(TN'_N) If you want *A*, and you are in situation *B*, then you ought to do X.[7]

---

[6] cf Meyer et al. (1999).

[7] Note that one may also achieve *A* by *satisficing* some design criteria. Herbert Simon's notion does not change our approach fundamentally.

Let us turn to a more detailed formal analysis of practical inferences and technical norms.

3.3 PDL and Niiniluoto's Technical Norms

With respect to technical norms we saw, in the introduction, that Niiniluoto claims that the (slightly rephrased) technical directive 'if you want *A*, and you are in situation *B*, then you should do α' is true if and only if

(*) Doing α is a necessary cause of *A* in situation *B*.

Our first task is to explicate the definiens (*) of Niiniluoto's proposal. To do so we need at least three types of elements: (*i*) individual states S_i of the world; (*ii*) (performances of) actions α, β etc., which transform one state of the world into another; and (*iii*) sentences φ, ψ etc. that may have different truth-values in different states of the world. With these three types of elements we are able to model the fact that some actions may change the truth-value of a statement and therefore can be seen as being done for *ends*. The three elements are the main ingredients of *dynamic logic*. We use propositional dynamic logic (PDL) to model technical norms and DDeL (a deontic extension of PDL) for the modelling of practical inference.[8] Basing ourselves mainly on Meyer (1988), we present here only those parts of PDL and DDeL that are relevant for explaining technical norms and practical inference.

Within the syntax of PDL, α, β etc., signify the executions or performances of actions,[9] which are recursively built from performances of atomic actions *a*, *b*, etc. Actions can be combined in three ways. One is *sequential composition*, expressed as α; β, which means that action α is followed by action β. (Rubbing down wood with sandpaper and then painting it.) Another is *joint action*, expressed by α & β, which means that α and β are performed simultaneously (for instance: lubricating a camshaft while moving the pushrods up and down). Finally there is *choice*, expressed by $\alpha \underline{\cup} \beta$, which means that one chooses between doing either α or β or possibly both (for example reading a book or listening to music (or both) may turn your evening in a happy one). An action is $\underline{\cup}$-free if it does not contain a choice aspect.

Secondly, to define obligations we also need the *negation* of an action, expressed as ~α. This notion is fraught with philosophical difficulties, but here we use it in the intuitive sense of refraining from an action, that is, of doing anything (not necessarily nothing) as long as it is not or does not include that action.[10]

[8] Because we look upon *actions* as a logical category of its own, we do not opt for a STIT logic.

[9] In PDL an action α is always successful – it changes a state in a predefined way.

[10] A comprehensive discussion of action negation is Broersen (2004).

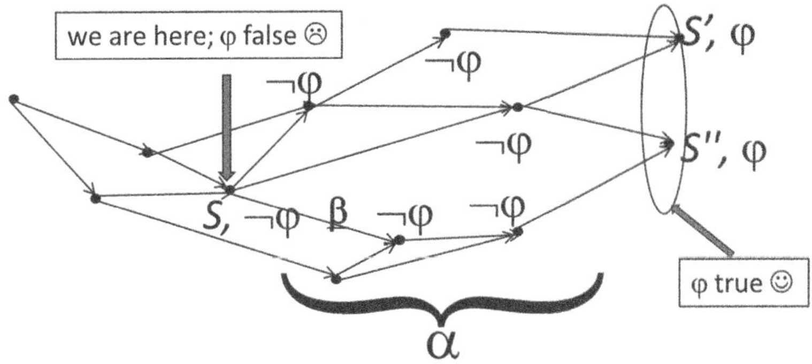

Fig. 3.1 Being in state S, complex action α in S achieves φ in S' or S''

Thirdly, PDL contains the expressions $[\alpha]\varphi$ and $\langle\alpha\rangle\varphi$.[11] $[\alpha]$ extends the semantics of the necessity operator $\Box$ from modal logic: $[\alpha]\varphi$ is true in S – formally $S \vDash [\alpha]\varphi$ – iff φ is true in all states S' which are the result of performing α in state S, or α cannot be performed in S. In the same way, $\langle\alpha\rangle$ extends the possibility operator $\diamond$. $\langle\alpha\rangle\varphi$ is true in S – formally $S \vDash \langle\alpha\rangle\varphi$ – iff, as Meyer (1988, p.114) phrases it, "there is some way by doing action α to achieve φ", that is, α is performable in S, and its performance does not guarantee the truth of φ in the resulting state S'. As in traditional modal logic, $[\alpha]\varphi$ is equivalent to $\neg\langle\alpha\rangle\neg\varphi$.

The Fig. 3.1 illustrates the framework. We adopt the convention that single arrows represent *atomic actions*, which change a state of the world into another one so that the truth values of some sentences change. Different arrows starting from the same state of the world then indicate a nondeterministic choice between atomic actions, that is, one or several of these actions are actually performed but it is left open which. A complex action α, which produces several subsequent state changes, is a recursive construction of sequential compositions, joint actions and (non)deterministic choices of atomic actions and corresponds to a complex directed graph of arrows.[12]

PDL takes the weakest form of a normal modal logic. Besides modus ponens and the rule of necessitation ($\vdash \varphi$ implies $\vdash \Box\varphi$, or in other words, logical truths are necessary) it only uses the Kripke (distribution) axiom:

(Kα) $\quad [\alpha](\varphi \rightarrow \psi) \rightarrow ([\alpha]\varphi \rightarrow [\alpha]\psi)$.

[11] PDL harbors still more operators, but these do not play a role in our analysis.

[12] When Meyer, van de Hoek and van Linder apply PDL to formalize the dynamics of beliefs, desires, intentions, commitments, (Meyer et al. 1999) they use a deterministic version of PDL and even in a deterministic context this fine-tuned psychological dynamics turns out to be fairly complicated. To paraphrase engineering knowledge we think as a first approximation we may dispense with these agent specific notions. Engineering means-end knowledge strives to be as least as possible subject-dependent or subjective. We cannot however avoid the nondeterministic version of PDL since engineers often need to choose between different in principle possible actions to achieve their goal.

(Kα) says that if action α necessary leads to states where φ cannot be true without ψ being true, then if the performance of α will always lead to φ it will also always lead to ψ. So if α makes φ always sufficient for ψ, then we only need perform α to achieve φ in order to achieve ψ, which is not insignificant for engineers. We do not impose any other property on the action operator.

3.3.1 Sufficient Means

In order to apply PDL to engineering means-end knowledge, we follow (Hughes, Kroes, and Zwart 2007) and paraphrase the *means* to achieve something with the *actions* that agents can perform to realize a particular state of the world or, in the language of the adopted framework, make a possible state of the world the actual state. Accordingly, we take *ends* to be the *possible states* in which certain desired states of affairs are instantiated.[13] With these definitions, a *weakly sufficient means* can be defined as follows:

Definition 3.1 Action α in state S is a *weakly sufficient means* for realizing φ iff there is some way by doing action α to achieve φ: $S \models \langle\alpha\rangle\varphi$.

Weakly sufficient means play an important role in forensic and reversed engineering. It may be helpful for formalizing reasoning about which processes led to an accident, or for functionally decomposing an artifact. In contrast, $[\alpha]\varphi$ cannot be used to define a strongly sufficient means in the same straightforward way. We could have done so if $[\alpha]\varphi$ were false when it is impossible to perform α in state S, which corresponds to $M_{\alpha,S}$ being empty.[14] However, it is a consequence of the adopted definition of the $[\alpha]$ operator that $S \models [\alpha]\varphi$ is (trivially) satisfied if performing α is impossible in that state of the world. Consequently, the definition of (strongly) sufficient additionally requires that α can be performed or equivalently, that $M_{\alpha,S}$ is non-empty. A condition that serves this purpose is $S \models \langle\alpha\rangle\top$, with $\top$ being the tautology.

Definition 3.2 Action α in state S is a *(strongly) sufficient means* for realizing φ iff the performance of α in state S always leads to a state in which φ is true and it is possible to perform α: $S \models [\alpha]\varphi \wedge \langle\alpha\rangle\top$.[15]

[13] We recognize that in natural language means are often identified with objects or instruments and not actions, but do not think this poses an important objection to our approach. We may always translate the object-as-means language into to the actions the means-object performs. If an ignition coil is the means to cause an ignition in the combustion chamber at exactly the right time, we may rephrase this means as the action of igniting.

[14] $M_{\alpha,S}$ denotes the set of states of the world that is the result of performing α in state S.

[15] Although this definition seems intuitively plausible, it lacks important relevance conditions. Typically we do not call an action a sufficient means to some end if this end is inevitable anyway,

3.3.2 Necessary Means

In principle an action β can be defined as *a necessary means* to achieve φ if all weakly sufficient means α for φ involve β. However, this requires a formal definition of the intuitive notion of involvement, which complicates things considerably. The possible complexity of compound actions enables the involved actions to be performed 'in chunks along the line' of other complex actions. If α is an action that is sufficient for achieving one's overall goal, which is being a writer, then writing a book, action β, certainly will be a necessary part of α. But obviously action β is not an atomic action. A complex action α may therefore subsume many other actions in the form of α-parts that together amount to β.

To accommodate the idea of the performance of an action along the way of another, in Hughes et al. (2007) the involvement relation has been introduced and axiomatized. The intuitive idea of an action α involving an action β is that one cannot do α without doing β 'along the way'. If, for instance, α consists of performing the sequence of sub-actions α_1; β_1; α_2; β_2; α_3, (in this order), and $\beta{:} = \beta_1; \beta_2$, then α involves β, although α does not involve the action $\beta_2; \beta_1$ because the order of the sub-actions counts. We write $\alpha \rhd_S \beta$ to indicate that α involves β. The subscript index S is required because whether or not an action involves another action may depend on the state of the world. Note that the involvement relation between actions α and β comes close to the idea that doing β is necessary for doing α.

With the involvement relation in place we are able to construct our definition of β being a necessary means for achieving φ in state S. We take this to correspond to the situation where there is a path σ such that (a) the initial state of σ is S; (b) the final state of σ is a φ-state; and (c) σ does β along the way, and where additionally every path σ that satisfies (1a) and (1b) also satisfies (1c).

The first clause guarantees that some part of actions from the current state to a desired state exists and that arriving at the latter requires doing α along the line. The second clause makes α a necessary condition since all paths require that α is being done along the way. The formal definition reads as follows (Hughes et al. 2007, definition 3.2):

Definition 3.3 An action β is a *necessary means* to φ in state S iff

1. For some action α: $S \models \langle\alpha\rangle\varphi$ and $\alpha \rhd_S \beta$
2. For every $\underline{\cup}$-free action α: if $S \models \langle\alpha\rangle\varphi$ then $\alpha \rhd_S \beta$

---

such as, for example, the setting of the sun. According to definition 3.2 all actions which start before sunset and end after sunset are sufficient means for the occurrence of the sunset. There are several attempts to repair this, but for the purposes of the present paper we can ignore this problem (cf. Hughes et al. 2007, p. 215/6).

To see that the second clause, which its restriction to $\sqcup$-free actions, is strong enough, consider an action α that achieves φ but is not $\sqcup$-free. Then we can rewrite α into $\alpha_1 \sqcup \ldots \alpha_i \ldots \sqcup \alpha_n$ with all α_i being $\sqcup$-free. If all α_i involve β, β is necessary.

3.3.3 Technical Norms

With these three definitions in place, we are now able to sketch the semantics of technical norms of the type

(TN'_N) 'If you want *A* and you are in situation *B*, then you ought to do X.'

As stated above, Niiniluoto's claim is that a norm of this type is true if and only if it is true that 'Doing X is a necessary cause of *A* in situation *B*'.

First, however, we must deal with a problem of this truth condition for technical norms. The 'and only if' part implies that not wanting *A* in situation *B* is sufficient for X being a necessary cause of *A* in situation *B*. This is a consequence we do not accept. We only subscribe to the right-to-left direction, i.e., *A* being necessary for *B* is sufficient for the truth of (*) but not necessary. This still has the consequence that if (*) is the case, then you do not want *A* or you are not in situation *B* or you ought to do X, which is perhaps somewhat odd. More adequately one could define: if you are in situation *B* and you want *A*, then you ought to do X iff doing X is a necessary cause of *A* in situation *B*.

With Niiniluoto's proposal modified in this way, we come up with the following semantics for it, which basically just amounts to reading 'necessary cause of …' as 'necessary means to …':

(Necessary reading) 'If you want φ, and you are in situation *S*, then you should do β' is true in *S* if $\exists\alpha$: $S \models \langle\alpha\rangle\varphi$ and for every action $\sqcup$-free action α in *S*: if $S \models \langle\alpha\rangle\varphi$ then $\alpha \rhd_s \beta$

Of course in the practice of engineering and design technical norms with this definition of their truth-value are almost always false. Many roads lead to Rome and not all pass the Rubicon. But as stated in the introduction, empirical adequacy is not our primary goal here; we want to show how technical norms, or means-end statements, can get truth-values, by using dynamic logic.

As mentioned above in footnote 3, Niiniluoto also considered weaker variants of (TN'_N). Our semantics generalizes naturally to the variant of the norm that states that 'it is sufficient for you to do α'. This variant we take to be true if doing α is a (strongly) sufficient means to φ in situation *S*.

(Sufficient reading) 'If you want φ, and you are in situation *S*, then it suffices to do β' is true in *S* if $S \models [\beta]\varphi \wedge \langle\beta\rangle\top$.

This definition achieves our goal of showing how technical norms can get a truth-value, but again one may wonder to what extent this definition of the second reading is empirically adequate. We will briefly discuss the complexities involved in this issue in the Discussion. We complete this section by addressing the remaining issue of the formal validity of practical inference.

3.4 Deontic PDL and Practical Inferences

Just as the semantics of propositional dynamic logic (PDL) suffices to formulate the truth conditions for technical norms, it suffices also to define the truth condition of (PI)'s second premise, the anankastic statement, which expresses that something is (or is not) a necessary condition for something else. We assume this truth condition to be the same as the sufficient truth condition of (TN'_N), or rather we assume the second premise of a practical inference simply to mean that given the state of the world a particular action is a necessary means to a given end:

(Anankastic premise) 'Unless ε is performed, it will not be the case that φ' is true if $\exists\alpha$: $S \models \langle\alpha\rangle\varphi$ and for every $\underline{\cup}$-free action α in S: if $S \models \langle\alpha\rangle\varphi$ then $\alpha \rhd_s \varepsilon$

Thus, if all paths from S in M that make the hut habitable involve heating the hut, then 'unless the hut is heated, it will not become habitable' is true in S.

This is how far we can proceed using just dynamic logic, which mainly describes how actions may change states of the world. To formulate the semantics of the first premise of (PI), which contains the notion of 'wanting', and its conclusion, which contains the 'must' of practical necessity, we need a language that is minimally equipped with the deontic 'ought to' operator. To that end we turn to Dynamic Deontic Logic as formulated by John Jules Meyer in his seminal (1988). In this paper, Meyer sets out to circumvent the well-known paradoxes of traditional deontic logic by letting the 'Forbidden', 'Ought to' and 'Permissible' operators have actions as their arguments instead of states of the world. To make this work, he introduces a special deontic proposition V, which "is a sentential constant denoting the so-called 'undesirable state-of-affairs', e.g., sanction, (liable to) punishment, trouble (with conscience, for example)". He then defines the notion that some action α is forbidden (F) in a state S by:

Definition F $S \models F\alpha$ iff $S \models [\alpha]V$.

This means that α in state S is *forbidden* iff performing α in S always leads to the undesirable state-of-affairs. Likewise an action α is *obligatory* in S iff failing to perform α in S always brings one to the undesirable state:

Definition O $S \models O\alpha$ iff $S \models [{\sim}\alpha]V$

Finally, an action α is permitted in S if it is not forbidden:

Definition P $S \models P\alpha$ iff $S \models \neg F\alpha$.

These definitions seem to enable us to formalize the conclusion of the Action version of (PI) – 'Therefore one must heat the hut' – in DDeL as $S \models O\varepsilon$, where ε is the action of heating the hut. We can define this statement to be true in *S* iff all performances of ε in *S* lead to states *S′* in which V is false, provided ε can be performed in *S*.[16]

When formalizing the deontic aspect of practical inference this way, we ignore the issue of individuality. The conclusion would state that the action of heating the hut must be performed, without specifying by whom it must be performed, although supposedly all actions are performed by someone. We cannot likewise ignore this issue for the first premise, however. Its 'want' must be formalized as the wanting of someone if the formal treatment is to make sense. Our semantics must therefore be able to manage obligations for specific individuals.[17] For this purpose, and thus extending DDeL to DDeL*, we introduce a personalized deontic operator $O_A(\alpha)$ which expresses that person *A* has an obligation to perform α. Individual obligations, however, need not be imposed. Agents can also voluntarily take obligations upon themselves. Thus when an agent *A* has obliged herself to perform action α, we also write this as $O_A(\alpha)$.[18]

With individualized 'oughts' we are now fully equipped to formalize (PI)'s conclusion in its 'true' form '*A* ought to heat the hut' as $S \models O_A(\varepsilon)$, where ε is the action 'heating the hut'.[19] For the first premise, we take seriously Von Wright's (1972: p. 55) remark that "action cannot possibly follow logically from premises about intentions and epistemic attitudes" and interpret the statement '*A* wants α' to mean or stand for '*A* has obliged herself to perform α', or formally $S \models O_A(\alpha)$. Of course one may object that wanting is not always a form of obliging oneself. However, we take the wanting in the first premise not to be some casual wanting like in 'I want it to be nice weather'. From such wanting indeed little could follow in the strict deductive sense. Instead we look upon the wanting as signifying a top priority in agreement with Von Wright who explains the first premise as follows: "we want to attain

[16] Note that ε need not be atomic but may consist for instance of two subsequent subactions: $\varepsilon := \varepsilon_1; \varepsilon_2$. In such a case $O[\varepsilon_1; \varepsilon_2] \equiv O\varepsilon_1 \wedge [\varepsilon_1]O\varepsilon_2$. (Note that in contrast $F[\varepsilon_1; \varepsilon_2] \equiv [\varepsilon_1]F\varepsilon_2$.) This means that V is true in all states along the paths towards the state where the hut is finally heated; only in that final state V is false.

[17] Individual deontic operators may also be found in for instance work of Wieringa and Meyer, (1993), Herzig and Lorini, (2010), and other STIT logic publications.

[18] There are ways for DDeL* to bring out the distinction between voluntary obligations and imposed obligations. Voluntary obligations result from a previous action of the person who has the obligation, namely the action of taking an obligation upon oneself, whereas for imposed obligations the action of imposing the obligation is performed by another person than the person who has the obligation.

[19] It is here that the step from the *Dynamic-States version* to the *Action version* of PI is actually made.

the end as a result or consequence of something which we do" (PI, p. 160). Such wanting involves a form of commitment that we see as at least close to obliging oneself to bring about the stated end.[20] We return to this issue in the Discussion section.

Finally we are ready to investigate whether practical inference, as introduced by Von Wright's and made more explicate by us, is logically valid in DDeL*. In our adapted, individualized Action version the inference is:

(PI′) A has obliged herself to make the hut habitable.
Unless the hut is heated, it will not become habitable.

A is under the obligation to heat the hut.

With φ being 'the hut is habitable', α being 'making the hut habitable' and ε being 'heating the hut' the semantical counterpart of this inference reads:

(PI′) $S \models O_A(\alpha) \wedge [\alpha]\varphi \wedge \neg\varphi$
$S \models \langle\alpha\rangle\top \wedge \forall\alpha\,(\langle\alpha\rangle\varphi \rightarrow \alpha \rhd_S \varepsilon)$

$S \models O_A\,(\varepsilon)$

To show that the truth of the premises implies the truth of the conclusion, let us first assume that 'making the hut habitable' is $\underline{\cup}$-free, which means that it is a linear order of actions; that it is possible for *A* to perform it, and that it involves 'heating the hut'. As mentioned before, ε, heating the hut, need not be an atomic action. We let ε refer to that string of actions starting from *S* to *S′* in which the hut is finally heated. We let β refer to the remainder of the string of actions, between *S′* and *S″* in which α is completed and the hut is habitable. So we have $\alpha = \varepsilon;\, \beta$ (and of course β may be non-existent). Consequently, more syntactically, the DDeL* representation of (PI) is:

(PI″) $O_A(\varepsilon;\, \beta) \wedge [\varepsilon;\beta]\varphi \wedge \neg\varphi$
$\langle\varepsilon;\, \beta\rangle\top$

$O_A\,(\varepsilon)$

That (PI″) is valid follows very easily from one of the theorems of DDeL which states that $O(\varepsilon;\, \beta) \equiv O\varepsilon \wedge [\varepsilon]O\beta$. In case α is not $\underline{\cup}$-free, it may be rewritten in a

[20] Meyer and others similarly claim that 'being committed to α' intuitively corresponds to 'having promised (to oneself) to perform α next (or at least a.s.a.p.)' (Meyer et al. 1999, p16). Note, further, that although one may want unattainable things, here we do assume that the things wanted are at least empirically attainable. This is relevant for engineering and will be taken up again in the final Discussion section.

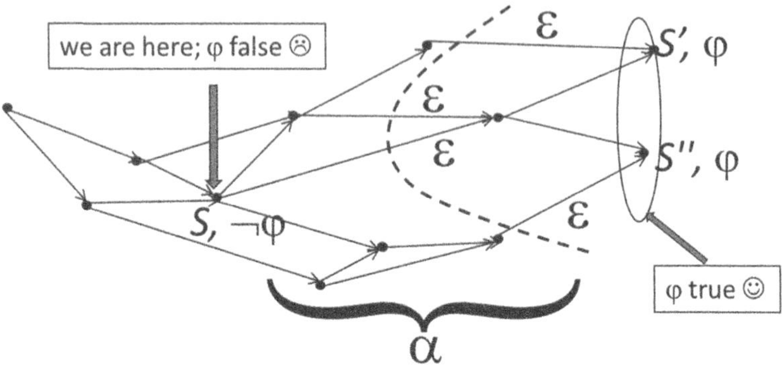

Fig. 3.2 Being in S and wanting to achieve φ in S' or S'' one cannot escape to perform ε somewhere along the line

disjunctive normal form $\alpha := \alpha_1 \cup \ldots \alpha_i \ldots \cup \alpha_n$ in which all the α_i are $\cup$-free. Because α is finite, all its branches end up in states where φ is true (since $[\varepsilon;\beta]\varphi$), and all these branches involve the activity of heating the hut. Thus we can identify ε with a structure of branches where each branch starts in state *S* and ends in a state S' in which the heating has finished, and take β to be the rest of α, and then we can again apply $O(\varepsilon;\beta) \equiv O\varepsilon \wedge [\varepsilon]O\beta$ to show that that (PI″) is valid (Fig. 3.2).

Let us summarize, then, why (PI″) is valid. An agent *A* in some state of the world 'really' wants something, which is not yet the case in that state and which is her goal φ, and therefore obliges herself to achieve φ by performing an action α (the first premise). By empirical necessity, by the laws of nature jointly, so to speak, all action paths that bring her from the initial state of the world to a state where her goal is realized involve the performance of another action β (second premise). Then we may validly conclude that *A* is obliged to perform β.

3.5 Discussion

To recap the starting point of our paper, in a practical inference (PI) as defined by Von Wright, an obligation to act is deduced from a want or an intention to realize an end and from an empirical constraint on how this end can be achieved. Technical norms (TN), considered as 'contracted' PIs, are conditional statements with the setting of an end as the antecedent and an obligation to act as the consequent. In this paper we used dynamic logic to determine formal truth conditions for technical norms and to define validity for practical inferences. But by doing so, we adjusted them considerably to make them fit into our DDeL* framework. The main question that remains to be discussed here is to what extent the results regarding our adapted PI and TN versions still say something substantial about the original ones. To what extent do our paraphrased versions of TN and PI do justice to the intuitions that underlie our uses of these statements and forms of reasoning?

A reason to doubt that they do them justice is the observation that, in practice, technical norms and practical inferences are always defeasible or non-monotonic. Every engineer would agree that for hardly any design problem, a necessary condition exists that is part of all solutions. No engineering task seems to be absolutely determined in this way. Thus in real-life situations a technical norm will have the form: 'If you want φ, and you are in situation S, then *normally* you should (or: it suffices to) do α'. This is so because in real-life situations the underlying anankastic statement is seldom deterministic and almost always has the form: 'unless α is performed, *normally* it will not be the case that φ'. Accordingly, the pertaining conclusion is that someone who wants φ and is in situation S *normally* should perform α.

Formally, our semantics can cope with non-monotonicity by introducing a relation r that for every state determines the set of normal states S_n that are, up to a certain extent, similar to S. This r enables us to define a defeasible conditional $\psi \Rightarrow \varphi$, which is already true in S if only the *S-normal* ψ-states satisfy φ (but not necessarily all ψ-states). Then, we can use ψ as a sufficient precondition for the necessary means-end relation involving α and φ by applying the conditions 1 and 2 of Definition 3.3 not only to the unique state S but to all normal states S_n and in this way arrive at a defeasible TN version. With the same mechanism we may render the formal version of PI non-monotonic.[21]

Where would this leave us with regard to the adequacy of our analysis? On the one hand we have engineers and other professional practitioners who, on a daily basis, draw defeasible conclusions about what should be done, and on the other we have constructed a non-monotonic framework for drawing indefeasible conclusions from self-obligations and anankastic statements. The proposed introduction of defeasibility into the framework, although opening a way to bring our approach more in line with reasoning as it actually occurs, requires a similarity-between-states-of-the-world notion, which is barely recognizable in real-life means-end reasoning. More in-depth research should find out how in practice professionals cope with the defeasibility of their conclusions and whether our similarity-based mechanism can accommodate this reasoning in practice.

A second, seemingly even more profound departure from real-life practical reasoning of our formal semantics is the assumption that an agent can only oblige herself to perform *successful* actions. In every state S and for every action α, the semantics determines whether $S \models \langle\alpha\rangle\top$ holds, that is, whether α can be performed in S. Moreover, for all S and performable α the semantics fixes the resulting world states. The framework therefore does not allow for actions with undetermined end results.[22] In real life, agents often initiate actions without being sure whether they will result in the intended end. In the course of their work, engineers regularly change their course of action to adapt to changing circumstances or changing requirements. In addition, research engineers, who are engaged in extending the limits of what is technically possible, often start on projects of which they are not

[21] See for more technical details Hughes et al. (2007), pp. 225–227.

sure that they are realizable. Take for instance, the first transatlantic communications cable or keeping heavier-than-air contraptions airborne.

Whether the success assumption of α poses a profound or only a superficial obstacle to the adequacy of our model, and how much it takes to repair this, depends heavily on fundamental philosophical questions. Our model comes close to a form of technical realism according to which all technical possibilities are fixed once and for all. From this point of view one may argue that taking up an unattainable engineering task is not performing an action at all and therefore does not need to be represented in the framework. Technical constructivism, according to which the technically possible is a realm of human construction, is much more difficult to conciliate with our framework, especially for engineering tasks that have never been performed before. The quest for pushing technological boundaries will often involve the performance of actions that fail to achieve the projected ends. Technical realism is closely related to scientific realism according to which the laws and theories of physics lie hidden 'out there', to be discovered by science. Constructivism, in contrast, is related to scientific empiricism or instrumentalism, which acknowledges a good deal of human construction in the development of scientific theories. Furthermore, the two points of views differ markedly in that technical realism is primarily backward-looking and technical constructivism primarily forward-looking. It is not our task here to decide between the two. We only note that our framework fits the former perspective better than the latter.

The last and perhaps most salient difference between our formal account and the intuitive notions of practical inference and technical norm is the interpretation of the 'want' or intention-statement. In our formal version of PI we paraphrase '*A* wants φ' as '*A* obliges herself to (bring about) φ'. Even if we restrict 'wants' to a person's top priorities, for which it is plausible that they involve some form of self-obligation, we have to acknowledge important differences between the two. Wants are akin to desires (Irvine 2006). The wants in practical inference appear to be much more similar to desires than to self-obligations, and as the distinction between desires and self-obligations is obvious, so must the wants be different from self-obligations. To oblige oneself, is the performance of an action, whereas ending up wanting something does not appear to be related to the performance of a prior explicit action. Wants are typically generated beyond the grasp of one's own will; they happen to you and may fluctuate, whereas self-obligations seem rather to be decisions taken because one wills them, at a precise moment, and do not fluctuate.

Whereas we acknowledge that in general wants do not imply self-obligations, in our formal analysis we have interpreted wanting in its strongest form, which we take to be close enough to self-obligation to be allowed to treat them as equivalent. In doing so, we are convinced that we have excavated and modeled all the logical

[22] This differs from the claim that for $\langle\alpha\rangle$ and $[\beta]$, α can and β possibly cannot be performed. The reader interested in the logic of effectiveness may consult Harz (2007) (in German).

structure and the validity there is to practical inference. This is a strong claim, which may be underpinned by the following robustness argument. None of the authors who so far have seriously studied the validity of practical inference have found a valid way to deduce an (obligation to) an action from a want or intention and an empirical necessity. All of them, when establishing a logically valid connection between the premises and the conclusion of a practical inference, have, more or less independently, used forms of the same strategy, which we have applied, and which can be likened to a reduction-to-the-same-denominator. Similar to the process of reducing fractions to the same denominator, this strategy puts the 'want'-premise and the 'must'-conclusion on the same footing. For instance, the form in which Von Wright ultimately considered practical inference to be valid is (1972, p.47):

(PI^I) *X* intends to make it true that E
He thinks that, unless he does *A*, he will not achieve this

X intends to do *A*.

In Von Wright's own words this form of practical inference exhibits a *transmission of intentions* from premises to the conclusion. It is exactly this form that Meyer et al. (1999, p. 15) paraphrase in their framework. In the same sense we may call our valid version of the practical inference *transmission of obligations*. Also Broome, in his analysis of practical reasoning (2002), applies the same reduction-to-the-same-denominator strategy. He paraphrases the 'want'-statement and the 'must'-conclusion in the future tense and infers 'Chris will borrow money' from 'Chris will buy a boat' and 'for Chris to buy a boat a necessary means is for Chris to borrow money'. Like Von Wright before him, however, Broome (ibid, section 2) claims the inference to be valid without further ado. Finally we may even turn to Immanuel Kant, who claims in his *Groundwork* that 'whoever wills the end, wills also (according to the dictate of reason necessarily) the indispensable means thereto which are in his power'. Again, this, according to Kant, analytically true statement may be seen as an instance of the same strategy. Answering Segerberg's (1980) call to apply dynamic logic to philosophical problems, we also came to the same result, and which we now in retrospective consider to be a sound philosophical (and not technical) norm: If you want a practical inference to be logically valid, you should apply something similar to the reduction-to-the-same-denominator strategy.

References

Broersen, J. (2004). Action negation and alternative reductions for dynamic deontic logics. *Journal of Applied Logic, 2*(1), 153–168.

Broome, J. (2002). Practical reasoning. In J. Bermúdez & A. Millar (Eds.), *Reason and nature: Essays in the theory of rationality* (pp. 85–111). Oxford: Oxford University Press.

Harz, M. (2007). *Zur Logik der technologischen Effektivität*. Cottbus: Brandenburgischen Technischen Universität. https://opus4.kobv.de/opus4-UBICO/frontdoor/index/index/docId/3738.

Herzig, A., & Lorini, E. (2010). A dynamic logic of agency I: STIT, capabilities and powers. *Journal of Logic, Language and Information, 19*(1), 89–121. https://doi.org/10.1007/s10849-009-9105-x.

Hughes, J., Kroes, P., & Zwart, S. (2007). A semantics for means-end relations. *Synthese, 158*(2), 207–231.

Irvine, W. B. (2006). *On desire: Why we want what we want*. Oxford: Oxford University Press.

Meyer, J.-J. (1988). A different approach to deontic logic: Deontic logic viewed as a variant of dynamic logic. *Notre Dame Journal of Formal Logic, 29*(1), 109–136.

Meyer, J.-J. C., van der Hoek, W., & van Linder, B. (1999). A logical approach to the dynamics of commitments. *Artificial Intelligence, 113*(1), 1–40.

Niiniluoto, I. (1993). The aim and structure of applied research. *Erkenntnis, 38*(1), 1–21. https://doi.org/10.1007/BF01129020.

Segerberg, K. (1980). Applying modal logic. *Studia Logica, 39*(2–3), 275–295.

Von Wright, G. H. (1963a). *Norm and action: A logical enquiry*. London: Routledge & Kegan Paul.

Von Wright, G. H. (1963b). Practical inference. *The Philosophical Review, 72*(2), 159–179.

Von Wright, G. H. (1963c). *The varieties of goodness*. London: Routledge & Kegan Paul.

Von Wright, G. H. (1972). On so-called practical inference. *Acta Sociologica, 15*(1), 39–53.

Von Wright, G. H. (1999). Deontic logic-as I see it. In P. McNamara & H. Prakken (Eds.), *Norms, logics and information systems: New studies in deontic logic and computer science, Artificial intelligence and applications* (Vol. 49, pp. 15–25). Amsterdam: IOS Press.

Wang, Y. (2016). A logic of goal-directed knowing how. Synthese, 1–21. https://doi.org/10.1007/s11229-016-1272-0

Wieringa, R. J., & Meyer, J.-J. C. (1993). Actors, actions, and initiative in normative system specification. *Annals of Mathematics and Artificial Intelligence, 7*(1–4), 289–346.

Chapter 4
Intuition and Ineffability: Tacit Knowledge and Engineering Design

Mark Thomas Young

Abstract My goal in this paper is to show how the study of intuition in engineering design allows a fresh perspective from which to approach the issue of tacit knowledge, and one which may even help us gain some traction on stubborn philosophical problems. The first section of this paper seeks to outline the nature and role of intuition by examining the limitations of attempts to formalize the practice of engineering design. As an element of engineering practice that is commonly understood to resist codification, and be acquired exclusively through practice and experience, intuition not only shares a number of characteristics with philosophical accounts of tacit knowledge, but it also reveals promising new directions for its analysis in both historical and philosophical contexts. The second section of this article aims to draw out correlations between characteristics of intuition in engineering design and phenomenological aspects of the accounts of tacit knowledge provided by Michael Polanyi and Harry Collins. It will be shown how both thinkers emphasize the use of judgment stemming from a cultivated receptivity to relevant features of a task, as central to their accounts of tacit knowledge. Finally, I aim to show how a phenomenological understanding of tacit knowledge provides us with a solution to what I call the "ineffability problem"; the idea that because tacit knowledge resists codification it is ineffable and therefore possesses little explanatory value.

Keywords Tacit knowledge · Intuition · Engineering · Design · Phenomenology

The distinction between tacit and explicit knowledge represents a central and fundamental concept employed by scholars for the classification of knowledge and expertise. Disciplines employing the term "tacit knowledge" span a wide spectrum, from management studies and psychology to history and philosophy. However, despite the wide appeal of the idea of tacit knowledge, an increasing number of

---

M. T. Young (✉)
Department of Philosophy, University of Bergen, Bergen, Norway
e-mail: Mark.Young@fof.uib.no

A. Fritzsche, S. J. Oks (eds.), *The Future of Engineering*, Philosophy of Engineering and Technology 31, https://doi.org/10.1007/978-3-319-91029-1_4

scholars consider the concept to be overly ambiguous or inadequately conceptualized. In the first place, many are uncomfortable with the fact that 'most of the literature that uses the concept of tacit knowledge does little more than report that there are features of engineers' knowledge that are difficult, if not impossible to articulate' (Nightingale 2009). Furthermore, this resistance to articulation, which is often taken to be the defining feature of tacit knowledge is also the source of concerns surrounding the epistemological viability of the concept itself. As Tim Thornton asks in a recent work, 'how can states that cannot be articulated count nevertheless as knowledge? What is the content of such knowledge, for example? What is it that is known?' (Gasciogne & Thornton 2013, p. 6)

This general and persistent uncertainty surrounding the nature of tacit knowledge has led a number of recent philosophers and historians to question the explanatory value of the concept. According to Peter Dear, 'One reason to be wary of the "tacit knowledge" approach to expertise and skill is that…it appears to require the attribution by the analyst of additional, hidden properties possessed by the people who are credited by others with a particular skill or expertise' (Dear 2004). Kathyrn Olesko, on the other hand, suggests that 'the broader historiographic implications of tacit knowledge – not in what it implies *should* be investigated, but in what *cannot* be examined under its aegis – are most problematic and troublesome' (Olesko 1993).

Both thinkers voice a complaint concerning the concept of tacit knowledge that is becoming increasingly common – what I call the 'ineffability problem'. Put simply, the problem represents the concern that because tacit knowledge resists a particular form of articulation – codification in rules or principles, it is ineffable and therefore must fail to carry explanatory weight. Understanding tacit knowledge as ineffable in this way, has led some historians to argue that 'tacit knowledge' should be used only as an explanatory 'last resort' (Olesko 1993), as a placeholder term encompassing something which escapes the current reach of historical explanations. However, defining tacit knowledge negatively – as a mysterious 'other', which defies historical explanation, raises its own set of problems in that it risks rendering the concept too broad. As Paul Nightingale notes 'the ability of tacit knowledge to explain a host of very diverse phenomena, which on closer inspection actually turns out to involve something else, suggests a substantial weakness in how the concept is used' (Nightingale 2009).

In response to these concerns, this paper aims to lay groundwork for a more robust approach towards dealing with tacit knowledge in the history and philosophy of technology that will permit it to play a more substantial explanatory role. In order to do so, we will turn our attention to the role intuition plays in engineering design. My goal in this paper is to show how the study of intuition in engineering design allows a fresh perspective from which to approach the issue of tacit knowledge, and one which may even help us gain some traction on stubborn philosophical problems.

The first section of this paper seeks to outline the nature and role of intuition by examining the limitations of attempts to formalize the practice of engineering design. This analysis will be shown to yield a conception of intuition as an ability to recognize appropriate solutions, relevant factors or respond to contextual features of an engineering task. Insofar as this ability is understood to be acquired exclusively through practice and experience, and resist codification, intuition not only shares a number of characteristics with philosophical accounts of tacit knowledge, but it also reveals promising new directions for its analysis in both historical and philosophical contexts.

The second section of this article aims to draw out correlations between this account of intuition in engineering design and phenomenological aspects of the accounts of tacit knowledge provided by Michael Polanyi and Harry Collins. It will be shown how both thinkers emphasize the use of judgment stemming from a cultivated receptivity to discern relevant features of a task as central to their accounts of tacit knowledge. Finally, it will be shown how a phenomenological conception of tacit knowledge; as a form of cultivated receptivity, provides a solution to the ineffability problem by revealing how tacit knowledge can be understood to be uncodifiable without being ineffable. But first however, we must turn our attention to the role intuition plays in engineering.

4.1 The Controversial Role of Intuition in Engineering Design

Historians and philosophers interested in the nature of engineering have long looked towards the practice of design as the essential component. However, the nature of design itself presents its own host of thorny philosophical difficulties and for this reason is often the subject of dispute. A central point of dispute concerns explaining how engineers design fundamentally *new* kinds of artifacts (Love 2002). While it is common to understand innovative artifacts, such as the Large Hadron Collider in these terms, we also have reason to consider the design of more familiar forms of technology in the same way. For the design of each and every suspension bridge also reflects the unique environment from which it emerged; the local geography and availability of materials, for example, or the specific economic and political constraints affecting design and construction. Indeed, it is this uniqueness characterizing the designs of artifacts that lead some theorists such as architect Juhani Pallasmaa, to argue that 'design is *always* a search for something that is unknown in advance' (Pallasmaa 2009, p. 110; emphasis added). For this reason however, philosophical accounts of engineering design face a peculiar epistemological challenge; to explain how it is that engineers court uncertainty in the practice of designing and constructing artifacts. In this section we will review different aspects of engineering practice which are commonly cited as explaining how engineers cope with this uncertainty. It will be my suggestion that it is by looking at the different ways in which engineering practices have sought to meet this challenge that we can gain insight into the nature of tacit knowledge essential to engineering design.

Since the enlightenment, one of the most common ways of conceptualizing the nature of engineering design has been in terms of the practical application of scientific knowledge. Understanding engineering in this way, as a form of 'applied science' inevitably casts the practice of engineering design in terms of the extrapolation of plans for technical artifacts from known scientific laws. As Edwin Layton Jr. notes, 'the theory that engineering, and technology generally are no more than applied science, in its extreme form, holds that technology consists of applying the fruits of the basic sciences in a more or less mechanical, deductive manner' (Layton 1991). This ideal conception of the practice of design as determined by the content of the laws of natural science, rose to dominance in the U.S.A throughout the 1950s, where it helped encourage increased funding in basic research and influenced a shift in focus in engineering education away from practice based learning and towards a curriculum centered primarily around physical science and mathematics.[1]

However, for many theorists, such efforts to make engineering more scientific failed to address the continued presence of unformalizable factors in design, such as intuition, which many considered to ensure the status of engineering design as epistemologically inferior in relation to the natural sciences. The design science movement, which gained prominence in the 1960s, emerged in response to a widespread desire to render the methods of engineering design more scientific. Proponents of this movement, such as Herbert Simon, suggested that it was 'the methods of the cookbook that originally put design into disrepute and drove it from the engineering curriculum' and argued for its rehabilitation by formalizing the process of design through the use of 'computer programs, optimizing algorithms, search procedures, and special-purpose programs' in order to prevent 'the design process hiding behind the cloak of "judgment" or "experience"' (Simon 1996, p. 135). Today, various formal methods employed in contemporary engineering design, such as optimization theory, or axiomatic design, are often advocated for the same reasons. In a recent work on analytic design methodology, Gyung-Jin Park for example, argues that 'the usage of (scientific) analysis results depends on the designer's personal judgment and it is obvious that this process is inefficient. Therefore design engineers are trying to find design principles and establish scientific ways for design' (Park 2007, p. 6).

In this paper, I will use the term "formal approaches to design" to refer to any approach which aims to replace the use of personal judgment and intuition in the process of design with procedures governed by explicit rules, or the application of universal methodological principles. Such approaches to design share a number of features; because they are based on explicit rules, they help to determine aspects of the activity of design in advance and can therefore be learned outside of practical

[1] See Ferguson, Eugene, S. *Engineering and the Mind's Eye*. (U.S.A: M.I.T Press, 1993), 58 & Trevelyan, James. "Towards a Theoretical Framework for Engineering Practice" in Williams, Bill & Figueiredo, José & Trevelyan, James. (eds) *Engineering Practice in a Global Context: Understanding the Technical and the Social* (U.K: CRC Press, 2014) pp., 33–61.

contexts, for example, in classroom settings. Formal approaches to engineering design thus adopt a particular stance towards the epistemological challenge that we noted above by contending that uncertainty in design is best countered through a rule governed application of general principles of design methodology and natural science. For such general principles to be effective however, proponents of formal approaches to design must resist understanding individual cases of design as inherently unique and maintain instead 'that there are sufficient uniformities in problems and in devices for solving them to qualify the solvers as professionals…. (who) apply very general principles, *standardized* knowledge, to concrete problems' (Moore 1970, p. 55).

Yet while formal approaches to the practice of engineering design have exerted a strong influence on education, policy and practice, they have not gone unchallenged. At least since the late 1960s, critics have emphasized the limitations such approaches face in attempting to replace the role of personal judgment and intuition in engineering design.

The first line of criticism we will consider concerns a central premise underlying many analytic approaches to engineering design; the idea that, given enough information about a particular system, decision-making processes may be performed on the basis of mathematical and theoretical analysis rather than personal judgment. Critics question the feasibility of this idea by drawing attention to the limitations of theoretical techniques to yield accurate predictions concerning proposed design systems. For despite the increasing sophistication of computational methods, such as modeling and prediction, alongside corresponding techniques for quantifying and measuring uncertainty, complex technologies continue to yield unexpected outcomes (Rosenberg 1982, p. 122). This is most commonly observed to apply to the design of socio-technical systems, where agents 'may redesign parts of the system from within in unforeseen ways (precluding a) single vantage point from which complex systems can be designed and controlled' (Vermaas, Kroes, Light & Moore 2008). However, it is often noted that these limitations appear not only as a consequence of unpredictable human behavior – similar limits also characterize our ability to predict the physical function of complex technologies. As Nathan Rosenberg notes;

> For a range of products involving complex, interdependent components or materials that will be subject to varied or prolonged stress in extreme environments, the outcome of the interaction of these parts cannot be precisely predicted. In this sense, we are dealing with performance characteristics that scientific knowledge or techniques cannot predict very accurately. The performance of these products, therefore, is highly uncertain (Rosenberg 1982, p. 122).

Furthermore, in cases where it is possible to subject design alternatives to the various regimes of testing employed in modern engineering practice, we must remain aware of the fact that knowledge gained concerning the behavior of an artifact in a testing environment does not necessarily transfer directly to products in use. An apt illustration of these limitations is found in the use of wind tunnels in aeronautical engineering which have long been observed to subject models to different conditions than those experienced in atmospheric flight. Not only have wind tunnels in the past been known to generate small scale turbulence that is not present in the atmosphere (Vincenti 1990, p. 35), but they have also been observed to exaggerate the increase in drag at high speeds (Rosenberg 1982, p. 125). The use and limitations of wind tunnels testing throughout the twentieth century therefore reflect how the quality of information that can be gained from testing regimes is limited not only by the quality of our testing technology but also by the presence of scale effects and 'latent physical and other phenomena that, while bearing little relevance for the initial artifact, can prove to dominate the behavior of scaled up or pared down descendants' (Petroski 1994, p. 94). Together, these limitations characterizing the capacity of theoretical techniques and experimentation to yield accurate predictions pose an inherent challenge to the notion that decision-making processes in engineering design can be governed by formal design methods alone.

The second major line of critique advanced towards formal approaches to design concerns the extent to which these methods fail to account for the creative work involved in formulating design problems. Analytic methods such as optimization and axiomatic design represent techniques designed to provide solutions for clearly defined design problems and therefore do not address the question of how design problems themselves emerge. Furthermore, unlike the kinds of clearly defined problems which are often addressed by mathematics and natural science, the problems engineers face are commonly understood to be 'ill-defined' (Rittel & Webber 1973). In *The Reflective Practitioner*, Donald Schön outlines the way engineering design problems must be constructed from situations which are characterized by conflicting demands while at the same time taking into account resources which are initially uncertain and liable to change throughout the process of design and construction. According to Schön;

> When we set the problem, we select what we will treat as the "things" of the situation, we set the boundaries of our attention to it, and impose upon it a coherence which allows us to say what is wrong and in what directions the situation needs to be changed. Problem setting is a process in which, interactively, we *name* the things to which we will attend and *frame* the context in which we will attend to them (Schön 1983, p. 40).

Furthermore, not only must practitioners "frame" design problems at the outset 'from the materials of problematic situations which are puzzling, troubling and uncertain' (Schön 1983, p. 40), but they are also often required to "reframe" design problems in the face of unpredictable developments during the design process itself (Schön 1983, p. 131).

Schön's analysis of the process by which design problems are formulated complement similar claims made by design theorists Rittel and Webber, who introduced the term 'wicked problems' in their seminal article "Dilemmas of a General Theory of Planning". Rittel and Webber drew attention to the way that standard procedures in systems engineering, which urge practitioners to first understand the problem and then search for a solution, failed to account for the creative and active process by which practitioners formulate problems *throughout* the process of design;

> one cannot understand the problem without knowing about its context; one cannot meaningfully search for information without the orientation of a solution concept; one cannot first understand, then solve. The "systems-approach" of the first generation is inadequate for dealing with wicked-problems. Approaches of the "second generation" should be based on a model of planning as an argumentative process in the course of which an image of the problem and of the solution emerges gradually amongst the participants, as a product of incessant judgment, subjected to critical argument (Rittel & Webber 1973).

Rather than replacing judgment then, Rittel, Webber and Schön argue that formal design methods are best understood as *complementing* the use of judgment throughout the entire process of engineering design.

The final line of critique advanced towards formal approaches to design that we will consider draws attention to the way in which the problems engineering design aim to solve can be understood to be inherently unique. The previous section emphasized the extent to which engineering design problems are formulated within specific technological, political and economic constraints. Here we find an implicit challenge to the central presupposition underlying analytic approaches to design; that it is sufficient uniformities in problems which allow them to be solved by the application of standardized methods and knowledge. In contrast, critics maintain that it is because 'engineers have to deal with situations that are always different from situations they have previously encountered – different clients, with different needs and different tolerances for risk and uncertainty, based in different geographical situations (that) the experience (they) acquire cannot simply be reduced to a set of rules for decision making' (McCarthy 2010). If we are to recognize the inherent uniqueness of individual cases of engineering design, then we must also understand the application of general principles of design to require the exercise of judgment. Successfully applying standardized forms of knowledge in engineering design therefore requires modifying standardized knowledge to accommodate the idiosyncratic nature of individual cases of design. As we will see in the next section, it is the failure to exercise these forms of judgment successfully, rather than a failure to follow methodological principles of design or perform calculations correctly, that is often cited as a central cause of failure in engineering design practice.

Together, these concerns have led theorists from a variety of disciplines to argue that appeal to formal approaches to design alone fails to account for the way in which engineers tackle the epistemological challenges posed by design. Such arguments, which seek to highlight the limitations of analytic approaches to engineering design are often made in tandem with suggestions for changes to the nature of engineering education itself. Since the late 1960s, concerns have been voiced that the traditional engineering curriculum does not adequately prepare engineers for practice. An increasing recognition amongst some engineers that 'the principles and techniques of engineering design can never be fully articulated' (Ferguson 1993, p. 9), for example, has led many to argue that engineers should follow the model of medical training which combines theoretical study with clinical practice, by supplementing courses in engineering or design theory with on-site training.[2] For many, the central feature of design practice which the traditional engineering curriculum fails to account for is the role played by judgment. According to Henry Petroski;

> Engineering judgment, by whatever name it may be called, is what from the very beginning of a conceptual design identifies the key elements that go to make up an analytical or experimental model for exploration and development. It is judgment that separates the significant from the insignificant details, and it is judgment that catches analysis going astray. Engineering judgment is the quality factor among those countless quantities that have come to dominate design in our post computer age. Judgment tells the designer what to check on the back of an envelope and what to measure at the construction site. Judgment, in short, is what avoids mistakes, what catches errors, what detects flaws, and what anticipates and obviates failure (Petroski 1994, p. 121).

Furthermore, a number of authors suggest that the priority accorded to numerical and analytical techniques over experience and judgment in engineering education can be understood to be indirectly responsible for a variety of historical accidents. In their studies of engineering failure, both Ferguson and Petroski identify judgment and intuition as an indispensible capacity which allows engineers to recognize contextual features relevant to the structural integrity of buildings and bridges.

The importance of judgment in successful engineering design is reflected clearly in the analysis of a number of high profile engineering failures. The report produced for the Federal Works Administration after the collapse of the Tacoma Narrows bridge in 1940, for example, noted how designers had regarded stiffness under traffic loads, rather than stiffness in suppressing the oscillations of the bridge in high winds (Petroski 1994, p. 158), as the most relevant factor in their design;

> Evidently the designers had relied on the weight of the long span of the bridge and on the relatively small cable sag to provide the necessary vertical rigidity. There is every evidence that insofar as rigidity under moving loads is concerned their judgment was justified. They had not expected that aerodynamic action could become so powerful as to put the great suspended mass of the structure into dangerous or objectionable motion (Ammon, Karman & Woodruff 1941, p. 47).

Likewise, for the case of the Hartford Coliseum, a sports arena in Connecticut which collapsed in 1978, it appears that engineers considered only pure compression in their calculations for the cruciform members from which the roof was to be constructed.[3] The report issued a year after the collapse by Lev Zetlin Associates, Inc. cited design deficiencies as the cause of collapse, noting in particular how;

> The choice of the typical member as a cruciform, a section that is weak in bending and torsion, was based on the design assumption that such bending and torsion would be negligible in the space truss (Lev Zetlin Associates 1978).

---

[2] See, for example Brooks, Harvey. "Dilemmas of Engineering Education" IEEE Spectrum 4, (1967) No. 2 Feb & Ferguson, Eugene, S. *Engineering and the Mind's Eye*. (U.S.A: M.I.T Press, 1993), & Trevelyan, James. "Towards a Theoretical Framework for Engineering Practice" in Williams, Bill & Figueiredo, José & Trevelyan, James. (eds) *Engineering Practice in a Global Context: Understanding the Technical and the Social* (U.K: CRC Press, 2014) pp., 33–61.

[3] Ferguson, Eugene, S. *Engineering and the Mind's Eye*. (U.S.A: M.I.T Press, 1993), 39.

A separate investigation of the collapse by Loomis and Loomis Inc. identified torsional buckling triggered by the added weight of snow as the primary cause of the collapse, observing that 'torsional buckling is so uncommon a problem that it is not covered by building codes and only by a few textbooks, so most designers don't bother to look for it' (ENR 1979).

As is well known, both the Tacoma Narrows Bridge and the Hartford Coliseum collapsed in extreme weather conditions. Interestingly, for each of these cases, warning signs were observed but dismissed as irrelevant; the roof of the Hartford Coliseum, for example, exhibited unexpected amounts of deflection upon assembly, a fact dismissed by engineers who maintained that the deviations reflected the simplifying assumptions of theoretical calculations and were not therefore a cause for concern (Levy & Salvadori 1992, p. 72). Similarly, prior to the catastrophic collapse of the Tacoma Narrows bridge in 1940, engineers had been aware of high winds causing suspension bridges to oscillate, but did not regard this to be significant enough to warrant reinforcement (Petroski 1994, p. 61). As Ferguson emphasizes, 'numerical calculations always embody human judgment' (Ferguson 1993, p. 39) and in both these cases, like many other engineering failures, it was not faulty calculations but faulty *judgments* (i.e. the choice of which factors to regard as relevant), which were to blame for the eventual collapse of the structures (Ferguson 1993, p. 183).

We have already seen how, in different ways, Ritell, Webber, and Schön drew attention to the fact that before engineers set about attempting to solve a problem, they must first formulate the problem itself; a process which involves using judgment to determine what elements of a situation will constitute relevant factors in the attempt to find a solution. As Schön emphasizes however, 'although problem setting is a necessary condition for technical problem solving, it is not itself a technical problem' (Schön 1983, p. 40). In other words, because this process of formulating design problems exists outside the grasp of formal design methods, these techniques cannot provide us with guidance.

This uncertainty surrounding the practice of problem formulation reflects limitations facing attempts to formalize a wide range of practices, from psychotherapy to management. Rittel and Webber employ examples from social planning to illustrate the essential role played by judgment in determining not only the boundaries of a problem, but also by implication, the nature of potential solutions;

> Consider, for example, what would be necessary in identifying the nature of the poverty problem. Does poverty mean low income? Yes, in part. But what are the determinants of determinants of low income? Is it deficiency of the national and regional economies, or is it deficiencies of cognitive and occupational skills within a labor force?....Or does the poverty problem reside in deficient physical and mental health?....Does it include cultural deprivation? Spatial dislocation? Problems of ego identity? Deficient political and social skills – and so on (Rittel & Webber 1973).

Furthermore, argue Rittel and Webber, unlike the problems of mathematics, we have no absolute criteria for determining which of these judgments are right and which are wrong (Rittel & Webber 1973). Eugene Ferguson observes a similar quandary to exist in engineering design where, 'making wrong choices is the same kind of game as making right choices; there is often not *a priori* reason to do one thing rather than another (Ferguson 1993, p. 9). This difficulty is further exacerbated by the problems of using past practice as a guide; the fact that certain structures have proved successful in the past fails to provide us with decisive confirmation that the judgments embodied in a particular design are sound. This insight constitutes, in the words of Petroski 'an essential rule of design *logic:* the absence of failure does not prove that a design is flawless, for a latent failure mode may be triggered by yet-unexperienced conditions' (Petroski 1994, p. 160).

So if the structured reasoning of formal design methodology is unable to provide us with guidance in these cases, then what exactly do engineers base such judgments on? The typical way of answering this question is through appeal to intuition; a capacity typically described as an ability to recognize appropriate solutions, relevant factors or respond to contextual features of an engineering task. Intuition in engineering bears many of the hallmarks of conceptions of tacit knowledge in philosophy; it is commonly understood to resist codification, for example, and is therefore considered to be acquired exclusively through practice and experience rather than formal education. In the next section, we will examine more closely the parallels between conceptions of tacit knowledge in philosophy and the role of intuition in engineering design by examining the phenomenological dimension of the accounts of tacit knowledge found in the work of both Michael Polanyi and Harry Collins.

4.2 Intuition As Tacit Knowing: Connoisseurship and Language Use

Among philosophers, language use is often cited as a prime example of the exercise of tacit knowledge. Harry Collins, one of the central figures in the current debates surrounding tacit knowledge, for example, argues that language constitutes a 'repository of tacit knowledge'.[4] In doing so, Collins aligns himself with the work of philosopher Michael Polanyi, who introduced the term "tacit knowledge" in part through a sustained examination of the practice of language use. For both thinkers, the characterization of language use as an example of tacit knowledge derives from a perception of meaning and experience as existing in a state of constant change. Polanyi, for example, argued that;

> Since every occasion on which a word is used is in some degree different from every previous occasion, we should expect that the meaning of a word will be modified in some degree on every such occasion. For example, since no owl is exactly like any other, to say 'This is an owl', a statement which ostensibly says something about the bird in front of us, also says something new about the term 'owl', that is, about owls in general (Polanyi 2005, p. 116).

[4] See Collins, Harry M. "Language as a Repository of Tacit Knowledge" in Schilhab, T. & Sternfelt, F. & Deacon, T. (eds) *The Symbolic Species Evolved* (Dordrecht: Springer, 2012) also Collins, Harry M. & Evans, Robert. *Rethinking Expertise* (Chicago: University of Chicago Press, 2007) & Collins, Harry M. *Tacit and Explicit Knowledge* (U.S.A: University of Chicago Press, 2010).

Polanyi's insistence on the dynamic nature of meaning forms the crux of his argument against the possibility of explaining language use as an entirely rule governed process. Because the world exists in a state of flux, never repeating any previous situation (Polanyi 2005, p. 83), Polanyi argued, the act of applying the general terms of our language requires speakers to use judgment in deciding whether or not a particular experience is to be captured by a particular concept. Furthermore, this process resists formalization because 'the outcome cannot be predicted from the previous use of language, for it may involve a decision to correct, or otherwise modify, the use of language' (Polanyi 2005, p. 99). Indeed, it is through the active decisions that language users must constantly make in deciding how their terms are to relate to the world, that language achieves its sense of consistency;

> we can achieve consistency only by identifying manifestly different situations in respect to some particular feature, and this requires a series of personal judgments. First, we must decide what variations of our experience are irrelevant to the identification of this recurrent feature, as forming no part of it, i.e. we must discriminate against its random background. Secondly, we must decide what variations should be accepted as normal changes in the appearance of this identifiable feature, or should be taken, on the contrary, to discredit this feature altogether as a recurrent element of experience (Polanyi 2005, p. 83).

According to Polanyi, the cognitive capacity that allows proficient language users to discern relevant from irrelevant features of our experience when applying general terms is connoisseurship; an ability which Polanyi argues is communicated by example rather than precept (Polanyi 2005, p. 56);

> To classify things in terms of features for which we have names, as we do in talking about things, requires the same kind of connoisseurship as the naturalist must have for identifying specimens of plants or animals. Thus the art of speaking precisely, by applying a rich vocabulary exactly, resembles the delicate discrimination practised by the expert taxonomist (Polanyi 2005, p. 84).

Furthermore, connoisseurship has a distinct phenomenological dimension. As Polanyi notes, 'a word in use *looks different* from the way it does to someone who meets it as a totally foreign word' (Polanyi & Prosch 1975, p. 35). In other words, the acquisition of language skills is accompanied by changes to the way practitioners experience the world. To capture this point, Polanyi employs the metaphor of learning to use a stick to feel one's way in the dark, a process which he argues involves a '*Phenomenal transformation.* The sensation of the probe pressing on fingers and palm, and of the muscles guiding the probe, is lost, and instead we feel the point of the probe as it touches an object' (Polanyi & Prosch 1975, p. 36). In other words, as we develop expertise, the nature of what we experience is modified; features irrelevant to the task at hand recede from view, allowing more significant features to occupy our attention.

The conception of language use that Polanyi describes reflects many of the features of engineering design that we discussed in the previous section; the inherent uniqueness of every design problem, the limitations of using past practice as a guide, and the irreducible role of judgment in facilitating successful performance. Furthermore, the analogy draws our attention to the importance of relevance for both language use and engineering design; the ability to recognize the significance of tonal shifts, or a shrug of the shoulders, to the meaning of a speaker's phrase is not unlike the capacity of an experienced engineer to respond to the particularities of certain building sites and materials with an appropriate design. In both cases, we can make mistakes. However, there are reasons why we should consider those whose education combines formal training with practical experience as our best bet. For in both cases, our ability to discern the relevance of significant features of our environment is not determined solely by rules of grammar or laws governing the structural integrity of materials. Instead, it is determined in large part through a honing of our skills of interpretation through practice.

The analogy with Polanyi's account of language use therefore implies a concrete suggestion for a way of conceptualizing the nature of tacit knowledge that avoids the ineffability problem. In particular, it provides a novel way of explaining *why* tacit knowledge resists codification. One common answer to this problem involves the idea that there are simply too many rules determining skillful practices to allow for codification, or that the rules used are, for some reason or another, not consciously available to practitioners. The perspective I am outlining here, on the other hand, suggests a very different answer; that tacit knowledge is inarticulable not because it rests on something inexpressible that is *known* in advance, but rather upon a cultivated receptivity to relevant features of a particular environment. It is by expanding in this way, what we allow as a source of skillful expertise beyond the possession of information, that we can subvert the ineffability problem. For insofar as tacit knowledge is understood phenomenologically; as an openness to the world that allows one to *see* things that go unnoticed by a novice, codification no longer needs to be considered a relevant factor to its explanation.

This phenomenological perspective also forms an important element of the account of tacit knowledge provided by Harry Collins in his seminal work, *Changing Order*. In a case study examining how individuals develop expertise in building TEA lasers for experimental research, Collins showed how learning to build a laser occurred through practice, as individuals acquired the ability to discern aspects of the device significant to the goal of constructing a functioning laser. Collins describes this expertise as visible in the way, Bob Harrison, the subject of the study;

> confidently ignored certain parameters that were once thought vital (flatness of tubes, anode marking) and in the way he overlooked differences between the lasers that I noticed and would have thought important: for example, the thickness of the bottom electrode, the glue and the wires. It was even evident in certain displays of more ordinary perceptual skill such as his ability to hear the quality of the sound made by an arc discharge and thereby know the discharge's characteristics (Collins 1985, p. 73).

Drawing on perspectives from the later philosophy of Wittgenstein, Collins suggests that the ability to perceive what is relevant in practical situations is a central feature of tacit knowledge. So central, in fact, that Collins even claims that 'one might say that learning tacit knowledge, or acquiring culture, is a matter of learning this indefinitely long list of what is insignificant, and *inter alia*, learning what is significant' (Collins 1985, p. 71).

As I noted above, the central benefit of understanding intuition in this way, as a form of cultivated receptivity, is that it allows us to avoid the problem of ineffability.

However, it also requires us to relinquish a number of commonly held assumptions about the nature of intuition and knowledge;

Firstly, it encourages us to resist the common tendency to think that because intuition cannot be codified we should regard it as 'related to unconscious and subconscious processes'.[5] To the contrary, insofar as an experienced practitioner is more able than a novice to discern features of an environment that are most relevant to the task at hand, the exercise of tacit knowledge should be understood to require *heightened* levels of awareness.

Secondly, this perspective also requires us to oppose the common tendency to regard knowledge in terms of what Harry Collins calls the algorithmical model; 'a notion of knowledge as a set of formal instructions or pieces of 'information' about what to do in a variety of circumstances, (the) sort of information that enables a computer to carry out its programmer's intentions' (Collins 1985, p. 57). To quote Donald Schön, understanding knowledge in this way binds us to an epistemology of practice 'which leaves us at a loss to explain, or even describe, practical competency' (Schön 1983, p. 20). It is precisely because knowledge is commonly conceived in this way, that practical competence is so often understood as governed by mysterious, ineffable determinants of practice. As Collins notes 'if we did not feel pulled toward trying to say what we do, and if we did not make the mistake of thinking this is central to the understanding of knowledge, we would find nothing strange about our brains' and bodies' abilities to do the things we call tacit' (Collins 2010, p. 117). Collins makes an important point; if we can resist the tendency to restrict knowledge to only what can be articulated in the form of propositions, then we might see that what makes skillful practices possible has been right in front of our eyes all along. For after all, nothing is more familiar and ubiquitous, to our everyday lives than the forms of cultivated receptivity which make the acquisition and maintenance of practical competencies possible.

4.3 Conclusion

Studies of engineering design provide philosophers with a rich resource to explore a variety of fundamental questions concerning the nature of tacit knowledge. Longstanding controversies surrounding the extent to which the process of design can or should be formalized reflect, after all, conflicting views on the nature and function of the forms of knowledge which make skilled practice possible. In this paper, I have argued that studies of engineering design reveal a conception of

[5] See, for example, Badke-Schaub, Petra & Eris, Ozgur "A Theoretical Approach to Intuition in Design: Does Design Methodology Need to Account for Unconscious Processes?" in Chakrabarti, Amaresh & Blessing, Lucienne T. M. (eds) *An Anthology of Theories and Models of Design: Philosophy, Approaches and Empirical Explorations* (London: Springer-Verlag, 2014) & Pahl, G. & Beitz, W. & Feldhusen, J. & Grote, K.-H. *Engineering Design: A Systematic Approach (3rd ed)* Ken Wallace & Luciënne Blessing trans. (London: Springer-Verlag, 2007), 9.

intuition as a form of cultivated receptivity which resonates with aspects of the philosophical analyses of tacit knowledge that we find in the work of Michael Polanyi and Harry Collins. Understanding tacit knowledge in this way, not only provides philosophers and historians with a more explanatorily robust conception of tacit knowledge, but it also expands the kinds of resources relevant to its analysis. In this respect, studies in anthropology and psychology[6] which explore the acquisition of skillful expertise as a process by which a perceptual system becomes attuned to different aspects of the environment, rather than a process of acquiring information, indicate promising avenues for future philosophical studies of tacit knowledge. The extent to which this constitutes a solution to the ineffability problem however, will depend on our willingness to resist the persistent tendency to conceive of knowledge exclusively in terms of propositions and follow Collins in understanding knowledge in the way Wittgenstein suggested: broadly, as that which allows us to perform certain practices competently.[7]

Acknowledgements I would like to thank Harald Johanessen and Sorin Bangu for their helpful comments on an earlier draft of this paper.

References

Ammon, C. H., von Karman, T., & Woodruff, G. B. (1941). The failure of the Tacoma narrows bridge. Report to the Federal Works Agency

Badke-Schaub, P., & Eris, O. (2014). A theoretical approach to intuition in design: Does design methodology need to account for unconscious processes? In A. Chakrabarti & L. T. M. Blessing (Eds.), *An anthology of theories and models of design: Philosophy, approaches and empirical explorations*. London: Springer-Verlag.

Collins, H. M. (1985). *Changing order: Replication and induction in scientific practice*. London: Sage Publications.

Collins, H. M. (2010). *Tacit and Explicit Knowledge*. Chicago: University of Chicago Press.

Collins, H. M. (2012). Language as a repository of tacit knowledge. In T. Schilhab, F. Sternfelt, & T. Deacon (Eds.), *The symbolic species evolved*. Dordrecht: Springer.

Collins, H. M., & Evans, R. (2007). *Rethinking expertise*. Chicago: University of Chicago Press.

Collins, H. M., & Reber, A. (2013). Ships that pass in the night: Tacit knowledge in psychology and sociology. *Philosophia Scientiæ, 17*(3), 135–154.

Dear, P. (2004). Mysteries of state, mysteries of nature: Authority, knowledge and expertise in the seventeenth century. In S. Jasanoff (Ed.), *States of knowledge London: Routledge* (pp. 206–224).

ENR. (1979). "New theory on why Hartford Roof fell", 14.

Ferguson, E. S. (1993). *Engineering and the mind's eye*. Cambridge: M.I.T Press.

Gasciogne, N., & Thornton, T. (2013). *Tacit knowledge*. Durham: Acumen Publishing.

Gibson, J. J. (1986). *The ecological approach to visual perception*. New York: Psychology Press.

[6]See for example, Gibson, James J. The Ecological Approach to Visual Perception (New York: Psychology Press, 1986) & Ingold, Tim. Making: Anthropology, Archaeology, Art and Architecture (New York: Routledge, 2013).

[7]See Collins, Harry M. & Reber, Arthur. "Ships that Pass in the Night: Tacit Knowledge in Psychology and Sociology" *Philosophia Scientiæ* Vol 17, No. 3, 2013 pp., 135–154 & Collins, Harry M. *Tacit and Explicit Knowledge* (U.S.A: University of Chicago Press, 2010).

Ingold, T. (2013). *Making: Anthropology, archaeology, art and architecture*. New York: Routledge.
Layton, E. T., Jr. (1991). A historical definition of engineering. In P. T. Durbin (Ed.), *Critical perspectives on nonacademic science and engineering, Research in technology studies* (Vol. 4, pp. 60–80). Cranbury: Associated University Press.
Lev Zetlin Associates. (1978). "Report of the engineering investigation concerning the causes of the collapse of the Hartford Coliseum Space Truss Roof on January 18, 1978".
Levy, M., & Salvadori, M. (1992). *Why buildings fall down: How structures fail*. New York: W.W Norton & Co.
Love, T. (2002). Constructing a coherent cross disciplinary body of theory about designing and designs: Some philosophical issues. *Design Studies, 23*, 345–361.
McCarthy, N. (2010). A world of things not facts. In I. van de Poel & D. E. Goldberg (Eds.), *Philosophy and engineering: an emerging agenda*. Dordrecht: Springer.
Moore, W. (1970). *The Professions: Roles and Rules*. New York: Russell Sage Foundation.
Nightingale, P. (2009). Tacit knowledge and engineering design. In A. Meijers (Ed.), *Handbook of the philosophy of science, Philosophy of technology and engineering sciences* (pp. 351–375). Amsterdam: Elsevier.
Olesko, K. M. (1993). Tacit knowledge and school formation Osiris, 2nd series, Vol. 8, pp. 16–29
Pahl, G., Beitz, W., Feldhusen, J., Grote, K.-H. (2007). Engineering design: A systematic approach 3rd ed., K. Wallace & L. Blessing, trans, London: Springer.
Pallasmaa, J. (2009). *The thinking hand: Existential and embodied wisdom in architecture*. New York: Wiley.
Park, G.-J. (2007). *Analytic methods for design practice*. London: Springer-Verlag.
Petroski, H. (1994). *Design paradigms: Case histories of error and judgment in engineering*. New York: Cambridge University Press.
Polanyi, M. (2005). *Personal knowledge: Towards a post-critical philosophy*. London: Routledge.
Polanyi, M., & Prosch, H. (1975). *Meaning*. Chicago: University of Chicago Press.
Rittel, H. W. J., & Webber, M. M. (1973). Dilemmas in a general theory of planning. *Policy Sciences, 4*, 155–169.
Rosenberg, N. (1982). *Inside the black box: technology and economics*. Cambridge, MA: Cambridge University Press.
Schön, D. A. (1983). *The reflective practitioner: how professionals think in action*. New York: Basic Books.
Simon, H. A. (1996). *The sciences of the artificial* (3rd ed.). Cambridge: MIT Press.
Trevelyan, J. (2014). Towards a theoretical framework for engineering practice. In B. Williams, J. Figueiredo, & J. Trevelyan (Eds.), *Engineering practice in a global context: understanding the technical and the social* (pp. 60–80). London: CRC Press.
Vermaas, P. E., Kroes, P., Light, A., & Moore, S. A. (2008). Design in Engineering and Architecture: Towards and integrated philosophical understanding. In P. E. Vermaas, P. Kroes, A. Light, & S. A. Moore (Eds.), *Philosophy and design: From engineering to architecture*. Dordrecht: Springer.
Vincenti, W. G. (1990). *What engineers know and how they know it: Analytical studies from aeronautical history*. Baltimore: John Hopkins University Press.

Chapter 5
The Engineering Knowledge Research Program

Terry Bristol

Abstract The engineering knowledge research program is part of a larger effort to articulate a philosophy of engineering and an engineering worldview. Engineering is not 'merely' applied science. Engineering knowledge requires a more comprehensive conceptual framework than scientific knowledge.

The paradigm shift from the limited scientific epistemology to the broader engineering epistemology involves a problem shift. The scientific inquirer seeks the 'facts' of 'objective' reality – out there. The participant engineer, embodied in reality, seeks 'methods' regarding how to work in the world to bring about a more desirable future.

The demonstration of complementarity and uncertainty in the new physics established the limits of the classical scientific concept of detached, observer-independent inquiry and knowledge. Quantum uncertainty entails that the inquirer is an active participant in the irreversible historical development of reality. Participant engineering inquiry and knowledge are concerned with 'practical reason'. As Kant pointed out practical reason is about 'how we should live' – the defining question of morality. Participant engineering research and development is concerned with 'the construction of the good.'

In the design of the modern k12 STEM curriculum it is crucial that developers recognize that the engineering representation of inquiry and knowledge requires a broader conceptual framework than is possible in the scientific representation. Efforts to integrate engineering into liberal arts and sciences colleges have stalled due to a failure to properly understanding of the relationship between engineering and science.

The scientific representation of inquiry isn't self-referentially coherent within the deterministic scientific worldview. The engineering representation of embodied inquiry is self-referentially coherent within a progressively emerging reality.

T. Bristol (✉)
Institute for Science, Engineering and Public Policy, Portland State University, Portland, OR, USA
e-mail: bristol@isepp.org

A. Fritzsche, S. J. Oks (eds.), *The Future of Engineering*, Philosophy of Engineering and Technology 31, https://doi.org/10.1007/978-3-319-91029-1_5

Keywords Engineering knowledge · Engineering philosophy · Embodied inquiry · Constructive emergence · Moral intelligence

5.1 Introduction

At end of World War II, President Truman asked Vannevar Bush, an engineer and inventor, to clarify the relationship between science and engineering (Bush 1945). Bush proposed that science should be understood as 'pure' research, the acquisition of knowledge, and engineering as 'applied science'. Not everyone agreed.

Dr. Walter Vincenti, Stanford professor of aeronautical engineering, presented a challenge in his book, *What Engineers Know and How They Know It* (1993) (Vincenti 1993). Vincenti notes: "Modern engineers are seen as taking over their knowledge from scientists and, by some occasionally dramatic but probably intellectually uninteresting process, using this knowledge to fashion material artifacts. From this point of view, studying the epistemology of science should automatically subsume the knowledge content of engineering." He then adds: "Engineers know from experience that this view is untrue" (Vincenti 1993). Vincenti illustrates his point by noting that scientific knowledge, albeit a useful tool, doesn't tell you how to build an airplane.

The engineering knowledge research program aims to more clearly articulate an engineering theory of knowledge. The core hypothesis is that, properly understood, engineering knowledge presents a more comprehensive understanding of knowledge, one that formally subsumes and supersedes the inherently limited scientific representations.

5.2 The Relevance

Two decades ago the high school technical curriculum was focused on science and math. Science Fairs were about demonstrating scientific phenomena. More recently Science and Technology Fairs have incorporated technology and engineering design challenges. The new STEM curriculum (viz. Science, Technology, Engineering and Math), having included engineering, needs a better understanding of engineering practice and an up to date understanding of the relationship between scientific and engineering knowledge. Most faculty and curriculum designers were educated primarily within the classical, 'pure' scientific tradition. They are ill-prepared to teach the E portion of the STEM curriculum (Felder and Brent 2016).

In 2015 MIT engineering professor Louis Bucciarelli under the sponsorship of MIT and the National Academy of Engineering organized a conference to reconsider the place of engineering in higher education (Downey 2015; Bucciarelli et al. 2015). Engineering began to enter higher education in the U.S. in the second half of the nineteenth century in response to the increasing industrialization. The founding colleges of engineering (e.g. MIT, RIT, et al.) followed the European polytechnic university model. With the strong emphasis on liberal arts education in the U.S. most colleges of engineering were created as semi-autonomous adjuncts to the established university structure. In keeping with the 'pure-applied' distinction, curriculum for engineers was severely limited largely excluding key elements of the liberal arts and human sciences. Bucciarelli was proposing a new integration of engineering into the traditional liberal arts and sciences.

At the 2015 conference, aside from the unsurprising resistance of engineering deans to a possible loss of their relative financial autonomy, there was an unanticipated intellectual pushback. The engineering educators pointed out a fundamental intellectual and cultural disjunction. The representation of inquiry and knowledge in the liberal arts and sciences was 'detached' from the practical problem solving world of engineering inquiry and knowledge (Bucciarelli 2000, 2003). The scientific community tended to represent their theory of inquiry and knowledge as seeking 'knowledge for knowledge sake' with no sense of possible application or expectation of beneficial value.

What emerged was recognition of a fundamental conceptual discontinuity between the engineering tradition and the liberal arts and sciences tradition (Bucciarelli et al. 2015). For instance, the practical engineering enterprise, aside from providing tools, was viewed, in a crucial sense, as irrelevant to the scientific agenda of acquiring 'pure knowledge'. Indeed the dominant, deterministic scientific worldview seems to entail that the engineering enterprise to develop reality is either misguided, or simply illusory. Articulating the reverse position, Vincenti emphasizes that demonstrable 'engineering knowledge' (e.g. how to build an airplane) simply cannot be made sense of in the limited scientific theory of knowledge.

My takeaway from the intellectual explorations at the conference was that the need for educational reform was reciprocal. The current image of engineering in higher education definitely needs to be broadened from the 'merely applied science'. Engineers need to appreciate their central historical role in the progressive unfolding of human civilization. Reciprocally, the liberal arts and sciences need to abandon their 'ivory tower' detachment to recognize their role in the 'real world' development enterprise. Per hypothesis, a successful integration of liberal arts and sciences and engineering calls for a new, more general, understanding, one that incorporates both the limited 'mere application' and 'pure research' representations, understanding them in a larger context, in a new way.

For a more general engineering theory of knowledge to subsume (viz. incorporate) the scientific it must be able to explain all the apparent successes of the scientific representation, albeit, in a new way. The more comprehensive engineering understanding of knowledge cannot be derived from the limited scientific understanding. Per hypothesis, the transition from the classical scientific epistemology to the engineering epistemology is logically and conceptually discontinuous. What is involved is a 'revolutionary' paradigm shift, an expansive, qualitative shift to a new type of conceptual framework, to a new way of understanding knowledge.

One of the simplest and best examples of a paradigm shift is the shift from the flat Earth theory to the spherical Earth theory. From the spherical Earth perspective the successes of the earlier flat Earth way of understanding can be subsumed, can be made sense of and explained as limited special cases with the realization that human observer-actors are very small on a very large sphere. The flat Earth theory is not simply, completely false, but it does not correctly understand the reasons for its successes. In superseding the flat Earth theory the spherical Earth theory understands the successes and the failures of the flat Earth theory *in a new way*, in terms of a more sophisticated type of conceptual framework.

5.3 The Strategy: The Heuristic of Paradigm Shifts

The situation prior to a paradigm shift is that there are observed phenomena that don't make sense in terms of the current theory. You can experience phenomena that demonstrate the limits of your current way of understanding reality without being able to make sense of those phenomena. You are 'stuck'. The dilemma is that you can't simply reason from the empirical demonstrations of the inadequacies of your current understanding to a new better understanding. How then is one to proceed? How might one come to discover a new, more general, paradigm?

I will briefly review two ways in which the limits of the scientific representation of inquiry, knowledge and reality were demonstrated in the twentieth century. It will not be possible to reason from these demonstrations to the new participant engineering paradigm. Nonetheless, it appears to me that these demonstrations offer 'clues to the new direction'.

5.4 The Limits of Scientific the Representation of Knowledge

There were two parallel paths in the twentieth century to the recognition of the inherent inadequacies of the scientific representation inquiry and knowledge – to the representation of inquiry and knowledge as scientific. One arose unexpectedly in the new philosophy of science and the other, in a flare of enigmas and paradoxes in new physics.

In the philosophy of science the dominant logical empiricist school argued that scientific inquiry advanced systematically, in particular, by means of one universal, logico-mathematical scientific method. This position 'stood to reason' if one accepted that objective reality was governed by ONE universal mechanical order. Galileo, one of the main developers of the scientific worldview strongly promoted the idea that the language of the universe was logico-mathematical.

Genuine scientific knowledge was characterized by its being repeatable over changes in time and location. For example, I can repeat Galileo's experiments at Pisa in the seventeenth century in Oregon in 2017. Generalizing, if the ultimate complete scientific understanding of reality is consistently repeatable over changes in time and location then all the phenomena in the universe must be governed by ONE universal (viz. time-space invariant), causal mechanical order (viz. he scientific hypothesis).

Thomas Kuhn, Paul Feyerabend, Imre Lakatos and Karl Popper rebelled against the logical empiricist's representation of both inquiry and knowledge. In his famous book, *The Structure of Scientific Revolutions* (1962), Kuhn argued that advances in knowledge were not systematic, but 'revolutionary', logico-mathematically discontinuous (Kuhn 1962). Advances in knowledge involved qualitatively new ways of understanding. Feyerabend and Lakatos were rigorous in showing that you couldn't reason from *within* the prior conceptual framework to the more advanced, more comprehensive, conceptual framework (Lakatos and Feyerabend 1991). A good analogy is that you can't logico-mathematically reason from an earlier invention to the next new more advanced invention. Numerous historical case studies clearly demonstrated that the historical advances in knowledge did not proceed, indeed, could not have proceeded logico-mathematically. These studies established the inadequacy of the logical empiricist's representation of inquiry and knowledge.

Another recognition of the limits of the scientific epistemology developed gradually through Popper's insistence that all meaningful 'scientific' theories must be falsifiable (Popper 2002). Popper was bothered by the seemingly endless defenses of both pseudo-scientific and scientific theories by 'after-the-fact' excuses for why the theory had 'only apparently' failed a test. What I refer to as Popper's Question challenges the proponents of any theory to specify ahead of time, before testing: what evidence if it were to occur would lead them to abandon their core hypothesis? The surprising realization was that neither pseudo-scientists nor scientists could answer Popper's quite reasonable sounding question. The unexpected lesson was that the conceptual machinery of a theory is always inadequate to describe a phenomenon that would falsify it (Bristol 2016).

What Popper's Question revealed was that the falsifying evidence must be 'conceptually outside' the theory being questioned. The falsifying evidence must be something that, by its very nature, cannot be made sense of in terms of the conceptual framework of the proponent's theory. The evidence, demonstrating an inherent incompleteness, must be qualitatively different from anything that can possibly be made sense of or explained in terms of the proponent's theory.

The scientific conception of objective reality is that there is ONE right way to understand reality. However, falsificationism insists that there must always be more than one successful way to understand reality. Popper's Question led us to recognize that for any meaningful theory there must be another successful, opposite *type* of theory. A simple way to express this is to say that reality is 'more ample' than can be captured by any one way of understanding. Formally, there must be another successful, conceptually discontinuous, complementary theory (Bristol 2016). The structure of human knowledge could no longer be understood in terms of the logical empiricist's representation. The body of meaningful, falsifiable knowledge could never, even in principle, be understood as ONE unified, conceptually coherent, logico-mathematical consistent, theory of everything.

With this philosophical background it is fairly straightforward to layout how, in the first half of the twentieth century, the recognition of the limits of the representation of knowledge as scientific arose in the new physics. In brief it was the empirical demonstrations of quantum theory and relativity that undermined the classical scientific, mechanical representation of knowledge of reality. It wasn't, as one might have expected, one simple experiment that conflicted with the presuppositions of classical physics. What happened was appropriately mysterious and enigmatic. Quantum theory demonstrated that there was more than one type of phenomena, more than one type of order and consequently more than one type of knowledge of reality. In a curious sense one might argue, at least initially, that science had too much success – too many successful theories. The problem is that these successful theories don't fit together, don't make sense in terms of each other. They are incommensurable. There is no conceptual common denominator.

Particles are the definitive phenomena of Newtonian physics and waves are the definitive phenomena of Maxwellian physics. Particle phenomena and wave phenomena are co-defined oppositely. Particles are local in space and time. Waves (fields) are non-local, distributed in space and time. A particle is a non-wave and a wave (field) is a non-particle. Particle phenomena and wave phenomena are formally complementary. In the new physics it was recognized that these two entire highly successful research programs, Newtonian physics and Maxwellian physics, were conceptually complementary (Bristol 2016).

If knowledge of reality is not conceptually unified as expected by the scientific epistemology, how are we to understand reality, in the new way? Einstein captured the dilemma succinctly: "[Traditionally] Physics is an attempt conceptually to grasp reality as it is thought independently of its being observed. In this sense one speaks of 'physical reality.' In pre-quantum physics there was no doubt as to how this was to be understood. In Newton's theory reality was determined by a material point [completely localized] in space and time; in Maxwell's theory, by the field [completely distributed] in space and time. In quantum mechanics it is not so easily seen" (Schilpp 1951).

The new physics demonstrated the inadequacy, the limits, of the scientific representation of knowledge of reality. Richard Feynman put it bluntly: "I think I can safely say that nobody understands quantum mechanics" (Feynmann 1964). In parallel, the new physics demonstrated the limits of the scientific worldview, the scientific understanding reality, the understanding of reality as scientific.

However, realizing that these highly successful complementary physical theories undermine the classical Scientific Hypothesis that all phenomena are govern by ONE universal order doesn't by itself automatically generate a new better theory of knowledge and understanding of reality and our place in it.

Quantum cosmologist Lee Smolin reflected that when his generation entered physics in the 1960s they were excited, hoping to resolve the question of the nature of quantum reality, a question, they felt, had been left in confusion by the founders. Smolin recently told a group of incoming graduate students: "It is now 2009 and it has become rather Kafkaesque that we have made no progress whatsoever" (Smolin 2009).

5.5 Clues to the New Direction

The curious and often enigmatic ways in which the limits of the scientific representations inquiry, knowledge and reality were recognized provide us with what I like to refer to as 'clues to the new direction'. The characteristics of the demonstrations seem to outline features of and constraints on any possible superseding paradigm.

There are at least these three key 'clues' from the new philosophy of science and the new physics that point toward an engineering theory of knowledge.

First, since complementarity entails that the universe is not governed by ONE time-space invariant order, if you choose to observe in one way you experience one type of order, if you choose to observe in the opposite way you experience the opposite type of order. The observer is necessarily continually choosing.[1]

Second, in order for the quantum choice, that 'collapses the possibility space (wave function potential)', to occur there must be a chooser, a participant actor-observer embodied in reality.

Third, the quantum choice actualizes a historically unique, mechanically irreversible future. The evolution of the universe is mechanically discontinuous and historically cumulative.

5.6 The Initiative from the Engineering Community

In addition to the developments in the new philosophy of science and the new physics, the third twentieth century initiative calling for a new theory of knowledge arose independently from within the engineering community. That the new engineering paradigm didn't arise from within the scientific community should not be surprising.

Vincenti's challenge was born of a frustration over the general failure of the scientific and philosophy of science communities to appreciate that engineering knowledge was different, indeed, that engineering knowledge was something more. Vincenti also found the scientific ontology of particles and waves to be inadequate to account for the engineering world of airplanes and cell phones. Vincenti reasons simply that if the (creative) products of engineering are not 'merely' the result of 'applied science' then the scientific approach to understanding the engineering ontology must be inherently inadequate. At best scientific approach can provide only limited insights as to how the engineering world works.

In the twenty-first century engineering community there has been a developing self-examination of the place of the engineer in reality (Petroski 2010). In his recent

[1] At a 2013 meeting of the American Physical Society John Heilbron, Berkeley professor of history of science, reexamined the broader worldview of quantum physicist Niels Bohr: "The primary payoff of his engagement with quantum physics for his wider philosophy was the discovery that multiple truths come… in complementary pairs" (Heilbron 2013).

book, *The Essential Engineer: Why Science Alone Will Not Solve Our Global Problems* (2010), Duke University engineer professor Henry Petroski offers a bold expansion of Vincenti's themes. According to Petroski, what we have been characterizing as scientific inquiry and knowledge can only be properly understood, in a new way, as a subroutine within the more comprehensive engineering knowledge research, development and deployment enterprise.

For Petroski it isn't a matter of science versus engineering. There aren't scientists and engineers. Petroski's engineering way of understanding doesn't reject or simply replace the scientific view. Petroski's engineering way of understanding inquiry and knowledge incorporates the incomplete scientific idealizations, albeit understanding them in a new way. All meaningful inquiry and all meaningful knowledge can only make sense in the practical engineering framework. All meaningful inquirers must be engineering inquirers.

I recall my own moment of revelation. I had casually imagined that I was a scientist seeking the universal laws governing objective reality. Then I realized that, since the new physics established that there were no universal laws and that 'objective' reality was an idealization, I couldn't have been doing what I imagined myself to have been doing. What many scientists thought they were doing, on reflection and closer examination, never really made sense. For instance, if all human actions are fully determined the 'scientific inquirer' is not 'actually' in a position to learn, to freely design and run experiments to test and potentially falsify his theories. To put it succinctly the scientific representation of inquiry is not self-referentially coherent.

In Petroski's representation of engineering, we learn by a problematic process that is always somewhat blind and uncertain. Learning always involves experimentation and exploration. Nobel laureate economist Herbert Simon characterized engineering inquiry and problem solving as the attempt to move from a current state of affairs to a more desirable future state of affairs (Simon 1981/1996). Petroski emphasizes the uncertainty. We learn by doing, by trying. We learn from our failures. Advances in engineering knowledge are recursively enabling, opening qualitatively new uncertainties, qualitatively new opportunities for novel inquiry and experimental explorations.

5.7 The Paradigm Shift Is a Problem Shift

One unexpected lesson of Kuhn's rigorous investigation of the real history of advances in knowledge was that even when the inadequacies of an otherwise successful theory might be clear to both its proponents and critics a new better theory doesn't just automatically appear. Major advances are conceptually revolutionary and cannot be derived or otherwise reasoned from within the current theory even with the help of considerable demonstrated counter-evidence. Proponents and critics may be frustratingly 'stuck' for long periods unable to abandon the current, reasonably successful theory. Even when the new advanced framework begins to appear the transition to the new way of understanding is often difficult and gradual.

I have tried to represent each of the three twentieth-twenty-first century lines of critical reasoning as being 'stuck'. First, Popper, Lakatos, Feyerabend and Kuhn detailed the inadequacies of the logical empiricist representations of the history and philosophy of science. Second, Bohr, Heisenberg (Heisenberg 2007), Planck and de Broglie (De Broglie 1958), argued that the empirical demonstrations of quantum theory established the inherent inadequacies of the classical scientific representation of knowledge and the scientific understanding of reality. Third, Vincenti's challenge details the fundamental inability of the scientific theory of knowledge to make sense of engineering knowledge. At best each of these three lineages provided only clues to the revolutionary paradigm shift needed to bring forth the new better understanding of knowledge, the new better understanding of reality and our place in it.

American pragmatist John Dewey offered a helpful characterization of what I will defend as the proper frameworks involved in the paradigm shift (Dewey 1929/1980). Dewey noted two representations of inquiry and knowledge, which he referred to as the spectator and the participant. The spectator representation corresponds to the classical scientific portrayal. The spectator is a detached inquirer seeking the universal order (laws) governing objective reality – 'out there'. The spectator's initial ignorance declines as knowledge advances. Successful inquiry is a convergence toward a complete and consistent scientific knowledge of objective reality. The spectator's criterion of genuine scientific knowledge is that it is 'repeatable over changes in time and location'. Consequently the overall order governing all phenomena must be time-space invariant. A crucial feature of the spectator representation is that the actions of the inquirer must not interfere with the natural, 'objective' course of events or alter the structure of reality. If the inquirer interferes with the natural course of events while investigating he would be unable to separate the effects of his interference from the natural, inquirer-independent 'objective' course from events. The opportunity would be lost to converge to an inquirer-independent objective reality.

In the participant representation the inquirer is embodied in reality, a natural component of reality. The participant inquirer naturally alters on the course of events. Participant inquiry doesn't converge to full knowledge of a fixed, inquirer-independent 'objective' reality. When the participant learns reality develops. The participant's knowledge is not *about* reality like some sort of detached scientific description. Participant knowledge is something that develops *within* reality, an irreducible aspect of the nature of a developing reality.

Quantum theory requires a participant framework. Reality does not actualize without a participant observer. The embodied observer's choice (action) selectively brings forth one unique historical path from among the possibilities. The participant's choices are mechanically discontinuous, irreversible and historically cumulative. The essentially 'arbitrary' (viz. symmetry-breaking) character of the choice cannot possibly, by its very nature, be made sense of within any possible mechanical (viz. symmetric, time-space invariant) scientific worldview. Quantum physics provides a clue to 'the participant nature' of the new knowledge paradigm.

Perhaps the single most important implication of the paradigm shift from a spectator to a participant representation is that the transition involves a problem shift. For the spectator the problem of inquiry is to gain knowledge of 'how the universe [out there] works'. Einstein states the classical view plainly: "Physics is an attempt conceptually to grasp reality as it is thought independently of its being observed. In this sense one speaks of 'physical reality'" (Schilpp 1951).

For the participant, embodied in reality, the problem of inquiry is something like 'how to work in the world'. Participant engineering inquiry is not undirected and, not concerned merely with acquiring instrumental knowledge of 'how to work in the world'. Engineering inquiry has an agenda focused on selectively acquiring 'important knowledge' that empowers the engineer to bring about a more desirable future. William James, in his 1906–07 Lowell Lectures commenting on what I believe to be an earlier attempt to articulate an engineering theory of inquiry and knowledge, referred to 'pragmatism as a new name for an old way of thinking' (James 2015).

The spectator's answers to scientific questions are 'facts', about what is fixed, regular, and repeatable. In contrast, answers to the participant engineer's 'how to work in the world' questions are 'methods'. The 'methods' are about how to do things, about how to do things that bring forth a more desirable future. The participant problem of inquiry incorporates the limited scientific representation of the problem of inquiry (viz. the inductive sciences). Learning the current 'facts' about the relatively fixed and regular, is a step, a limited subroutine in the broader participant agenda to develop reality.

In the participant's framework what had been represented as fixed scientific 'facts' are newly understood. All facts (regularities, repeatables) and their associated theories are falsifiable, stable only within certain boundary conditions. Knowledge of the relatively stable facts is preliminary to developing them.

Whether the electron is a Newtonian-like particle or a Maxwellian-like wave is undecidable when posed as an objective question. Whether reality is completely and consistently particle-like or wave-like is undecidable. Participant inquiry is always facing an indeterminate reality requiring a 'free' choice of how to engage. In the participant understanding, the nature of reality is unfolding historically in a mechanically discontinuous (symmetry-breaking) manner.

As a thought experiment it is helpful to try to distinguish two stages of participant learning. In the first stage the participant inquirer can learn *about* how the currently evolved engineered world around him works. To learn how the world works in this engineering sense is to learn, for instance how the current socio-economic system works, or how to fly an airplane. Engineering reality is composed of dynamic embodied methods (viz. engineering and engineered processes). The participant himself is a dynamic embodied method in the world of dynamic embodied methods.

In the second idealized stage of learning the participant learns how to develop reality. When he learns a new advanced method he learns how to work in the world in a new way to bring forth a more desirable future. Implementing a new engineering process alters the participant's relationships *within* reality. Per hypothesis, acquisition of engineering knowledge involves the formation of novel synergistic relationships. The picture of an engineering ontology composed of dynamic embodied methodological learning processes suggests a process reality, reminiscent of Whitehead's Process Philosophy (Whitehead 1969).

In developing the broader philosophy of engineering research program I had expected to be able to keep questions concerning engineering knowledge (viz. epistemology) separate from questions concerning the composition of engineering reality (viz. ontology). With the recognition that the engineering inquirer must be a participant, must be an irreducible aspect of the composition of reality, these two lines of research, the epistemic and the ontic merge. If, as expected, the participant framework naturally embraces quantum complementarity, engineering reality can no longer be completely and consistently characterized in only one 'objective' way. Engineering knowledge is not 'conceptually unified' in the time-space invariant manner envisioned in the scientific characterization. There is not, cannot be, just one 'everywhere and always' solution of how to bring forth a more desirable future. The interrelationships of the embodied processes are not like in a Newtonian clockwork, they develop. Plausibly, the interrelationships are better understood along the lines of the new post-scientific ecological paradigm (Ulanowicz 2009). One might imagine the biosphere developing through an inherently symbiotic, cumulative, emergent, synergistic learning process. Despite this inevitable epistemic-ontic merger, in this essay I will continue to focus on the epistemological, on the nature and characteristics of engineering knowledge.

Engineering knowledge must reflect complementarity. There should always be opposite and alternative ways to bring about a more desirable future. Equivalent there should always be complementary research programs, complementary ways of learning. The overall architectonic of engineering knowledge must incorporate and reflect complementarity. Consequently, engineering knowledge cannot be represented as a consistent increase in only one conceptual type of knowledge, in only one type of research program. Advances in engineering knowledge involve a qualitative (viz. heterogeneous) expansion of *types* of methods, of *types* of capabilities, of *types* of potential. Thought of as creative inventions (e.g. how to build an airplane; aerodynamics), advances in engineering knowledge are, by their very nature, surprising, unexpected and unpredictable. Later inventions are conceptually advanced. They are not derivations from earlier inventions. They are not logical deductions from earlier engineering knowledge. How a new invention works cannot be understood in terms of the limited prior knowledge of how earlier inventions worked.

The question of how these new methodological inventions are discovered, how they become manifest, how they are developed and become actualized in reality, and how reality transforms as a result, is at the leading edge of research in the philosophy of engineering.

5.8 The Participant Inquirer as a Natural Inquirer

The participant engineer is an embodied actor and a natural inquirer within reality. The uncertainty of the new physics is embraced in the participant engineering framework, but newly understood. The engineer is always facing uncertainty and, consequently, of necessity, he is continually choosing. In quantum theory the participant's choice is referred to as 'arbitrary', reflecting that it cannot be made sense of within any possible scientific framework. In the detached scientific theory of knowledge the choice is concerned with 'what to believe about the objective world – out there'. In the engineering theory of knowledge the participant's choice is concerned with how to act in the real world.

Heisenberg realized that not only are the wave and particle phenomena complementary, but that the experimental setups used to demonstrate these phenomena must also be complementary. Taking this reasoning one step further the series of actions needed to create the alternative experimental setups must also be, in some irreducible aspect, complementary. Popper's falsificationism likewise requires that for any successful research program (viz. line of inquiry) there must be an equally successful complementary research program that can demonstrate a complementary type of phenomena. For every situation facing an engineer there must always be a range of qualitatively distinct ways to bring about a more desirable future.

The participant's choice of how to act in the world is always between (viz. in a range between) qualitatively distinct alternatives like particle and wave. All choice in the world involves such a qualitative bias. All action-choices in the world are value-laden.

The more desirable future that the participant engineer's choice of action and inquiry is hoping to bring forth, may be heavily constrained, but is also inherently under-determined. The outcome of an intended action, a doing, a method, can never be fully assured. All purported experimental repetitions actually have an irreducible uniqueness in time and place. The reasoning of the inductive sciences presupposes that repeated experimental trials are literally identical.

The quantum uncertainty newly understood in the participant framework constitutes a tacit question: how should I act? The participant is of necessity constantly choosing how to act in the world. The uncertainty is newly understood as an irreducible element of freedom in all action. The participant engineer is a natural explorer, a natural inquirer. Every action is to some irreducible extent experimental and exploratory, to some irreducible extent a questioning. Every action moves into a somewhat indeterminate future. Every action, every experimental trial, occurs at a somewhat different historical time in a somewhat different, historical circumstance. The Continental Existentialists characterized the participant (1) as instantiated in reality, (2) as having the ability to act, (3) as necessarily choosing, and yet (4) as lacking a script. There is nothing to tell him how he should act, or, with certainty, what the outcome his different actions might be. The Continental Existentialists often found these realizations of freedom to be anxiety provoking. The modern engineer understands this openness, this freedom, as representing the opportunity to be a truly creative problem solver, as representing the opportunity to bring about a more desirable future (Florman 1977).

The participant's choice doesn't decline with advances in engineering knowledge. The uncertainty simply transforms and develops. New types of questions and opportunities emerge with each advance. The participant's 'ignorance' is natural, necessary and ongoing. By contrast, in the scientific representation inquiry is convergent. As scientific knowledge advances scientific ignorance and uncertainty are supposed to decline. The engineering knowledge research enterprise is expansive, emergent and recursively enabling.

Participant engineers are 'opportunity actualizers' focused on discovering, developing and bringing forth value. What I have referred to elsewhere as Carnot's Epiphany (Bristol 2015) is that participants are engineers in a world of engineering. Alternatively expressed participants are doers in a world of doings, choosers in a world of choosing, learners in a world of learning, developers participating in the development of reality.

5.9 The Self-Referential Coherence of Engineering Inquiry and Knowledge

Because engineering knowledge is embodied in reality engineering inquiry is naturally self-reflexive. The question of the nature of reality is inseparable from the question of the nature of inquiry and knowledge. Inquiry into the nature of reality is inquiry into the nature of inquiry. Inquiry into the nature of inquiry is inquiry into the nature of reality. American Pragmatist Josiah Royce proposed what I refer to as Royce's Criterion of Self-Referential Coherence (Bristol 2016). Royce argued that whatever theory of the nature of reality that one proposes it must be able to account for itself, it must be able, self-inclusively, self-referentially, to account for itself as part of reality. Furthermore the theory must be able to provide an account of how it was learned. This seems to entail that 'learning' must be an irreducible aspect of the nature of reality. Knowledge of reality must be a self-referentially coherent aspect of reality.

Royce begins by arguing that whatever theory you propose must be able to account for the existence of problems. He referred to this as 'the problem of problems' (Royce 1965). Two fundamental types of problems concerned Royce: the problem of ignorance (viz. knowledge) and the problem of evil (viz. value). In the engineering knowledge research program with its agenda to bring forth a more desirable future these two problems merge.[2] In the scientific representation these two types of problem are essentially separate, incommensurable *types* of problems.

[2]This is the enfolding of the sciences and the humanities into the more general engineering framework.

The scientific worldview is unable to make sense of, let alone explain the origin or nature of our 'beginning ignorance'. The scientific representation of knowledge as the result of 'pure' research, as 'knowledge for knowledge sake', is unable to make sense of the potential benefit of knowledge in bringing forth a more desirable world. In a fully deterministic scientific worldview there is simply no way to make sense of inquiry. The scientist is not a natural inquirer. Inquiry is not a part of or contained in the ontology of the scientific worldview. The scientific representation of inquiry is as occurring from the position of a detached spectator. The scientific representation of inquiry and knowledge cannot make sense of itself. It is not self-referentially coherent. In contrast, engineering inquiry and knowledge are natural coherent aspects of engineering reality. The engineer is a natural inquirer. The engineering knowledge framework presented in this essay is able, self-referentially, to make sense of itself, thereby satisfying the test of Royce's Criterion of Self-Referential Coherence.

A second self-reflexive characteristic that differentiates engineering and scientific representations is that in the engineering process when we learn we also learn how to learn. When we acquire engineering knowledge we also acquire engineering knowledge of how to improve our methods of acquiring engineering knowledge. With advances in engineering knowledge we learn how to inquire in qualitatively new, better ways. We become better at how, when, where and about what to inquire in order to discover and develop increasingly valuable knowledge.

The logical empiricists assumed that there was just one universal, unchanging 'scientific method'. Their method paralleled what they supposed was the time-space invariant mechanical order governing all phenomena. It 'stood to reason', that there could be only one, correspondingly unchanging, logico-mathematical scientific method. In the engineering knowledge research program the method of inquiry itself develops and improves. In parallel, 'it stands to reason' that if embodied engineering knowledge and method of inquiry develop then reality, understood as a learning system, must also be developing. Engineering intelligence, in this sense, emerges recursively.

In his book, *What Technology Wants* (2010), Kevin Kelly, founder of *Wired Magazine*, emphasizes the meta-reflexive theme that when we learn we also concomitantly learn how to learn – better. Kelly argues that, "Not only is the aggregate process of evolution evolving, but it is evolving more ability to evolve, or greater evolvability… The evolution of evolution is change squared" (Kelly 2011).

5.10 Modern Examples of the Paradigm Shift

The paradigm shift from the scientific worldview to the engineering worldview, from the limited scientific representation of knowledge and reality to the engineering representation knowledge and reality is already well underway in a number of traditional disciplines.

University of Victoria biologist Robert G.B. Reid proposes a participant engineering view of biological evolution in his book, *Biological Emergences: evolution by natural experiment* (2007) (Reid 2007). Reid understands organisms as natural inquirers, as creative agents characteristically exploring and experimenting with new ways interrelating. Reid maintains that ecosystems are natural engineering enterprises, naturally seeking new better ways of developing the world, naturally seeking win-win symbiotic interrelationships that recursively increase the capacity of the ecosystem to perform engineering work.

Reid offers a new way of understanding Darwinian 'random mutation'. In the Darwinian framework there is a tacit presupposition that when organisms are reproducing they are attempting to produce identical copies of themselves. Variation can only arise 'accidentally' from 'copying errors'. The natural expectation of the scientific worldview, of Newton's clockwork reality, is for a conservative mechanical Steady State system that does not 'evolve'. Darwin's nineteenth century introduction of 'random errors' was scientifically enigmatic. By contrast Reid's engineering model incorporates uncertainty, newly understood from an engineering perspective. Biological organisms are actor-agents in an uncertain world, naturally generating experimental, exploratory variations (viz. potentially valuable inventions). In the participant engineering learning process variation is natural and expected.

In Darwinian evolution populations are supposed to be 'adapting', a sort of convergence, to their local fixed environment.[3] By contrast Reid observes that as populations acquire emergent engineering knowledge they gain increasing 'adapt-ability'. They acquire a qualitative, broadening capacity to live and work in more diverse environments. Life isn't adapting *to*, it is constructively emerging *from*. As the biosphere engineers its emergence it develops an increasing ability to perform (engineering) work.

Ecologist Robert Ulanowicz in his book, *A Third Window: Natural Life beyond Newton and Darwin* (2009), argues in concert with the themes of this essay for a post-scientific understanding of the emergence of the biosphere (Ulanowicz 2009). Ecologist Eric Schneider and Dorion Sagan, in their book, *Into the Cool: Energy Flow, Thermodynamics and Life*, represent the biosphere as an engine (viz. an agent able to perform engineering work). The engine of the biosphere evolves in a way that is analogous to the historical evolution of steam engines through successive (critically recursive) engineering design improvements (Schneider and Sagan 2005). As with Reid, they argue that with these improvements the biosphere increases its capacity to perform work (both qualitatively and quantitatively).

In support of the participant engineering way of understanding reality, master engineer George Bugliarello argued that modern engineers should be taught that what they are doing is a natural extension of biological evolution (Bugliarello 2003). The tacit implication is that biological evolution, the evolution of the biosphere leading to human civilization is the result of a recursively enabling engineering

[3]As with the inability of scientific models of inquiry to explain our 'beginning ignorance', the Darwinian model is unable to explain how it is that organisms (life), as part of a mechanically deterministic reality, are initially non-adapted.

enterprise. Organisms and ecological systems are themselves naturally inquiring, progressively developing their embodied methodologies.

There has been a paradigm shift in economics that parallel the paradigm shift in the understanding of the evolution of biological systems. From the classical scientific equilibrium economics of the last century and half, the new participant engineering approach is referred to as New Growth Economics (Warsh 2006). In scientific equilibrium economics supply and demand always work toward a fixed, Steady State equilibrium. If the economy develops there is no way to explain it scientifically in terms of the naturally conservative (symmetric) economic processes. Despite the enormous observed impact on economic growth of engineering advances (viz. tools and rules) these engineering advances are scientifically unpredictable. Engineering inventions cannot be made sense of, cannot be explained as the result of zero-sum equilibrium seeking supply-demand mechanics. Consequently, for most of the twentieth century economics engineering advances were referred to as due to unpredictable (viz. 'arbitrary'), 'exogenous' influences. In 1990, Stanford economist Paul Romer published an article entitled "Endogenous Technological Change" (Romer 1990) where he argued for a new post-scientific understanding of economic systems. Romer argued that economic systems are engineering enterprises, naturally net productive (viz. not zero-sum; non-conservative), naturally seeking to learn qualitatively new, better ways of doing things. Economic systems are, by their very nature, engaged in seeking, developing and utilizing engineering knowledge. By analogy with Kuhn's reconsideration of the history of science, Romer proposes to understand the historical advance the civilization as, in large measure, the result of engineering design advances. Real, meaningful economic actors and systems are historically emergent engineering knowledge research and development enterprises. For Romer, the economic engineering enterprise is recursively enabling and self-reflexively accelerating. New Growth Economics is congruent with Reid's biological emergence theme. Both maintain that advances in engineering knowledge increase and broaden reality's capacity to perform engineering work.

5.11 Knowledge and Value

The second most striking feature of the paradigm shift from a scientific spectator to a participant engineering theory of knowledge is the reintroduction of value into epistemology. Scientific inquiry was represented as seeking an observer-independent, value-free, objective description and explanation of how the supposed mechanical world works. Inside the reality of the deterministic scientific worldview there is no intrinsic value or meaning. Scientific knowledge is for knowledge sake.

Unlike the supposed scientific search for a complete and consistent understanding of reality, the engineering enterprise is not seeking all possible knowledge, all possible methods for working in the world. In engineering research and development, Royce's two types of problem ignorance (viz. knowledge) and evil (viz. value) merge. The engineering enterprise is not seeking just any knowledge. The engineering knowledge research and development enterprise seeks to discover and develop methods that serve to bring forth a more desirable future. The enterprise seeks methods that are able to recursively enable the emergence of the value of reality.

It is crucial to appreciate that the engineering enterprise doesn't begin with any understanding of what is valuable, doesn't begin with any pre-conception of what a more desirable future should look like. The question of how to bring something about (viz. instrumentally) and the question of what is of value, what is better, what is more desirable, have merged. Always to some irreducible extent the engineering enterprise is blind (viz. uncertain) concerning how to bring about a more desirable future. And to emphasize, it isn't simply that the engineering enterprise doesn't know how (viz. instrumentally) to bring about a more desirable future, the concomitant problem is that it doesn't know ahead of time what a more desirable future is, what it should look like. What constitutes the more desirable future cannot, by its very nature, be pre-conceived. This can be expressed by saying that the 'plan' of the engineering enterprise is itself qualitatively emergent. The more advanced opportunities of the plan emerge unexpected ways. The plan unfolds as it is actualized.

If the engineering enterprise had a firm pre-conception of value, of 'the good', at the onset of research, then the sought-after knowledge would be simply instrumental, about how to bring about what we know ahead of time to be valuable. But value, by its very nature, *emerges* qualitatively in a logically and mechanically discontinuous manner. The more valuable invention, the more valuable method cannot be predicted, cannot be derived from prior understanding of what has been valuable. Consequently the engineering knowledge research and development enterprise can only learn through experimentation and exploration, by trying and learning from failures.[4] Our understanding of what is valuable does advance, but can never fully prescribe the more desirable path to the future. Creative engineering advances are conceptually revolutionary. Advances in our understanding of what is more valuable, of what is more desirable, are revolutionary inventive discoveries.

In his *Critique of Practical Reason* (1788, 2004) Immanuel Kant pointed out that the engineering choices of practical reason are concerned with how we should live (Kant 2004). Kant then notes that the question of 'how should we live' is the defining question of morality. Accordingly the context of engineering research and development is concerned with an advancing morality. But what is morality? What might a more moral future look like? (Fesmire 2003) Kant argues that 'the good' cannot be 'conceived', cannot be pre-conceived, pre-specified. The engineering knowledge research, development and deployment program is properly understood as naturally concerned with the unfolding of a more desirable moral reality. Recognition that the engineering enterprise is concerned with the progressive manifestations of moral value suggests that the emergence of reality is driven by an embodied recursively unfolding moral intelligence.

[4]"... the purification of the motive in the ground of our beseeching." Little Gidding, Four Quartets, T.S. Eliot.

American Pragmatist John Dewey characterized the embodied engineering knowledge research program as 'the construction of the good' (Dewey 1929/1960).

The specific nature of the goal and the specific path isn't pre-conceivable. The engineering isn't teleological because the 'goal' is always changing, developing in unexpected ways. As Lao Tzu expressed it in the ancient *Tao Te Chung*, "The path that can be spoken [pre-conceived] is not the [real] eternal path" (Tzu 2016). The path forward is always uncertain, is always in doubt. One ideology will advise you that whenever in doubt always turn to the right. Another ideology will advise you that whenever in doubt always turn to the left. Each of these pre-specifiable iterations spirals in on itself and self-destructs. Per hypothesis, the path forward is the under-determined middle way. The creative middle way path to a more desirable future must always involve an irreducible aspect of the contrary, complementary processes and perspectives (Bristol 2016).

In 1893, Pragmatist Charles Sanders Peirce, citing numerous ancient wisdom precedents, noted that progressive philosophies and theologies from the earliest times seemed to have a common vision of the nature of our developing reality. Peirce's famous speculative essay was entitled "Evolutionary Love" (Peirce 2011).

5.12 So What?

The engineering knowledge research program lays the conceptual groundwork for Bucciarelli's proposed vision to merge the colleges of engineering with the colleges of liberal arts and sciences. Engineering, previously characterized as 'applied science', is newly understood to involve its own broader type of research, experimentation and exploration within a context of value concerns. Both the hard and human sciences, still commonly characterized as undertaking 'pure' research, detached from practical concerns, are to be newly understood as part of 'engineering research', as subroutines within the more comprehensive engineering research and development enterprise. In the post-mechanical engineering understanding of the emergence of civilization the humanities, history and philosophy in particular, are 'rediscovered' as essential components of the self-reflexive engineering learning process. This broadened understanding of the engineering enterprise naturally re-integrates the sciences and the humanities. Similarly, the arts and aesthetics are newly understood as essential components in the development of our understanding of value, of our understanding of ourselves, of our understanding of the more desirable future.

In short, the mature, broadened understanding of the engineering knowledge research and development enterprise presents a more general, more advanced framework for a fully integrated, practically engaged higher education.

Per hypothesis, the new improved understanding of the engineering enterprise naturally provides a critical perspective on the developing design of k12 STEM curriculum. First of all it clarifies and critiques the superficial conflict between the traditional images of 'pure research' and 'applied science'. It should be a point of emphasis within the STEM curriculum that the deterministic scientific worldview is not self-referentially coherent. Engineering provides the more general self-referentially coherent framework for understanding inquiry and action. Similarly, the inadequacies of the representation of mathematics as conceptually complete and consistent should be an essential part of the framework of the mathematics curriculum. Per hypothesis, novel mathematical conceptions (e.g. turbulence) emerge with engineering advances. The role of technology in society, investigated both prospectively and retrospectively, should be understood and evaluated within both the historical and modern existential, opportunistic engineering contexts.

Particularly difficult, but immensely important, for the developing STEM curriculum will be incorporation of the understanding of biological evolution and economic development as engineering enterprises. As George Bugliarello emphasized, engineers should be taught to understand themselves and their work as natural extensions of biological evolution. The modern engineer needs to understand his creative and value actualizing role as meaningful in recursively developing moral reality.

References

Bristol, T. (2015). *What is engineering? What is the value framework of engineering practice?.* Linus Pauling Memorial Lecture Series. Video. https://youtu.be/08kfPUEnpGM

Bristol, T. (2016). *Give space my love: An intellectual odyssey with Dr. Stephen Hawking*. Institute for Science, Engineering and Public Policy, Portland.

Bucciarelli, L. (2000). *Designing engineers*. Cambridge, MA/London: The MIT Press.

Bucciarelli, L. (2003). *Engineering philosophy*. Delft: Delft University Press.

Bucciarelli, L., Drew, D., & Sheila, T. (2015). *Liberal studies in engineering – Workshop report*. http://hdl.handle.net/1721.1/96672. Accessed on 12 Mar 2015.

Bugliarello, G. (2003). *The BIOSOMA: Reflections on the synthesis of biology, society and machines*. New York: Polytechnic Press.

Bush, V. (1945). *Science, the endless frontier: A report to the President*. Ann Arbor: University of Michigan Library.

De Broglie, L. (1958). *The revolution in physics: A non-mathematical survey of quanta*. New York: Noonday Press.

Dewey, J. (1929/1960). *The quest for certainty: A study in the relation of knowledge and action*. Los Angeles: Capricorn Press.

Dewey, J. (1929/1980). *The quest for certainty: A study of the relation of knowledge and action*. New York: Perigee Books.

Downey, G. (Ed.) (2015). Special issue: Liberal studies of engineering – A design plan. *Engineering Studies*, *7*(2–3). Cambridge, MA: Taylor and Francis.

Felder, R. M., & Brent, R. (2016). *Teaching and learning STEM: A practical guide*. San Francisco: Jossey-Bass/Wiley.

Fesmire, S. (2003). *John Dewey and moral imagination: Pragmatism in ethics*. Bloomington: Indiana University Press.

Feynmann, R. (1964). *The character of physical law* (Chapter 6, p. 129). Cambridge, MA: MIT Press.

Florman, S. (1977). *The existential pleasures of engineering*. New York: St Martin's Press.

Heilbron, J. (2013). *Bohr's creation of his quantum atom*, (video) http://meetings.aps.org/link/BAPS.2013.APR.P1.1. See also Heilbron, J., & Aaserud, F. (2013). *Love, literature and the quantum atom: Niels Bohr's 1913 trilogy revisited*. Oxford: Oxford University Press.
Heisenberg, W. (2007). *Physics and philosophy: The revolution in modern science* (Harper perennial modern classics). New York: Harper Collins.
James, W. (2015). *Pragmatism: A new name for some old ways of thinking*. NewYork: CreateSpace Independent Publishing Platform.
Kant, I. (2004). *The critique of practical reason*. Mineola: Dover Publications.
Kelly, K. (2011). *What technology wants* (pp. 341–342). London: Penguin.
Kuhn, T. S. (1962). *The structure of scientific revolutions*. Chicago/London: Phoenix Books, University of Chicago Press.
Lakatos, I., & Feyerabend, P. (1991). *For and against method: Including Lakatos's lectures on scientific method and the Feyerabend-Lakatos correspondence*. Chicago: University Of Chicago Press.
Peirce, C. S. (2011). *The philosophical writings of Peirce*. Mineola: Dover Publications.
Petroski, H. (2010). *The essential engineer: Why science alone will not solve our global problems*. New York: Vintage, Random House.
Popper, K. (2002). *Conjectures and refutations: The growth of scientific knowledge*. London: Routledge Classics.
Reid, R. G. B. (2007). *Biological emergences: Evolution by natural experiment*. Cambridge MA: The MIT Press.
Romer, P. (1990). Endogenous technological change. *Journal of Political Economy, 98* (5), Part 2: The problem of development: A conference of the institute for the study of free enterprise systems, pp. 71–102. Chicago: University of Chicago Press.
Royce, J. (1965). *The religious aspect of philosophy*. Gloucester: Peter Smith.
Schilpp, A. (Ed.). (1951). *Albert Einstein: Philosopher-Scientist* (Vol. 1, p. 81). New York: Tudor Publishing.
Schneider, E., & Sagan, D. (2005). *Into the cool: Energy flow, thermodynamics and life*. Chicago: University of Chicago Press.
Simon, H. (1981/1996). *The sciences of the artificial*. Cambridge, MA: MIT Press.
Smolin, L. (2009). *Informal chalk and talk*. Perimeter Institute Recorded Seminar Archive (PIRSA). http://pirsa.org/09080032/. Accessed on 15 May 2010.
Tzu, L. (2016). *Tao Te Chung* (S. Mitchell, Trans.). Hollywood: Simon & Brown.
Ulanowicz, R. (2009). *A third window: Natural life beyond Newton and Darwin*. West Conshohocken: Templeton Press.
Vincenti, W. (1993). *What engineers know and how they know it*. Baltimore: Johns Hopkins University Press.
Warsh, D. (2006). *Knowledge and the wealth of nations: A story of economic discovery*. New York: W.W. Norton.
Whitehead, A. N. (1969). *Process and reality: An essay in cosmology*. New York: Free Press (Simon & Schuster).

Chapter 6
Philosophy of Engineering and the Quest for a Novel Notion of Experimentation

Viola Schiaffonati

Abstract Epistemological issues in engineering knowledge have traditionally played a central role in the debate over the assessment of the philosophy of engineering as a disciplinary field. However, only few works have explicitly focused on experimental methodology and attempted to systematically compare the traditional experimental method of the natural sciences to the kind of experimentation carried out in engineering research. In this paper, by investigating some areas of computer engineering, and in particular autonomous robotics, I claim that traditional experimentation cannot be always applied as such to computer engineering and that the notion of explorative experiment is a good candidate to be considered. Explorative experiments are a form of investigation of novel ideas or techniques without the typical constraints of rigorous experimental methodologies. They are driven by the desire of investigating the realm of possibilities pertaining to the functioning of a technical artefact and its interaction with the environment in the absence of a proper theory or theoretical background.

Keywords Experimental method · Experimental computer science and engineering · Technoscience · Explorative experiment · Autonomous robotics

6.1 A Methodological Investigation

Epistemological issues in engineering knowledge have traditionally played a central role in the debate over the assessment of the philosophy of engineering as a disciplinary field. Foundational works, such as the classical book by Walter Vincenti (1990), have addressed both the design and the growth of engineering knowledge, while refusing the characterization of technology as applied science and promoting engineering knowledge as an epistemological species. More recently edited

V. Schiaffonati (✉)
Dipartimento di Elettronica, Informazione e Bioingegneria, Politecnico di Milano, Milano, Italy
e-mail: viola.schiaffonati@polimi.it

A. Fritzsche, S. J. Oks (eds.), *The Future of Engineering*, Philosophy of Engineering and Technology 31, https://doi.org/10.1007/978-3-319-91029-1_6

collections (van de Poel and Goldberg 2010; Michelfelder et al. 2013) have also covered an important role in setting up the methodological agenda of the discipline and contributed to shape it right from its beginning. If in the first of these edited collections (van de Poel and Goldberg 2010) a philosophy of engineering was acknowledged as yet non–existent and the necessary steps toward it were taken, in the second one (Michelfelder et al. 2013) reflections on the philosophy of engineering are spelled out, including reflections on principles, not only of ethical nature. A difference (and an evolution) already emerging in the titles of the two volumes.

Some recent attempts (Staples 2014, 2015) have set out the epistemological debate around engineering knowledge on clear ontological commitments: even if engineering is quite often based on science, what distinguishes the two is a peculiar methodology, more precisely a systematic attitude of engineering for the use and growth of objective knowledge concerning the physical world and how this can be made to meet some desired requirements (Staples 2014). Despite similarities with science, engineering has its own kind of knowledge and a peculiar way in which knowledge grows. In particular, engineering theories are explicitly falsifiable claims used for predicting and analyzing artefacts with respect to requirements, and the growth of engineering knowledge deals with error elimination (Staples 2015).

Within this framework, however, only few works have explicitly focused on methodology from an experimental point of view and, as far as I know, no one has attempted to systematically compare the traditional experimental method of the natural sciences – that is considered at the core of modern science – with the kind of experimentation carried out in engineering research. In this paper, by investigating computer engineering and in particular the field of autonomous robotics, I claim that the traditional notion of experimentation cannot be always applied as such to computer engineering, and I propose that the notion of *explorative experiment* is a better candidate to be considered in some cases. Explorative experiments are a form of investigation of novel ideas or techniques without the rigorous constraints of experimental methodology. They are experiments driven by the desire of investigating the realm of possibilities pertaining to the functioning of a technical artefact and its interaction with the environment in the absence of a proper theory or theoretical background (Schiaffonati 2016).

Moreover, while recognizing a continuity of the engineering sciences with the natural ones, my attempt is to show that the philosophy of engineering needs not only an adaptation of the traditional frameworks already established in the philosophy of science, but also in some cases a radical shift from them. This is in line with what has been suggested by Nordmann in discussing the technological turn in the philosophies of science and technology (Nordmann 2016), where he shifts away from the analysis of technological practice exclusively made through the lenses of a traditional philosophy of science. His invitation to take technology seriously and to empirically investigate the relation between making and knowing is the reason why in this paper I advance the novel notion of explorative experimentation, capable of explaining an engineering experimental practice that cannot be analyzed only with the traditional tools of the philosophy of experimentation.

If the philosophy of technology has already looked into experimentation and exploration (see (Franssen et al. 2016) for different examples), in this paper I focus on the application perspective by considering how these issues are treated in the practice of experimental computer engineering. The case of experimental computer engineering is emblematic under many respects. Already from its beginning, the plea for adopting the traditional standards of experimental science has been constant and the worry not to be taken seriously is evident in the words of one of its founders (Denning 1980). The calling for experiments in the field is justified as a way to assess its scientific status (Morrison and Snodgrass 2011), even if in most of the cases the notion of experiment to deal with is oversimplified. Moreover, a full adequacy to the same standards of traditional experimental sciences is advocated (Tichy 1998), without any critical analysis on whether and how these standards are applicable to computer engineering.

In what follows I briefly review experimental computer engineering and the calling for a more rigorous experimental approach as a way to assess the scientific status of the discipline. On the basis of some recent analyses I criticize the attempt to fully adequate computer engineering to the same standards of traditional experimental sciences and I propose a more articulated framework. In my endeavor, I plan to move along three different but interconnected directions. The first one deals with the notion of directly action-guiding experiment, as characterizing a significant part of the experimental practice in computer engineering, in opposition to the one of epistemic experiment. The second direction concerns the debate around engineering epistemology, and whether adapting frameworks from the traditional philosophical debate can suffice to take into account the peculiarity of the discipline. Finally, the third direction acknowledges the empirical turn in the recent philosophy of technology, introduces the framework of technoscience as an engineering way of being in science, and invites philosophers of science to take this notion seriously in order to shed light on a range of questions that have been neglected so far.

6.2 Experimental Computer Engineering

Probably the foremost concept of computer science as an experimental science dates back to the 1976 paper by Newell and Simon published in the occasion of their acceptance of the Turing award: "*Computer science is an empirical discipline. We would have called it an experimental science, but like astronomy, economics, and geology, some of its unique forms of observation and experience do not fit a narrow stereotype of the experimental method. None the less, they are experiments. Each new machine that is built is an experiment. Actually constructing the machine poses a question to nature; and we listen for the answer by observing the machine in operation and analyzing it by all analytical and measurement means available. Each new program that is built is an experiment. It poses a question to nature, and its behavior offers clues to an answer. Neither machines nor programs are black boxes; they are artifacts that have been designed, both hardware and software, and we can open them up and look inside.*" (Newell and Simon 1976, 114). It is easy to recognize in these words the characterizing idea of experiment as a question to nature. Accordingly, machines and programs are experiments and the appeal to experience is fundamental in contrast with the view of computer science as a purely theoretical discipline. However, this rather ingenuous view of experiments may have contributed to spread an oversimplified conception of both the very notion of experiment and how the experimental method can be applied to computer engineering.

The quest for experiments in computing began to be treated systematically at the beginning of the1980s, following a crisis in what was then called experimental computer science and engineering. In an ACM report published in 1979 (Feldman and Sutherland 1979), experimental research in computer science is strongly related to the measurement and testing of computing algorithms and systems. In the same issue of the journal, where the ACM report was published, the call for experimentation is expressed in terms of the recognition of the possibility for major advantages in the field (McCracken et al. 1979). At the same time, a 'rejuvenation' of experimental computer science is advocated from very concrete perspectives: for example, by promoting experimental facilities for computer systems research. However, experimental computer science is seldom defined in a precise way in this context, and experiments are conceived mainly as explorations. Experimental computer science is to be rejuvenated also according to Denning, who proposes in a short article that the experimental work produced in computer science should be judged by traditional standards (Denning 1980). Denning advances the idea that to implement experimentally a computer system is not just to build the system and 'see what happens'. In a way, this approach tries to go beyond Newell and Simon's 'construct and test' paradigm, by proposing that experimental computer science has to deal with the process of supporting and testing a hypothesis, thus making computing closer to the standards of rigor and the practice of natural sciences. Unfortunately, experimental science is reduced to the process of classifying knowledge derived from observations.

More recently, a trend has once again emerged toward making the experimental scientific method take center stage in computer engineering. These efforts have shown a renewed need for an experimental methodology in this discipline (Freeman 2008; Morrison and Snodgrass 2011). Experiments are deemed to have an impact on several aspects of computing: their importance is recognized for assessing computing systems' performance and for triggering new developments, and experimentation with prototypes is considered essential in use-inspired research and product design (Snir 2011). Moreover, the use of the experimental scientific method is advocated to understand computations that are often too complex for mathematical analysis, to prove their correctness, to check consistency with hypotheses, to uncover performance constraints, and to show whether original goals are met (Denning and Freeman 2009). 'Experimental computer science' has become a quite common label to which (at least) three different meanings can be associated (see (Feitelson 2006) for further details).

Despite the increasing interest in a more rigorous approach to computer engineering, many lament that the current methodology is inadequate and that, in comparison with other fields (e.g., natural sciences), there should be more and better experimentation in the field (Denning 2005). Indeed, several articles describe demonstrations rather than real experiments (Feitelson 2006), and their sections on experimental results present just weak examples to show the superiority of the proposed solution over a limited amount of alternatives (Zelkowitz and Wallace 1997, 1998). The invitation to experiment more and better is not always based on convincing arguments. On the one side, it is generally claimed that without experiments, computer science is in danger of drying up and becoming an auxiliary discipline (Tichy 1998) and that experimentation can accelerate progress, but no precise analysis of how this experimentation should be carried out in the various fields of computing is provided. On the other side, the adoption of a scientific (in the sense of experimental) approach in computing is justified by the tremendous success of scientific reasoning. Still, no serious inquiry is carried out about the possibility and the consequences of plainly importing traditional scientific methods into computer engineering (Morrison and Snodgrass 2011).

The call for an increase in experimentation has often been made with an emphasis on making computer science more scientific. However, this 'scientificity' is taken as a positive goal per se, without any discussion on whether this discipline can really benefit from being more scientific, or even on what it really means 'to be more scientific'. Many recommendations present common traits: they stem from the acknowledgment of a crisis in computer engineering that is meant to be overcome with a greater maturity of the discipline, in terms of a more rigorous experimental method and a more scientific approach to the search for solutions. Taking inspiration from experimental principles adopted in traditional scientific disciplines has become a leitmotif in many analyses, which recognize, for example, the benefits of replicable results (Juristo and Gomez 2012) or the importance of negative results (Barni et al. 2007), two of the cornerstones of experimental scientific method. Still, many issues remain open and require, as argued in the next section, a shift from general statements to concrete discussions in order to clarify the advantages and the limits of taking inspiration from traditional scientific disciplines.

As it may be clear already from this short overview, the call for experiments to assess the scientific status of computer engineering has a propagandistic attitude. Moreover, some proposals are weakened by a lack of clarity when it comes to the core concepts of experimental methodology. For example, experiments are wrongly identified with empirical methods (see for instance (Morrison and Snodgrass 2011)), which are instead traditionally defined as based on the aggregation of naturally occurring data, without the strict rules that are supposed to guide experiments, and, even when a distinction between empirical and experimental is made, the notion of experiment is sometimes naïvely intended. Even works that are considered a milestone in the development of experimental methods, for example (Juristo and Moreno 2011) in software engineering, while discussing in detail how experiments should be intended in this field, present some ingenuous views on experimentation. There are of course cases in which experiments are defined within a more solid conceptual framework including observation, measurement, and analysis of results (Freeman 2008); however, in computer engineering, these processes are seldom illustrated in terms of concrete examples similar to what natural scientists deal with in their practice. As computer engineering has different goals and aims than experimental physics or biology, the full adequacy to the same standards of traditional natural sciences cannot be taken for granted without a critical analysis pointing out similarities, differences, and limits.

6.3 A Wider Framework

The investigation on the nature and role of experimental method in computer engineering requires widening the usual conceptual framework that, as I have shown previously, is mainly focused on simply adapting conceptual tools already adopted in the natural sciences. Even if the engineering sciences show continuity with the natural ones, they work in the context of technological applications with an interest in the development of technological devices, processes and materials. Accordingly, experiments are mainly carried out to check whether these technical artefacts met the desired specifications via the technological production of these phenomena, instead of theories (Boon 2012). Technological production has to be meant broadly in this discussion, including such activities as research, development, design, testing, patenting, maintenance, inspection, and so on, with an attention on what engineers actually do in their everyday practice. This suggests an agenda for philosophical reflection on engineering that is distinct from the traditional philosophy of technology (van de Poel 2010), mostly focused on the analysis of the social, cultural, and political impact of technologies. Moreover technical artefacts, because they are physical objects with a technical function and use plan designed and made by human beings (Vermaas et al. 2011), are evaluated with respect to their intended functions and requirements, thus introducing in the experimental context a normative character which is absent from the experimental context of the natural sciences. These aspects set the frame for a technoscientific perspective, more precisely that particular perspective that can be meaningfully summarized as an 'engineering way of being in science' (Nordmann 2010).

To distinguish science from technoscience is not only a matter of terms or of different objects of research, rather it is a matter of guiding ideals and research orientations that shape practice in significantly different ways. Using the words of a programmatic paper: "*Both science and techoscience involve an interplay of representing and intervening. Science is defined by its orientation to the epistemic ideal of purification* [...]. *Technoscience is defined by its neglect or abandonment of this work of purification.* [...] *Technoscience is therefore a kind of research where theoretical representation and technical intervention cannot be held apart even in thought.*" (Bensaude-Vincent et al. 2011, 368). As these authors emphasize, in experiments it is not always possible to distinguish in an observed effect between the contribution of the researchers and the contribution of nature. Both the observed phenomena and the effects appear to be engineered, and there is no interest in a work of purification such as, for instance, in the case of a pharmaceutical therapy where it is irrelevant to try to understand the contributions achieved by technology and by nature in order to assess the efficacy of the therapy.

The mention of biomedical research and of laboratory experimental practice is useful to introduce an example of a type of experimental process which cannot be easily accommodated with the traditional notion of experiment such as, for instance, the clinical trial of an analgesic. Here the intended outcome is the efficient pain reduction and the experimental intervention is the treatment that might be administered. According to Hansson (2016) this is a case of a *directly action-guiding experiment* that is such when satisfies two criteria: (a) the outcome looked for consists in the attainment of some desired goal of human action (e.g., pain reduction) and (b) the interventions studied are potential candidates for being performed in a non-experimental setting in order to achieve that goal (e.g., treatment administered in ordinary conditions). This definition frames the notion of experiment as a technological form of experimentation driven by practical needs. As Hansson carefully reconstructs, this form of experimentation has non-academic origin and was already employed in pre-scientific times, playing a major role in early Renaissance for the development of experimental methodology. The practical attitude of directly-action guiding experiments is contraposed by Hansson to *epistemic experiments*, that is experiments which are oriented toward the knowledge of the world we live in, where the desired outcome is the one providing this information. It is fair to say that historical and philosophical accounts of experimental method have principally focused on epistemic experiments, and little attention has been devoted to directly action-guiding ones.

The idea that different notions of experimentation can coexist without any radical contraposition has been only recently pointed out in computer science and engineering, and it still represents an exception to the mainstream framework depicted in the previous section. As it has been reconstructed in detail in (Tedre 2015), at least five different views of experiments can be recognized in the practice of the field. There are the so called *feasibility experiments* aimed at empirically demonstrating ('demonstration' and 'experiment' are terms commonly used as synonymous in computing) the proper development and working of a technology. There are *trial experiments*, evaluating some aspects of a system using predetermined variables in a laboratory, and *field experiments*, aimed at evaluating these aspects of a system outside the laboratory, in the real world. There are also *comparison experiments* devoted to compare different solutions to look for the best one for a specific problem. And, finally, there are *controlled experiments*, those more similar to the traditional notion of experimentation and aimed at achieving generalization and prediction. What is important in this account is not how the notion of experiment should be used, but how it is actually used: "*Many would object against calling, for instance, feasibility demonstrations 'experiments,' arguing that the term 'experiment' has a special meaning in science. They are right. But if one looks at how authors in computing have used the term—not how it should be used—those five uses are easily found.*" (Tedre 2015, 190).

On the basis of Tedre's analysis it is evident that the notion of control, usually an essential component in experimentation in the natural sciences, covers a different role in computer science and engineering: not only directly action-guiding experiments individuate a different focus of attention with respect to epistemic ones, but also a crisis of the traditional notion of control seems to come into life. If in the natural sciences it is prescribed that the experimenter should be an outsider of the phenomenon to be explained, it is not clear how a person working in computing, which is aimed at producing computation-based artefacts, could be an outsider with respect to a phenomenon (i.e., an artefact) that (s)he has created (Tedre 2011). Except from some significant examples, experiments in computer engineering are usually performed by the same person that has created the artefacts and, at the same time, needs to test them, losing the sort of independence of the experimenter prescribed in the classical experimental protocol.

The crisis of the traditional notion of experimental control has been recently evidenced by Peter Kroes (2016) in his analysis of experiments with new technologies in socio-technical systems, where he discusses the shift from a control paradigm based on two assumptions: the experimenter is not part of the system on which the experiment is performed and (s)he is in control of the independent variables and of the experimental set-up. Accordingly, the experimenter is able to intervene both by changing these variables to evaluate their influence on the dependent ones and by varying the experimental set-up. This traditional control paradigm becomes problematic, and a consequent shift in the notions of intervention and control is observed, when considering new technologies as *socio-technical systems*, namely as hybrid systems composed of natural objects, technical artifacts artefacts, human actors, and social entities. The idea of controlling the experimental system from a center of command and control that is outside the system becomes highly problematic because the distinction between the experimental system and its environment is critical and the environment is complex due to the co-presence of technical artifacts and natural and social elements. Hence the control of these systems not only involves controlling both technical artefacts and social elements, but also problematizes the drawing of the line between the experimental system and its environment.

6.4 Explorative Experiments

The scenario previously described delineates a complex framework where even an adaptation (following for example the one proposed in (Staples 2015) in which engineering epistemology is analyzed in a Popperian framework) of the traditional notion of controlled or epistemic experiment appears out of focus. As I have attempted to show, this is not only due to the coexistence of different notions of experimentation in computing, but also to a novel type of experimental intervention where the borders between the experimental setting and the general environment are blurred, as well the possibility of controlling from the beginning and from the outside the experimental procedure. This is precisely the reason why in this paper I introduce the notion of explorative experiment, rather than adapting the traditional one. Contrary to Staples, although I recognize a continuity between the natural sciences and the engineering ones, I claim that the differences just pointed out call for a stronger differentiation of computer engineering experimentation from traditional experimental methods. In my opinion this is another step in the direction of a philosophy of engineering that, even if deeply rooted in the philosophy of science, is progressively gaining autonomy and takes seriously the invitation of the analytical turn in the philosophy of technology of not considering engineering just as an applied science (Franssen et al. 2016).

What has been previously described in terms of changes in the practice of experimentation fits particularly well with the following quote, coming from a different field (political sciences), but helping to delineate this mutated scenario: "*Design experiments differ significantly from randomized controlled experiments. Design experimentation starts with the presumption that the world is a messy place and that experiments will not be able to isolate the effect of single variables.* [...] *The focus of a design experiment is not to definitively accept or reject a hypothesis, but rather to iteratively refine the intervention* [...]. *The ultimate purpose is not to test general theory, but to probe the possibility and limits of the intervention. Design experiments do not create a sharp distinction between researchers and subjects: instead, the practitioners often become experimenters.*" (Ansell 2012, 163–164). This perfectly describes a part of the experimental practice carried out in computer engineering that has been mostly neglected so far. An interesting example is represented by the application of the action research methods to research in Information Systems as a framework to relax some preconditions such as the assumed independence of the observer from the observations (Schwabe and Krcmar 2000). It describes, moreover, what elsewhere I have called explorative experiments (Schiaffonati 2016) to mark down both their difference with respect to exploratory experiments in the philosophy of biology and their specificity to computer engineering, a discipline that has been constantly struggling between a scientific and an engineering characterization.

In some recent philosophical research *exploratory experimentation* has been used to label those forms of experimentation in science which are not always guided by theories. One of the first authors to recognize the epistemic importance of exploratory experiments (Steinle 1997) defines them as driven by the desire to obtain empirical regularities when no well-formed theories or no conceptual framework are available. The same term is used with a slightly different meaning in another article in the same year, but in the context of some early research in protein synthesis (Burian 1997), where exploratory experimentation is seen as a style of inquiry not guided by theory. These and other similar works are mainly directed against the theory-driven approaches of most of the philosophy of science in the spirit of experimentation as having a life on its own (Hacking 1983). Even if they recognize that exploratory experimentation is typically not free of theory, they aim at showing the epistemic significance of those inquiries that are not primarily theory-driven by presenting several detailed case-studies. The idea that "*The aim of exploratory experiments is to generate significant findings about phenomena without appealing to a theory about these phenomena for the purpose of focusing experimental attention on a limited range of possible findings.*" (Waters 2007, 5) is probably that serving better as an inspiration for more recent works devoted to provide evidence of the exploratory shift observed in the methodology of some areas of biology (Franklin 2005).

To my purposes, however, this emphasis on theory (even when theory is relegated as background knowledge) is out of scope, because it is not even completely clear what a theory in computer engineering is or whether references to a theoretical background play a key role in experiments in computing. This is also the reason why I use the term 'explorative' instead of 'exploratory' to mark a difference from the philosophical work focused on accounting the distinction between exploratory and theory-driven experiments on the ways in which experiments depend on theory. In my attempt to characterize explorative experiments in computer engineering I am interested, rather, in the appeal to complexity that has been stressed in the philosophical literature (Burian 2007), where some systems are considered too complicated to be investigated by means of a theory-driven approach. If this appeal to complexity certainly applies to biology, there are good reasons to apply it to computing fields as well, in particular when considering that what is the subject of the experimentation are not just the technical artefacts per se, but also the ways these artefacts are able to interact with the surrounding physical and social world.

Moreover, the notion of exploration has been recently investigated in the context of the philosophy of technology, and in particular in the analysis of engineering knowledge (Zwart and de Vries 2016). Within a taxonomy of innovative engineering projects explorations may be part of descriptive projects or means-ends knowledge projects and, in this latter case, they produce outcomes that are not hypotheses based on dependent variables.

So said, *explorative experiments* are experiments that are driven by the desire of investigating the realm of possibilities pertaining to the functioning of a computation-based artefact and its interaction with the environment in absence of a proper theory or theoretical background. More precisely, explorative experiments are a special kind of directly action-guiding experiments which possess the following features:

- They are devoted to testing technical artefacts, meant as artificial entities purportedly built by humans to fulfill a purpose and, therefore, having a technical function.
- They are focused on iteratively refining the intervention, meant as the union of knowledge and action characterizing experimental practice, and their ultimate purpose is not to test a general theory, but to probe the possibility and limits of the intervention.
- They do not force a sharp distinction between designers and experimenters and, instead, the practitioners often become experimenters.
- The control of the experimental factors cannot be fully managed from the beginning, but is in part carried out after the artefact has been inserted into its environment.

The form of control characterizing explorative experiments can be properly labeled as *a posteriori*, since it is carried out while and after the introduction of technical artefacts within society (not before), thus maintaining the centrality of control in experimentation, but under a different perspective. In accordance with (Radder 2009), I wish to maintain that not any kind of intervention in the material world counts as a scientific experiment and, thus, I reject the broad notion of social experiment without any form of control as described in (Small 1921).

Moreover, in order to fully characterize the notion of explorative experiment as characteristic of computer engineering, it is important to spell out its difference from software testing, intended as the process of trying to discover every conceivable fault or weakness in a work product (Kit 1999). Testing is possibly the closest thing to a standardized empirical procedure in computer engineering and shares with explorative experimentation having technical artefacts as objects of investigation. Traditionally tests are procedures to verify if the desired requirements are met in engineering contexts and experiments are processes in which an object of study is subjected to interventions and manipulations in order to obtain knowledge about the object in a predictable way. As it should appear clear from the previous discussion, these distinctions are more blurred than ever if we look at computer science and engineering. For example, what is sometimes called trial experiment in computing (Tedre 2015) is precisely a form of test for evaluating how a novel solution meets the desired requirements. However, apart from the difficulty of demarcating between these two concepts as they lie at the opposite side of a continuum, it is important to stress that explorative experiments are more general than specific tests on software products. 'More general' means here that they offer more opportunities in terms of knowledge, since the kind of knowledge acquired in the experiment, in particular when devoted to forms of explorations, aims to be generalized in the direction of the idea of *learning-by-exploration* that has been recently conceptualized in (van de Poel forthcoming).

6.5 Forms of Explorations

To make the former analysis more concrete, I introduce now some forms of exploration observable in the experimental practice of autonomous robotics. This is an engineering field oriented to develop robot systems that are autonomous in the sense that they have the ability to operate without continuous human intervention, in order to work in places hardly accessible by humans or in cooperation with humans in common environments. The discussion on experiments and the effort in developing good experimental methodologies have gained attention in autonomous robotics in the very last years. Accordingly, a number of initiatives have been promoted, ranging from workshop series (Bonsignorio et al. 2015), to special issues of journals (Bonsignorio and del Pobil 2015), to European projects funded under different programs, to a generalized interest in experimental issues.

When analyzing the experimental trends within this community, two different tendencies emerge: on the one hand, the principles of experimental method (such as comparison, reproducibility and repeatability) inspire the community in defining a more rigorous approach to experiments; on the other hand, these rigorous approaches are not yet full part of the current research practice. For instance, from the systematic analysis presented in (Amigoni et al. 2014), it emerges that only few of the experiments conducted in a significant sample of autonomous robotics articles come close to controlled experiments in the sense employed by (Tedre 2015) for computing in general.

A systematic analysis of the articles that in the last years have been given awards at the *IEEE International Conference on Robotics and Automation* (ICRA) and at the *IEEE/RSJ International Conference on Intelligent Robots and Systems* (IROS) – two large and impacting robotics research conferences – offers some interesting insights. Here, besides the typologies of experiments easily classified according to the categories introduced by Tedre (2015), forms of experimentation labelled as explorative emerge. Some examples follow.

Investigating the Role of Parameters In this case experiments are explorative because they are used to elucidate the relationship between the values of the parameters and the behavior of robot systems that is often not very precise, as the designer of the system has only a rough *a priori* idea of it. In particular, some explorative experiments are set up to elucidate the qualitative and quantitative effects of different parameters values on some measurable quantities relative to the behavior and the performance of the robots (Thobbi et al. 2011; Daniel et al. 2012).

Confirmation of Expectations or Hypotheses In developing robot systems usually designers consider and build upon a set of expectations about the behavior of artefacts when inserted in their operating environments. However, to build reliable models of the interaction between robots and their environment is not easy, and hypotheses can be only empirically confirmed since they are not based on a solid theoretical ground. In this case experiments provide an easy feedback to the design phase, also in the sense of a posteriori justification on some assumptions. They have an explorative character that manifest itself in the continuous interplay between design and experimentation (Grollman and Billard 2011; Chu et al. 2013).

Getting Insights on the Behavior of the Robot Systems Another form of explorative experimentation is represented by those processes used to acquire knowledge on how robot systems perform their tasks. These experiments provide quantitative results used by the robot designers to inspect, and possibly modify, the design of the internal methods of the robot system, thus providing a richer knowledge than that acquired in investigating the role of parameters. In this case, actually, the achieved knowledge ranges from measuring values of internal variables to generating ideas for alternative design solutions (Grollman and Billard 2011; Thobbi et al. 2011; Fasola and Mataric 2013; Gemici and Saxena 2014).

Assessing the Generality of the Robot Systems Amongst the different form of explorative experimentation, the most elaborated one is that used to gain knowledge about the behavior of robot systems in different settings in order to evaluate their generality. A particular relevant situation is that of robot systems, designed for an environment characterized as error-free, which are then experimentally tested in a setting with noisy data or data that the robot system has never experienced before. This last aspect is particularly relevant in the case of learning systems, as those proposed in (Deisenroth et al. 2014) and (Gemici and Saxena 2014).

6.6 Conclusions

I argued about the necessity of enlarging the framework of analysis when considering the nature and role of experiments in computer engineering. In particular, I emphasized that the traditional notion of controlled or epistemic experiment does not fit with a significant experimental practice in autonomous robotics. For this reason I introduced the notion of explorative experiment and provided examples of explorative forms of experimentation in autonomous robotics. This analysis of experimentation is meant as a first step in the direction of a larger investigation about the experimental approaches of the engineering sciences, beyond computer engineering, to assess the philosophy of engineering as a new disciplinary field.

A methodological problem needs surely to be addressed in future analyses, namely how to strength and further investigate the idea of explorative experiment. A purely conceptual analysis is out of scope in the context of a philosophy of engineering attempting to move ahead from the philosophy of science and the philosophy of technology. On the other side a pure ethnographic approach, in the form adopted in qualitative research, is limited in the attempt to shape the concept of explorative experiment from a more general perspective. Moreover, the systematic analysis of publications, by considering the way in which the experimental activity is described, is sometimes misleading since a gap between the real practice and the way in which this is described can be found. It is likely that all these different approaches will need to be practically blended in a creative and innovative way.

References

Amigoni, F., Schiaffonati, V., & Verdicchio, M. (2014). Good experimental methodologies for autonomous robotics: From theory to practice. In F. Amigoni & V. Schiaffonati (Eds.), *Methods and experimental techniques in computer engineering, Springer Briefs in Applied Sciences and Technology* (pp. 37–53). Cham: Springer.

Ansell, C. (2012). What is a "democratic experiment"? *Contemporary Pragmatism, 9*(2), 159–180.

Barni, M., Perez-Gonzalez, F., Comesana, P., & Bartoli, G. (2007). Putting reproducible signal processing into practice: A case study in watermarking. In *Proceedings of IEEE international conference on acoustics speech and signal processing*.

Bensaude-Vincent, B., Loeve, S., Nordmann, A., & Schwarz, A. (2011). Matters of interest. The object of research in science and technoscience. *Journal of General Philosophy of Science, 43*, 365–383.
Bonsignorio, F., & del Pobil, A. (2015). Special issue on replicable and measurable robotics research. *IEEE Robotics and Automation Magazine, 22*(3), 32–154.
Bonsignorio, F., Hallam, J., & del Pobil, A. (2015). *Special interest group on good experimental methodologies*. http://www.heronrobots.com/EuronGEMSig/gem-sig-events. Accessed Sept 2016.
Boon, M. (2012). Scientific concepts in the engineering sciences: Epistemic tools for creating and intervening with phenomena. In U. Feest & F. Steinle (Eds.), *Scientific concepts and investigative practice* (pp. 219–243). Berlin: De Gruyter.
Burian, R. M. (1997). Exploratory experimentation and the role of histochemical techniques in the work of Jean Brachet, 1938–1952. *History and Philosophy of the Life Sciences, 19*, 27–45.
Burian, R. M. (2007). On MicroRNA and the need for exploratory experimentation in post-genomic molecular biology. *History and Philosophy of Life Sciences, 29*(3), 285–311.
Chu, V., McMahon, I., Riano, L., McDonald, C., He, Q., Martinez Perez-Tejada, J., Arrigo, M., Fitter, N., Nappo, J., Darrell, T., & Kuchenbecker, U. (2013). Using robotic exploratory procedures to learn the meaning of haptic adjectives. *Proceedings of ICRA*, 3048–3055.
Daniel, C., Neumann, G., & Peters, J. (2012). Learning concurrent motor skills in versatile solution spaces. *Proceedings of IROS*, 3591–3597.
Deisenroth, M., Englert, P., Peters, J., & Fox, D. (2014). Multi-task policy search for robotics. *Proceedings of ICRA*, 3876–3881.
Denning, P. J. (1980). What is experimental computer science. *Communications of the ACM, 23*(10), 543–544.
Denning, P. J. (2005). Is computer science science? *Communications of the ACM, 48*(4), 27–31.
Denning, P. J., & Freeman, P. (2009). Computing's paradigm. *Communications of the ACM, 52*(12), 28–30.
Fasola, J., & Mataric, M. (2013). Using semantic fields to model dynamic spatial relations in a robot architecture for natural language instruction of service robots. *Proceedings of IROS*, 143–150.
Feitelson, D. G. (2006). *Experimental computer science: The need for a cultural change*. Unpublished manuscript. http://www.cs.huji.ac.il/~feit/papers/exp05.pdf. Accessed Oct 2016.
Feldman, J. A., & Sutherland, W. R. (1979). Rejuvenating experimental computer science. *Communications of the ACM, 22*(9), 497–502.
Franklin, L. R. (2005). Exploratory experiments. *Philosophy of Science, 72*, 888–899.
Franssen, M., Vermaas, P. E., Kroes, P., & Meijers, A. W. M. (2016). *Philosophy of technology after the empirical turn*. Cham: Springer.
Freeman, P. (2008). Back to experimentation. *Communications of the ACM, 51*(1), 21–22.
Gemici, M., & Saxena, A. (2014). Learning haptic representation for manipulating deformable food objects. *Proceedings of IROS*, 638–645.
Grollman, D., & Billard, A. (2011). Donut as I do: Learning from failed demonstrations. *Proceedings of ICRA*, 3804–3809.
Hacking, I. (1983). *Representing and intervening*. New York: Cambridge University Press.
Hansson, S. O. (2016). Experiments: Why and how? *Science and Engineering Ethics, 22*(3), 613–632.
Juristo, N., & Gomez, O. (2012). Replication of software engineering experiments. In B. Mayer & M. Nordio (Eds.), *Empirical software engineering and verication* (pp. 60–88). Berlin: Springer.
Juristo, N., & Moreno, A. M. (2011). *Basics of software engineering experimentation*. Dordrecht: Kluwer.
Kit, E. (1999). *Software testing in the real world*. Harlow: Addison-Wesley.
Kroes, P. (2016). Experiments on socio-technical systems: The problem of control. *Science and Engineering Ethics, 22*(3), 633–645.

McCracken, D. D., Denning, P. J., & Brandin, D. H. (1979). An ACM executive committee position on the crisis in experimental computer science. *Communications of the ACM, 22*(9), 503–504.

Michelfelder, D. P., McCarthy, N., & Goldberg, D. E. (Eds.). (2013). *Philosophy and engineering. Reflections on practice, principles, and process*. Dordrecht: Springer.

Morrison, C. T., & Snodgrass, R. T. (2011). Computer science can use more science. *Communications of the ACM, 54*(6), 36–38.

Newell, A., & Simon, H. (1976). Computer science as empirical inquiry: Symbols and search. *Communications of the ACM, 19*(3), 113–126.

Nordmann, A. (2010). Science in the context of technology. In M. Carrier & A. Nordmann (Eds.), *Science in the context of application. Boston studies in the philosophy of science* (pp. 467–482). Dordrecht: Springer.

Nordmann, A. (2016). Changing perspectives: The technological turn in the philosophies of science and technology. In M. Frassen, P. E. Vermaas, P. Kroes, & A. W. Meijers (Eds.), *Philosophy of technology after the empirical turn* (pp. 107–125). Cham: Springer.

Radder, H. (2009). The philosophy of scientific experimentation: A review. *Automated Experimentation, 1*(2), 1–8. https://doi.org/10.1186/1759-4499-1-2.

Schiaffonati, V. (2016). Stretching the traditional notion of experiment in computing: Explorative experiments. *Science and Engineering Ethics, 22*(3), 647–665.

Schwabe, G., & Krcmar, H. (2000). Piloting socio-technical innovation. In *Proceedings of the European conference on information systems*, Paper 27.

Small, A. W. (1921). The future of sociology. *Publications of the American Sociological Society, 15*, 174–193.

Snir, M. (2011). Computer and information science and engineering: One discipline, many specialties. *Communications of the ACM, 54*(3), 38–43.

Staples, M. (2014). Critical rationalism and engineering: Ontology. *Synthese, 191*(10), 2255–2279.

Staples, M. (2015). Critical rationalism and engineering: Methodology. *Synthese, 192*(1), 337–362.

Steinle, F. (1997). Entering new fields: Exploratory uses of experimentation. *Philosophy of Science, 64*, S65–S67.

Tedre, M. (2011). Computing as a science: A survey of computing viewpoints. *Minds and Machines, 21*, 361–387.

Tedre, M. (2015). *The science of computing*. Boca Raton: CRC Press/Taylor & Francis Group.

Thobbi, A., Gu, Y., & Sheng, W. (2011). Using human motion estimation for human-robot cooperative manipulation. *Proceedings of IROS*, 2873–2878.

Tichy, W. F. (1998). Should computer scientists experiment more? *Computer, 31*(5), 32–40.

van de Poel, I. (2010). Philosophy and engineering: Setting the stage. In I. van de Poel & D. E. Goldberg (Eds.), *Philosophy and engineering: An emerging agenda* (pp. 1–11). Dordrecht: Springer.

Van de Poel, I. (forthcoming). Society as a laboratory to experiment with new technologies. In E. Stokes, D. Bowman, & A. Rip (Eds.), *Embedding and governing new technologies*. Singapore: Pan Stanford Publishing.

van de Poel, I., & Goldberg, D. E. (Eds.). (2010). *Philosophy and engineering: An emerging agenda*. Dordrecht: Springer.

Vermaas, P., Kroes, P., van de Poel, I., Franssen, M., & Houkes, W. (2011). *A philosophy of technology. From technical artefacts to sociotechnical systems*. San Rafael: Morgan & Claypool.

Vincenti, W. G. (1990). *What engineers know and how they know it*. Baltimore/London: John Hopkins University Press.

Waters, C. K. (2007). The nature and context of exploratory experimentation. *History and Philosophy of the Life Sciences, 19*, 275–284.

Zelkowitz, M. V., & Wallace, D. R. (1997). Experimental validation in software engineering. *Information and Software Technology, 39*(11), 735–743.

Zelkowitz, M. V., & Wallace, D. R. (1998). Experimental models for validating technology. *Computer, 31*(5), 23–31.

Zwart, S. D., & de Vries, M. J. (2016). Methodological classification of innovative engineering projects. In M. Franssen, P. E. Vermaas, P. Kroes, & A. W. M. Meijers (Eds.), *Philosophy of technology after the empirical turn* (pp. 219–248). Cham: Springer.

Chapter 7
Demarcating Simulation

Nico Formanek

Abstract Philosophy of computer simulation has for long tried to answer the question: What makes computer simulation special? Answers normally proceed by giving a demarcation criterion, a property which only holds for computer simulation. The definition by demarcation has also been attempted in different fields, maybe most famously by Popper for science. So it might not come as a surprise that philosophy of mathematics has tried to demarcate computer aided proof. What is remarkable is that the demarcation criteria for computer simulation and computer aided proof are rather similar. As the debate in philosophy of mathematics is more advanced, I will show how its arguments can be translated for the philosophy of computer simulation. I argue that even in the light of computer methods human justificatory capabilities still remain central. In closing I point out, that the focus on justificatory methods like verification/validation in software engineering challenges the traditional philosophical account of engineering.

Keywords Philosophy of computer simulation · Epistemology of computer simulation · Philosophy of software engineering

7.1 Definition by Demarcation

How do we distinguish between different fields of inquiry? How do we separate science from non-science? How do we account for the philosophical novelty of computer simulation?

Questions like these are best answered by giving a demarcation criterion. We thereby define the boundaries of its object and (in a more traditional parlance) say something about its nature. The defining property should be sufficient and necessary for demarcation. There should be no other objects, apart from the one we want to demarcate, instantiating that property.

N. Formanek (✉)
Höchstleistungsrechenzentrum, Stuttgart, Germany
e-mail: formanek@hlrs.de

A. Fritzsche, S. J. Oks (eds.), *The Future of Engineering*, Philosophy of Engineering and Technology 31, https://doi.org/10.1007/978-3-319-91029-1_7

Students of Aristotle will not be surprised that this sounds similar to the traditional *genus-differentia* theory of definition. Why then another name? In Aristotle's account in general only species (defined by genus) have essences and one is only able to define that which has an essence. Definition by demarcation differs that it ignores the requirement to specify a genus.

One famous example is Popper's definition of science by demarcation (see Popper 2005). He proposed the property of falsifiability as demarcating criterion. A statement is deemed scientific if and only if it is falsifiable. Of course such definitions are highly contended and counterexamples may come readily to mind. Popper's account is no exception and has had numerous critics. The necessary part of the definition has been attacked by giving examples of scientific statements which are not falsifiable, while the sufficient side has suffered blows from supposedly non-scientific but falsifiable propositions.

The historical problems of demarcation should not discourage from attempting those definitions again, for it is the task of philosophy giving adequate characterisations to make discourse possible and as I will show much is learned even from failure.

7.2 Opacity in Computer Simulation

Frigg and Reiss argued that a philosophy of computer methods is not needed because questions arising from that field are already dealt with either in the special sciences or in philosophy of science (see Frigg and Reiss 2009). Responding to that critique Humphreys introduced the term *epistemic opacity* (Humphreys 2009 p. 618). The concept's function is to characterize what makes computer simulation (or more broadly construed computer science) special by demarcating from other methods. In that way the critique would have been countered and the need for a specialized philosophy of computer simulation established.

Drawing on an analogous debate in philosophy of mathematics concerning computer aided proofs I will show that the introduction of this concept doesn't answer the critique fully.

In several spots Humphreys states that computers replace humans in terms of generating knowledge and justifying belief (Humphreys 2009 p. 616, 617). He calls this the *anthroprocentric predicament*, which is generated by epistemic opacity, semantics and temporal dynamics of computer simulation as well as in-practice-considerations. To be fair to his argumentation it has to be said, that he considers these points to be somehow connected and generating the predicament together.

I focus on epistemic opacity because it is the only point directly justifying a singular role for computers as epistemic agents. Also Humphreys says that „[m]any, perhaps all, of the features that are special to simulations are a result of this inability of human cognitive abilities to know and understand the details of the computational process. "(Humphreys 2009 p. 619). So the other three points might even be reducible to opacity.

There are two definitions of epistemic opacity in Humphreys (2009). Both have as subject a process and are relativized to an Agent X.

The first introduces opacity:

> [...] a process is epistemically opaque relative to a cognitive agent X at time t just in case X does not know at t all of the epistemically relevant elements of the process. (Humphreys 2009 p. 618)

An epistemically relevant element is an element that is necessary for the justification of other elements in the process. As an example consider a pupil who has memorized the rules for symbolic differentiation without knowing their derivation from limits. She is now able to differentiate sufficiently good-natured functions without being able the justify the result. The process of differentiation is epistemically opaque for her. It should be noted that epistemic relevancy is also an agent-relative concept. Justificatory steps might be evident for one agent while requiring further elucidation for another.

The second definition is about *essential* opacity:

> A process is essentially epistemically opaque to X if and only if it is impossible, given the nature of X, for X to know all of the epistemically relevant elements of the process. (Humphreys 2009 p. 618)

Humphreys uses these definitions to argue that the nature of humans makes computer simulations essentially epistemically opaque and that "[t]his feature is novel because, prior to the 1940s, theoretical science had not been able to automate the process from theory to applications in a way that made the details of parts of that process completely inaccessible to humans." (Humphreys 2009 p. 618). Simulation is essentially opaque because "[...] no human can examine and justify every element of the computational processes that produce the output of a computer simulation or other artifacts of computational science." (Humphreys 2009 p. 618).

By computational process Humphreys has in mind the workings of a computer at the basic hardware level. For example to numerically solve a differential equation the corresponding difference equation is computed. At hardware level the necessary arithmetic is carried out by logic gates, consisting of transistors, implementing boolean logic. One operation of a logical gate can be considered one step in the computational process. Even for a simple addition lots of gate operations happen in parallel, if difference equations are calculated lots of additions are carried out sequentially. It should come as no surprise that modern processors with billions of transistors capable of billions of operations per second generate so many steps in the computational process that humans can't know nor justify them in a lifetime. This is what compels Humphreys to talk about the essentially epistemic opacity of that process. Essential epistemic opacity hinges on the interpretation of impossibility in a rather non-standard way. The traditional story about modalities is that logical possibility encompasses metaphysical possibility which in turn contains physical possibility which includes human possibility. As the opacity is called essential one might think that this hints to logical or at least metaphysical impossibility in the definition. But not so, Humphreys explicitly states that he has human possibility in mind. He urges the reader to consider what humans can do in practice.

> A fourth novel feature of computational science is that it forces us to make a distinction between what is applicable in practice and what is applicable only in principle. (Humphreys 2009 p. 623)

A similar argument for novelty has been made in the philosophy of mathematics concerning a notion of proof which I will discuss in the course of this paper. Humphreys noted this point (Humphreys 2009 p. 617) and might have derived his concept of opacity from it. The similarity has also been noted by Grüne-Yanoff and Weirich in Grüne-Yanoff and Weirich (2010) – without them elaborating further on it.

7.3 Essential Opacity in Other Fields

If essential opacity is to be used as demarcating feature of computer methods the human possibility interpretation is prone to counterexamples of the following sort: For most of the processes involved in scientific instruments humans cannot in practice know every element in the process. The biologist using a microscope cannot (even if quantum mechanics would allow it) know the path of every photon through her instrument. So most scientific instruments are essentially epistemic opaque.

There even is an example from the foundation of maths where the impossibility of justifying something is a logical one. In his foundational program Hilbert tried to justify the infinitary methods of mathematics using only finitary methods (for an overview see Zach 2016) It has been proven that this can't be done the way he envisioned. As humans are arguably finite beings with finite capabilities the infinitary part of mathematics is essentially epistemically opaque for us.

But if already those two examples instantiate epistemic opacity without having something to do with computer methods, opacity obviously cannot be used to demarcate them.

In the following sections I will argue that this failure of demarcation should draw our attention to different and philosophical interesting modes of justification in software engineering.

7.4 Opacity in Computer Aided Proof

Proofs in mathematics are taken to consist of a finite list of premises, some (finite) inferential steps and a conclusion. Take for example Euclid's proof for the infinity of primes.

Premiss: Let $p_1, p_2, \ldots, p_n$ be a finite list of primes.
Premiss: Let $P = p_1 p_2 \ldots p_n + 1$.
Step: P is divisible by at least one prime p. (follows from the fundamental theorem of arithmetic)
Step: p cannot be a member of p1,p2,…,pn, because dividing P by any pn leaves a remainder of 1.
Conclusion: Any finite set of primes can be extended to a larger finite set of primes.

This is a traditional mode of *informal* proof in mathematics. Normally proofs like this aren't even given in list form like above but stated in prose. Mathematicians suppose that every informal proof has a corresponding *formal* proof. The formalization procedure involves a statement of the logical language, the rules of inference and possibly a semantic for that language. The general form of a formal proof is the same: List of premisses, steps, conclusion. Steps have to conform to one of the allowed rules of inference, for example modus ponens or reductio ad absurdum. In contrast to informal proofs it is not possible to skip steps, which makes formal proofs much longer. The upshot is, that every step in a formal proof is explicitly and directly justified by its corresponding rule of inference.

The advent of computer aided proof techniques in the 1970s (most famously the computer proof of the four colour theorem Appel and Haken 1976), lead Tymoczko questioning the traditional concept of mathematical proof (Tymoczko 1979). To mark the novelty of computer aided proofs he introduced the concept of *surveyability* which all non-computer proofs were supposed to have. He defined it implicitly as follows:

> A proof is a construction that can be looked over, reviewed, verified by a rational agent. (Tymoczko 1979 p. 59)

The classical concept of mathematical proof requires that the looking over, reviewing and verifying finishes in finite time. If we take as rational agent a human it becomes quite obvious that surveyability for proofs is similar to not being epistemically opaque for processes. On the side of non-opacity on should be able to examine and justify every step in a simulation while on the side of surveyablity on should be able to review and verify every step in the proof. Or to make the analogy even more striking: An unsurveyable proof is epistemically opaque for the human mathematician.

Tymoczko, using the computer aided proof of the four colour theorem as an example, alleged that the proof was unsurveyable (Tymoczko 1979 p. 68). The proof needed the help of a computer to rule out 1936 possible counterexamples, which were to intricate for any human mathematician to solve in any realistic time. Therefore, said Tymoczko, a new concept of proof is needed. Furthermore computer aided proofs wouldn't be a-priori anymore (see Tymoczko 1980). The ensueing debate in philosophy of mathematics focused much on this point (for example see Teller 1980). As of now this debate is still ongoing and the consensus seems to be that computer aided proofs are (contra Tymoczko) a-priori (see McEvoy 2008).

The mathematicians themselves were more concerned with developing techniques allowing humans to check computer generated proofs. This shows that they saw unsurveyable proofs as somewhat deficient and not worthy of the trust a well checked classical proof enjoys. Their remedy was quite ingenious. They started employing small programs called proof checkers, which were themselves surveyable (or epistemically transparent). Those proof checkers verified the computer generated proofs. The mathematicians trusted the proof checker (which they surveyed) and in turn trusted the computer aided proof. This was no easy work and the first verified proof of the four colour theorem came up only in 2005 (see Gonthier 2008).

What is the upshot of this debate for the philosophy of computer simulation? First of all surveyability would be a direct counterexample against the idea that computer simulation can be demarcated by epistemic opacity. There is at least computer aided proof in which it figures prominently. To be fair, this isn't exactly what Humphreys was claiming. He took computational science to be defined by the anthropocentric predicament, arguing that computer methods "push humans away from the centre of the epistemological enterprise" (Humphreys 2009 p. 616). And it can be contented that those methods include computer aided proof. However the debate from mathematics shows, that even if there is opacity mathematicians try to reduce it to point where it doesn't affect their justifications any more.[1] The same can be said about computer simulation. Software engineers developed techniques of verification and validation which serve the same end as proof checkers – namely to increase the trust in the results of a computer simulation. While the field of verification and validation is vast (see for example Winsberg 2010) it is by any means much messier than the deductive proof checking in mathematics. The trust generated by verification and validation follows a more inductive spirit (testing of singular cases and inferring the validity for all cases for example). Therein lies the main difference between the two. Contrary to Humphreys computers don't replace humans as epistemic arbiters, they rather supplement their justificatory capabilities while humans remain the ultimate judges.

7.5 Justification in Software Engineering

Engineering is generally taken to be about making stuff (work). So the philosophy of engineering is concerned with the relation of engineered things to the world (Franssen et al. 2015; Gabbay et al. 2009). Software engineering is about the rather abstract thing called software. Therefore the natural questions philosophers asked about software were questions of identity and existence (see Gruner 2011). I will argue that in light of the preceeding section the focus in software engineering shifts from mere creation to justification – a concept which seems foreign to engineering. Philosophy should follow suit.

First of all I take computer simulation to be part of the broader field of software engineering. I also consider scientists implementing a simulation involved in an engineering activity. In the traditional picture of engineering its theoretical tools and methods are grounded in some parent science (see Gruner 2010). So the justification for those methods arises from their parent science too. For example the statistical properties of any numerical algorithm are given by mathematical analysis. The application of the Ising model to magnetism is due to a physical interpretation, justified by physics. I am not disputing that engineers take part in all those different

[1] There is an argument to the effect that unsurveyability was accepted in mathematics prior to computer aided proofs (see Arkoudas and Bringsjord 2007). If this is correct it would even undermine Humphreys more general argument.

activities, this might even be a defining characteristic of the discipline. But in designing an artifact an engineer uses those methods rather than justifying them. Looking at computer simulation the engineered artifact itself – namely the simulation – is often the object under consideration. Simulation studies are conducted to find out about properties like parameter sensitiveness, convergence rates etc. of a particular simulation. Those and the procedures of verification and validation are in turn used to assess the epistemic power of the simulation result. Engineers are so involved a twofold activity: (1). Establishing and justifying the simulation methods. (2). Using those methods to arrive at a simulation result.

This, I contend, is new to engineering and needs to be reflected in the philosophy of engineering.

References

Appel, K., & Haken, W. (1976). Every planar map is four colourable. *Bulletin of the American Mathematical Society, 82*, 711–712.

Arkoudas, K., & Bringsjord, S. (2007). Computers, justification, and mathematical knowledge. *Minds and Machines, 17*, 185–202.

Franssen, M., Lokhorst, G.-J., & van de Poel, I. (2015). *Philosophy of technology, The Stanford encyclopedia of philosophy*. New York: New York City College of Technology.

Frigg, R., & Reiss, J. (2009). The philosophy of simulation: Hot new issues or same old stew? *Synthese, 169*, 593–613.

Gabbay, D. M., Meijers, A., & Woods, J. (2009). *Philosophy of technology and engineering sciences*. Amsterdam: North Holland Elsevier.

Gonthier, G. (2008). Formal proof—The four-color theorem. *Notices of the AMS, 55*, 1382–1393.

Gruner, S. (2010). Software engineering between technics and science. *Journal for General Philosophy of Science, 41*, 237–260.

Gruner, S. (2011). Problems for a philosophy of software engineering. *Minds and Machines, 21*, 275–299.

Grüne-Yanoff, T., & Weirich, P. (2010). The philosophy and epistemology of simulation: A review. *Simulation & Gaming, 41*, 20–50.

Humphreys, P. (2009). The philosophical novelty of computer simulation methods. *Synthese, 169*, 615–626.

McEvoy, M. (2008). The epistemological status of computer-assisted proofs. *Philosophia Mathematica, 16*, 374–387.

Popper, K. R. (2005). *Gesammelte Werke: Band 3: Logik der Forschung*. Tübingen: Mohr-Siebeck.

Teller, P. (1980). Computer proof. *The Journal of Philosophy, 77*, 797–803.

Tymoczko, T. (1979). The four-color problem and its philosophical significance. *The Journal of Philosophy, 76*, 57–83.

Tymoczko, T. (1980). Computers, proofs and mathematicians: A philosophical investigation of the four-color proof. *Mathematics Magazine, 53*, 131–138.

Winsberg, E. (2010). *Science in the age of computer simulation*. Chicago: The University of Chicago Press.

Zach, R. (2016). *Hilbert's Program*, The Stanford Encyclopedia of Philosophy (Spring 2016 Edition).

Chapter 8
Three Stages of Technical Artifacts' Life Cycle: Based on a Four Factors Theory

Nan Wang and Bocong Li

Abstract Technical artifacts had been a lost theme in philosophy for a long time until the emergence of philosophy of technology in the end of nineteenth century. In the twentieth century, Herbert Simon put forward sciences of the artificial, and Peter Kroes and Anthonie Meijers proposed a theory of the dual nature of technical artifacts. From the point of view of philosophy of technology, Aristotle's four causes theory is an explanation on the nature of technical artifacts in essence rather than that of the natural world. However, Aristotle, Simon, Kroes and Meijers did not pay attention to technical artifacts' life cycle, which inevitably leads to the failure of their theory of technical artifacts to some extent. The fact that technical artifacts, like living beings, are as mortal as human beings means that they also have a life cycle from birth to death. In order to interpret these phenomena, we plan to raise a theory of four factors of technical artifacts. However, this paper only aims to define the life cycle of technical artifacts as three stages, i.e. creation, vitality, and ending. And it also going to examine the creation stage and re-interpret Aristotle's four causes theory according to the four factors theory.

Keywords Technical artifact · Life cycle · Three stages · Four factors

8.1 Introduction

More than millions of years ago, there were neither human beings nor technical artifacts on the Earth. During the evolution of the human beings, the technical artifacts emerged with the beginning of the human beings. Exactly speaking, human beings begun to make technical artifacts and technical artifacts were created by human beings. With the development of history and society, the number and types of technical artifacts have risen steadily. Now human beings live in a world full of

N. Wang · B. Li (✉)
School of Humanities and Social Science, University of Chinese Academy of Sciences, Beijing, People's Republic of China
e-mail: wangnan@ucas.ac.cn; libocong@ucas.ac.cn

A. Fritzsche, S. J. Oks (eds.), *The Future of Engineering*, Philosophy of Engineering and Technology 31, https://doi.org/10.1007/978-3-319-91029-1_8

technical artifacts. On the one hand, human beings made various technical artifacts to achieve their own purposes, on the other hand technical artifacts affected on human beings in many ways as well.

Human nature has been one of the most important topics philosophers are concerned about. Different philosophers have different ideas on it. Benjamin Franklin defined man as a tool making animal. Karl Marx further developed Franklin's point of view. Marx stated that the making and using technical artifacts not only distinguished human beings themselves from animals, but also promoted the development the society. He pointed it out,

> Men can be distinguished from animals by consciousness, by religion or anything else you like. They themselves begin to distinguish themselves from animals as soon as they begin to produce their means of subsistence, a step which is conditioned by their physical organization. By producing their means of subsistence men are indirectly producing their actual material life. (Marx 2000, p.6)

There is no doubt that human beings cannot live without technical artifacts. Without technical artifacts human society would be no longer in existence. However, the technical artifacts that filled with the material life in which human beings lived did not become a core issue of philosophy until the emergence of philosophy of technology in the end of the nineteenth century. Historically, making and using technical artifacts is a largely unthinking activity (Mitcham 1994, p. 1) and a lost theme in the field of philosophy (Li 2002, p.2).

With the rapid growth of technology and engineering, the technical artifacts began to draw wide attention in the second half of the twentieth centuries. One of the important contributions is *The Sciences of the Artificial* (1969) by Herbert A. Simon, who was a Nobel Prize laureate, and an American political scientist, economist, sociologist, psychologist, and computer scientist. In this book, Simon identified the four indicia that distinguish the artificial things from the natural ones, and set the boundaries for sciences of the artificial. According to him, the artificial things

> [a]re synthesized (though not always or usually with full forethought) by human beings; may imitate appearances in natural things while lacking, in one or many respects, the reality of the latter; can be characterized in terms of functions, goals, adaptation; and are often discussed, particularly when they are being designed, in terms of imperatives as well as descriptives. (Simon 1996, p. 5)

Almost a half century later, some philosophers advanced a new research program "The Dual Nature of Technical Artifacts" at the beginning of the twenty-first century. The contributors were Dutch school of philosophy of technology, including Peter Kroes, Anthonie Meijers, Hybo Houkes, Pieter E. Vermaas. They argued that the technical artifacts "can be said to have a dual nature: they are (i) designed physical structures, which realize (ii) functions, which refer to human intentionality". The technical artifacts are the creations of mind and matter. (Kroes and Meijers 2002; Kroes 2012)

Simon and the Dutch school, in different times and with different forms of expression, both reveal the current conception of technical artifacts. It is believed that "artificiality extends not just to some physical objects but also to intellectual phenomena such as science, art, and technology—to the extent that they are characteristic of human life". (Negrotti 2015, p. 127) From these views, the issue of technical artifacts is certainly not trivial in philosophy, especially in philosophy of technology and engineering, and the nature of technical artifacts is the combination of physical and intellectual matters.

In the history of philosophy, Aristotle put forward his famous four causes theory. Although Aristotle did not regard his four causes theory as a theory of technical artifacts, contemporary scholars may find out that the essence of four causes theory is a theory of technical artifacts. Moreover, just as a man is a mortal being, who is an individual destined to die, a technical artifact is also the similar thing. Both their lives can be divided into a number of stages according to certain principles. This article is going to propose a theory on three stages of technical artifacts' life cycle, based on the comments on Aristotle's four causes theory, and Kroes and Meijers's dual nature theory.

8.2 Life Cycle of Technical Artifacts and Its Three Stages

When it comes to characteristics of technical artifacts, it is easy to see that all artifacts are necessary to go on a journey from being created to ending. During the process of being created and ending a technical artifact has its vitality. It means that a technical artifact is a physical and mortal being. The life journey a technical artifact goes on is similar to that of human beings, which consists of being born, growing, ageing and dying. Accordingly we can say a technical artifact also has its own life cycle.

The life cycle of a technical artifact (or product life cycle) can be divided into several stages characterized by the revenue generated by the product. But it seldom appears in philosophy. Schinzinger and Martin defined engineering as a process from concept to product, which includes initiation of task (idea, specific request, or market demand), design (preliminary analysis, simulation/prototyping, specifications for materials and components, detailed shop drawings), manufacture (assembly/construction, quality control/testing), implementation (advertising, sales and financing, use of the product), and final tasks (rebuilding, recycling, disposal of materials and wastes). (Schinzinger and Martin 2000, p.3) It is an idea close to life cycle of technical artifacts but under the name of "progression of engineering tasks".

8.2.1 Stage 1: The Creation Stage of Artifacts

The creation of technical artifacts is the first stage, in which the idea or design of a technical artifact can become a concrete product. In this sense some scholars regarded the technical artifacts as the creation of mind. This tendency seems to be obvious especially among the studies on engineering design, though it cannot be denied that the design is the premise of and the first step to create artifacts.

However, human beings cannot make technical artifacts from nothing. Technical artifacts have to be made of or from some kinds of materials, such as stone, bronze, timber, plastics and so on. Without design there would not be technical artifacts. Likewise, without materials there would not be technical artifacts.

Here it is necessary to take notice of the two different meanings of the word "material". Among the several meanings of material, there are two we have to pay attention to. One refers to the matter from which a thing is or can be made, while the other denotes the tangible substance that goes into the makeup of a physical object. According to the first meaning, only a small number of physical objects can be used as material to make technical artifacts, while according to the second one, all physical objects are materials. In this article the word "material" refers to the first one instead of the second one. For example, in the Stone Age, the stone could become an important material for human beings, while bronze was unlikely; similarly, in the Bronze Age, bronze could be an important material, while iron was impossible. Moreover, man can use brick, stone, wood and even ice, as materials, to build a house, but cannot use air and water to build a house, because air and water are not materials for building a house.

Because there would be no technical artifacts without designs and materials, the creation of technical artifacts involves both the conceptual creation and the matter one. Just as Peter Kroes says, technical artifacts are creations of mind and matter.

The creation of technical artifacts needs the factors of the design and the material, but also the agent and the tool. The tool is used by man to make technical artifacts, and the agent refers to the people who make artifacts with materials by tools on the basis of designs. According to Genesis in the Bible, God created the world out of nothing, and as long as we are nothing, he can make something out of us. However, the agents, no matter engineers, workers or other occupational staff, have to create artifacts with materials by tools. It is obvious that when the design is completed, agents are still unable to create technical artifacts if they do not have physical materials and tools.

It should be noted that the relationship between materials and products is much more complex than most commentary suggests. Firstly, the technical artifacts can be made of such a wide variety of materials. In ancient times, different kinds of materials, especially those which can be found in nature, were used to make artifacts; while in current times, more kinds of artificial things are invented and used to make artifacts by human beings, such as steel, plastic, nylon material, and so on.

Secondly, materials and non-materials all are things, but only the former can become technical artifacts. Because only materials have their own particular potentials to turning purposes into the specific artifacts, while non-materials do not have. Philosophy usually is concerned about all else with things, however, philosophy of technology and engineering only focus on materials, rather than non-materials, and its judgment on whether a thing can be a material is nothing else but meeting the purpose. According to this, a technical artifact can be made of or from more than one kind of materials for all their potentials are required to serve a purpose or purposes. In other instances, one kind of materials can become more than one kind of artifacts with increasing discovery and exploration on hidden potentials based on the advances in technology. The petroleum is one such example. It is mainly used for vehicles now, but it has been used for centuries to be the fuel of lighting and medical product before vehicles came out.

8.2.2 *Stage 2: The Vital Stage of Artifacts*

The second stage of artifact life cycle is the vitality of artifacts. According to the Kroes and Meijers's dual nature of technical artifacts, a conclusion can be drawn that technical artifacts, as mortal beings, have two kinds of vitality, i.e. the structural and functional vitality, or the physical and socio-functional vitality. The dual vitality is closely related to each other. In a certain sense, human beings are dual beings of body and mind, as well as the technical artifacts are those of structure and function. Because the dual vitality of artifacts has a close relationship with Kroes and Meijers's theory, this article will not discuss this issue further.

8.2.3 *Stage 3: The End Stage of Artifacts*

The third stage is the ending of artifacts for they cannot maintain their vitality forever. Most artifacts end in the form of wastes, when they do not have their original physical and socio-functional vitality. However, the ways that they end are varied. There is a special situation in which the artifacts with their original physical and functional vitality are still regarded as wastes, due to the fact that their economic and social value comes to an end. In modern society, when a new type of computers appears in the market, it is easy to regard the old type of computers as wastes. For example, although a 286 computer functioned well, it is still discarded as useless stuff or a white elephant when the consumers have purchased a new type.

In a different case, wastes are considered as the corpses, or called dead bodies of original products. It implies those wastes cannot function normally as original products do. It caused by the different structures of wastes compared with those of the original products. The artifacts originate from people's purposes initially; while wastes do not root in people's purposes, even the oppose ones.

It is interesting that there is no discussion on wastes in ancient time. However, this should not be a missing theme any more presently. From the philosophical point of view, this phenomenon often implies the alienation. In this context, wastes are not natural objects, but a special kind of artifacts, or rather, as remainders or dead bodies of the original artifacts. They also are different from the artificial materials and final products, because they are the results from the combination of natural change and human operation; furthermore, they usually do harm to natural environment and human beings. In modern society, wastes bring a lot of trouble to human being life. Engineers and philosophers should pay attention not only to the creation and the dual vitality of artifacts, but also to the end of artifacts.

The three stages of technical artifacts' life cycle have raised a lot of issues. But the following will only focus on those issues in the creation stage.

8.3 Re-interpreting Aristotle's Four Causes Theory

As one of the greatest philosophers, Aristotle's writings covered many subjects which consisted of a comprehensive system of Western philosophy. In this system it contains the views on technical artifacts in *Physics* and *Metaphysics*.

Following the previous philosophers, Aristotle involved in the causal investigation of the world in which human beings lived. The causal investigation aims to search for the Wisdom which is knowledge about certain principles and causes of things, and "men do not think they know a thing till they have grasped the 'why' of (which is to grasp its primary cause)" (*Physics* II 3). However, Aristotle made a critical comment on the results reached by his predecessors (*Metaphysics* I 3–7). He believed that the predecessors merely touched upon one or more of the material, formal, efficient and final causes, and was lack of a complete understanding of the range of possible causes and their systematic interrelations. (Falcon 2006) This leads to his general account of the four causes. According to him, the cause means

> (1) That from which, as immanent material, a thing comes into being, e.g. the bronze is the cause of the statue and the silver of the saucer, and so are the classes which include these. (2) The form or pattern, i.e. the definition of the essence, and the classes which include this (e.g. the ratio 2:1 and number in general are causes of the octave), and the parts included in the definition. (3) That from which the change or the resting from change first begins; e.g. the adviser is a cause of the action, and the father a cause of the child, and in general the maker a cause of the thing made and the change-producing of the changing. (4) The end, i.e. that for the sake of which a thing is; e.g. health is the cause of walking. For 'Why does one walk?' we say; 'that one may be healthy'; and in speaking thus we think we have given the cause. (*Physics* II 3 and *Metaphysics* V 2)

Among the four causes, the (2), (3) and (4) has a special relationship. An adequate explanation of the *formal cause*, e.g. the shape of a statue, requires a reference to the *efficient cause* or the principle that produces the statue. Similarly, a proper interpretation of the *efficient cause*, e.g. the art of bronze-casting, also needs to refer to the *final cause* or that for the sake of which everything in the production process is done.

When he tried to explain causes of things with the four causes theory, Aristotle actually provided an explanation that makes a reference to the *telos* or end of the process, without beliefs, desires, intentions or anything of this sort. Some have contended that Aristotle explains natural process on the basis of an inappropriately psychological teleological model; that is to say, a teleological model that involves a purposive agent who is somehow sensitive to the end. (Falcon 2006)

A contrast view on the "causes" of the technical artifacts can be found in a Chinese book named *Kaogongji* [The Records of Examination of Craftsmen], which is an official book of the *Qi* state on handicraft art during the Spring and Autumn period (770–476 BC) composed by an unknown author. This book is believed to be a late first millennium B.C. addition to the *Zhouli* [Rites of the Zhou], one of three ancient ritual texts (the "Three Rites") listed among the classics of Confucianism appeared in the middle of the second century B.C., a replacement for a lost portion from earlier Zhou times.

In Chinese opinion, the idea that the artificial things are invented and created by the sage has the historical allusions. *Yijing* [Book of Changes], one of the earliest classic books in China and even in the world, described how the ancient sages to do it based on their examination of natural phenomena. Following this tradition, *Kaogongji* proposed the principle of "four elements" (heaven, earth, material, human), which is the earliest systematic and all along the elementary Chinese views of technics (Fan 2005: 119), as the explanation of the technical artifacts. The following passage has further explanation to it.

> The heaven has the seasonal and climate change, the land has the geographical differences, the material has various properties, and craftsman has different types of creativity and skill. Gathering these four can produce good products. (Chapter of "The Six Jobs in China", Wen 1993, p. 5)

Here, *Kaogongji* offers an interesting Chinese contrast to Aristotle's four (material, formal, final, and efficient) causes. The Chinese views the artificial world by the way of the natural world, which means to establish and explain the technical artifacts in terms of the aspects of the natural world (human being is also regarded as a part of natural world or the human aspect of the natural world). The Chinese view can be exemplified by the making of the ancient carriage.

> The square of the cross board at the rear of a carriage is the symbol of the earth, the round of hood is heaven, the thirty-arm of wheel is the thirty days of each month, and the twenty-eight-bow-hood frame is twenty-eight constellations. (Wen 1993:53)

Whereas Aristotle projects his four causes, which are derived from reflection on human making, into the world of nature, the *Kaogongji* limits itself to aspects of the world that the human worker needs to take into consideration. There is no attempt to apply the Chinese analysis to reality as a whole or to give it metaphysical significance.

8.4 Interpreting the Creation Stage Based on the Four Factors

There is a paradoxical phenomenon in the history of philosophy. As the ordinary people, the philosophers could have nothing to do with the technical artifacts in every aspects of their life. However, they turned a blind eye to technical artifacts in their work or regarded this issue as a not fully valued one. Aristotle is one of this sort, though his four causes theory, which is originally an account of natural world, provides a kind of explanation of technical artifacts.

One reason for the phenomenon is that majority of philosophers belong to slave-owning class. Because of the classical preconception, the philosophers pay little attention to the technical work for which the slaves are responsible. Aristotle appeared to be an exception when he argued that "if the shuttle would weave and the plectrum touch the lyre without a hand to guide them", and "if every instrument could accomplish its own work, obeying or anticipating the will of others". (*Politics*, I 4) However, his arguments are certainly not the prediction of automation and cybernetics, but intend to defend and justify the slavery in terms of the context. (Jean-Yves Goffi 2000, p. 39)

Aristotle's four causes theory is developed in the context of his science of physics, but its application expands on the explanation of the human action and artistic production, besides that of the natural world. He does not limit the artificial to one or two causation, as other philosophers do, but offers the most general presentation of four causes. However, this theory contains the inductive reasoning, as well as the teleological argument, which signifies a view of devalued technical artifacts and action. He stated that "the smith fashions the Delphian knife for many uses", while the nature "makes each thing for a single use, and every instrument is best made when intended for one and not for many uses". (*Politics*, I 2)

Further, some philosophers argued that the meaning of the term "cause" that Aristotle used is different with that in modern times. One of the famous arguments is made by Martin Heidegger in "The Question Concerning Technology" (1954). According to him, people currently are accustomed to "representing cause as that which brings something about", i.e. "to obtain results, effects". Thus the four causes "of being responsible bring something into appearance. They let it come forth into presencing [An-wesen]". (Heidegger 1977)

Although Aristotle's four causes theory suffers criticism, it is still the source of proposing a new theory on the nature of technical artifacts, i.e. the four factors of artifacts. As the confusions and misunderstandings caused by the term "cause", the new theory will adopt the term "factor", which means one of the parts of something that makes it work, or a quality that makes someone or something effective, according to Cambridge Dictionary. The descriptions on technical artifacts in Chinese book *Kaogongji* can be considered as the embryo of this idea.

The four factors of artifacts contain the material, design, agent and tool factor. (1) The material factor, which is similar to the material cause, describes the material out of which something is composed. (2) The design factor will replace Aristotle's formal and final cause. The design "is the human power of conceiving, planning, and bringing to reality all of the products that serve human beings in the accomplishment of their individual and collective purposes". (Buchanan 2015) It indicates that it is the design that determines the form of a technical artifact, as well as achieves human purpose by bring a technical artifact into reality. (3) The agent factor substitutes for the efficient cause. The agent here refers to something that produces or is capable of producing an effect. The agent factor means the resource that from which the change of a technical artifact under consideration proceeds. (4) The tool factor is newly added. The tool factor denotes something (as an instrument or apparatus) used in performing an operation or necessary in the practice of a vocation or profession. The significance of tool factor in making and using a technical artifact is more important currently than in ancient time for the increasing complexity of technology and engineering.

There are still some important issues in vitality and ending stage, but more discussions on it will appear in other articles.

8.5 Conclusion

The making and using technical artifacts not only distinguishes human beings themselves from animals, but also produces a material life of human beings. Technical artifacts are created by human beings, and have a similar life cycle with their makers. But they have never been one of the philosophical concerns until the birth of philosophy of technology in the end of the nineteenth century.

Herbert A. Simon, Peter Kroes and Anthonie Meijers contributed to this topic in different ways. Simon created the discipline of artificial science and emphasized the value of design, and Krose and Meijers proposed the idea of the dual nature of technical artifacts, e.g. the structure and function, and technical artifacts are the creation of mind and matter.

However, none of them refers to technical artifacts' life cycle. The life of technical artifacts, which is as mortal as that of human beings, has a life cycle from birth to death with several stages. (1) The creation of technical artifacts is the first stage, in which the concept or idea of a technical artifact becomes a concrete product. (2) The second stage of artifacts life cycle is the vitality of artifacts. The dual vitality of technical artifacts consists of structural and functional one, or the physical and socio-functional one. (3) The third stage is the ending of artifacts for they cannot maintain their vitality forever. The artifacts end in the form of wastes, when they do not have their original physical and socio-functional vitality. However, the ways that they end are varied.

Differing from Aristotle, Simon, Kroes and Meijers, this article regards that technical artifacts' life cycle consists of three stages and the nature of artifacts is a unity of four factors, e.g. the material, design, agent and tool factor. In this article only the creation stage is briefly examined based on a new four factors theory which is inspired by Aristotle's four causes theory.

References

Buchanan, R. (2015). Design ethics. In J. B. Holbrook (Ed.), *Ethics, science, technology, and engineering: A global resource* (pp. 557–563). Farmington Hills: Macmillan Reference USA.

Falcon, A. (2006). *Aristotle on causality. Stanford encyclopedia of philosophy*. http://plato.stanford.edu/entries/aristotle-causality/#toc. Online stanford encyclopedia of philosophy. Accessed 31 Nov 2016.

Fan, Q. (2005). Kaogongji de gongyi sixiang [The crafts thoughts of *Yijing*]. *Shi xue yue kan* [Journal of Historical Sciences], *10*, 119–121.
Goffi, J. -Y. (2000). *Jishu zhexue* [Philosophy of technology] (D. Maoyong, Trans.). Beijing: Commercial Press.
Heidegger, M. (1977). The question concerning technology. In W. Lovitt (Ed.), *The question concerning technology and other essays* (pp. 3–35). New York: Harper & Row Publishers.
Kroes, P. (2012). *Technical artefacts: Creations of mind and matter*. Dordrecht: Springer.
Kroes, P., & Meijers, A. (2002). The dual nature of technical atrtifacts—Presentation of a new research programme. *Techne: Journal of the Society for Philosophy and Technology, 6*(2), 4–8.
Li, B. (2002). *Gongcheng zhexue yinlun* [An introduction to the philosophy of engineering]. Zhengzhou: Daxiang chubanshe [Daxiang Press].
Marx, K. (2000). *The German ideology*. Online Version: Marx/Engels Internet Archive. marxists.org.
Mitcham, C. (1994). *Thinking through technology: The path between engineering and philosophy*. Chicago: University of Chicago Press.
Negrotti, M. (2015). Artificiality. In J. B. Holbrook (Ed.), *Ethics, science, technology, and engineering: A global resource* (Vol. 1, pp. 127–129). Farmington Hills: Macmillan Reference USA.
Schinzinger, R., & Martin, M. W. (2000). *Introduction to engineering ethics*. Boston: McGraw Hill.
Simon, H. A. (1996). *The sciences of the artificial* (3rd ed.). Cambridge, MA: MIT Press.
Wen, R. (1993). *Kao gong ji yi zhu* [Translation and Annotation of Kao Gong Ji]. Shanghai: Shanghai Classic Publishing House.

Chapter 9
Métis: Reconfiguring the Philosophy of Engineering

José Aravena-Reyes

Abstract The philosophy of engineering often conceives engineering as subordinate to technology, that is, as a kind of technological problem solving through scientific knowledge. However, many engineering solutions have been criticized because of the particular technical objects produced. Philosophical arguments by such thinkers as Gilbert Simondon, Gilles Deleuze, and Bernard Stiegler challenge this approach and suggest new ways for understanding engineering. Using etymological arguments, Carl Mitcham has contributed to the debate, although he consistently points to a *lógos* built on the epistemological grounds of modern science. This work proposes a reconfiguration of his etymological analysis that rescues *métis*, a word used by the Greeks to describe this cunning and inventive mode of thinking, something which is strongly associated with ancient and contemporary technical production, but has yet to be explored philosophically. The inclusion of *métis* in philosophical discussions of engineering may open new perspectives on the role of engineering in society, especially in the face of problems with the contemporary production of technical objects and the implications in the so-called anthropocentric age.

Keywords Philosophy of engineering · Technology · Invention · Techné · Métis

9.1 Introduction

Even though the philosophy of engineering can be considered a recently constituted discipline (Jaramillo 2015; Silva 2011), it is not uncommon to see that engineering studies are developed from the perspective of the philosophy of technology. This can be largely attributed to the fact that it is almost intuitive to relate engineering and technology. However, Van de Poel and Goldberg (2010: 2) affirm that for many researchers, engineering and technology are different areas: primarily, engineering

J. Aravena-Reyes (✉)
Faculty of Engineering, Civil Construction Department, Federal University of Juiz de Fora, Juiz de Fora, MG, Brazil
e-mail: jose.aravena@ufjf.edu.br

A. Fritzsche, S. J. Oks (eds.), *The Future of Engineering*, Philosophy of Engineering and Technology 31, https://doi.org/10.1007/978-3-319-91029-1_9

should be an activity that produces technology, but a type of technology for which engineers themselves are specifically responsible, excluding other meanings given to this word, such as the notion of "technologies of the self" (Foucault 1988) or "technologies of the intelligence" (Levy 1990). The technology assigned to the engineers' work would be a depository of a concept of very delimited technicity, at times too narrow, which may well be the basis of the common sense that considers engineering to be a profession that deals with problem solving through the use of scientific knowledge. However, given recent socio-political and environmental problems of globalized and highly technologized production in modern society, such definition is limited and it is urgent to establish a solid foundation that can help to better understand the technical production process and the role engineering has in it (Mitcham 1998).

From this perspective, the question "what is engineering?" should not always be answered based on ontology with a positivist nature that puts science as a reference, because it runs the risk of tautologically reinforcing the dominant perspective.

On this subject, Gilbert Simondon, for example, is emphatic: scientific knowledge is not an absolute condition for the existence of technical objects.[1] The (pre) ontology of Simondon (1958a) refers to an object in constant change, in constant process of individuation. In this way, approaches such as Mario Bunge's (1980: 45) do not express clearly the inventive dimension of technical production and are not useful for this purpose, for they are unable to describe in universally demonstrable and stable terms the production of what is new, as a manifestation of the unstable or variable.

To Simondon, the development of the technical object is not found in the specialization of its functions according to externally defined intentionalities, but in the inner coherence of such objects. Therefore, to talk about technology as an advanced stage of technique (i.e., a technique that evolves in response to advances in science), more than describing the particular evolutionary condition of the existence of technical objects, might show the external intentionality that drives the current capitalist process of technical production, inside which the scientific knowledge serves as a guide for the production of contemporary technical objects, but hardly to explain – in philosophical terms – its productive-inventive nature or its evolutionary process.

The situation becomes critical with regard to the constant complaints made by several social actors who point out the harmful effects caused by the contemporary mode of technical production, questioning even the idea of technological progress. Although this has always been a controversial subject, for years the cultural pessimists have criticized the idea of a technical and scientific progress and, faced to the serious problems affecting the humankind by the harmful effects produced by the actual functioning of several technical objects, it is hard nowadays to categorically state that society has evolved by the support of the progress of science and technology. On the contrary, there is strong critical positioning. For example, the Brazilian anthropologist Eduardo Viveiros de Castro claims that several choices of human

[1] According to the author, scientific knowledge does not allow the prediction of all the effects of technical objects. (Simondon 1958b: 57).

technique have caused or may cause the extinction of mankind (2014: 133), reason by which there are a group of people on planet ("earthbounders") that simply are not expecting any global solution in the dominant model. They are preparing themselves to the critical future events: the anthropocentrics events. In effect, Viveiros de Castro do not disagrees with the social criticisms, he just appears to have the conviction that analyses made within the epistemological framework of the dominant model of life, does not offer answers to the problems of the anthropocentric era.

Such statements lead to serious reflection on the course that the contemporary mode of production is taking. In a sense, the approach of the dominant thought has failed to provide answers to social criticism: it seems that it is no longer possible to invent technical objects that avoid the harmful effects of contemporary technical production. All that remains is to think that there is no other possible world, and that therefore the only option is to improve current production methods, always taking as a basis, the epistemological ground of dominant thought.[2] Under these circumstances, returning to the etymological roots of the word technology can help elucidate the reasons that led engineering to become a passive agent that solves problems without questioning the constitutive order of the active agents that created these problems in the first place.

The genealogical work deserves full attention, but an additional effort is also required to elucidate the technical condition itself, to reposition the philosophical foundations of engineering in order to enhance the human inventive capacity towards new worlds that are less harmful for life on the planet. In this direction, this work intends to reconfigure the philosophy of engineering in relation to the dominant thought, giving back the importance of the inventive process in engineering through the rescue of "*métis*",[3] an ability of the living being associated with the word technique from its birth, in both Greek culture and philosophy.

9.2 *Techné* in the Philosophy of Technology

Paul Durbin (2006) wrote that Carl Mitcham, pioneer of the philosophy of technology, was who most clearly elaborated arguments "for" and "against" many philosophers of technology. He also pointed out that the main work of Mitcham[4] served as much as a basic reference as it did a suitable history of the field of philosophy of technology. Durbin seems to adhere to the idea that Mitcham is one of the leading researchers in the field of philosophy of technology and engineering. The main

[2]This is named by Simondon as the "hypertelic phenomenon": an object's unadaptation with respect to its environment, caused by the excessive specialization produced by an analytical more than synthetic thought, that, in its limit, it represents the total failure of the technical object.

[3]The neglect of *métis* in the history of ancient thought was reported by Detiene and Vernant (1974: 17) and also by Anagnostopulos & Chelidoni (2008: 436).

[4]Thinking Through Technology (1994) is seen as the principal work on the philosophy of technology by Carl Mitcham.

issue addressed by Mitcham is technology and, although in his early work he established a clear difference between technique and technology (1989), in subsequent works he used an etymological approach to the word "technology" (Mitcham and Schatzberg 2009: 28).

Indeed, in his main work, Mitcham devoted an entire chapter to address the philosophical issues of *techné* (1994: 114). In this study he highlighted the importance of etymological studies of the word 'technology', a subject later revisited in the book edited by Meijers (2009). In fact, to develop his concept of engineering sciences, Mitcham uses and guides himself with the term 'technology', for which he elaborates a complete etymological analysis, citing especially Socrates, Plato and Aristotle.

Mitcham (1994) tries to establish a history of ideas about technology, taking as a starting point the *techné* of Socrates and Plato described in Gorgias, Ion and Philebus' dialogues. In them, early Plato initially adopted the idea that *techné* referred to those human activities about which one could speak or reason (and teach), while later Plato would undertake a classification of *techné* according to the type of knowledge to be used. From this etymological (and historical) analysis, Mitcham argues that the first appearance of the word 'technology' in its Greek form (*techné* + *lógos*), comes from Aristotle (Mitcham 1994: 199).

To support this argument, the author cites extracts from the Metaphysics and from Rhetoric (Aristotle 2015, 2012) in order to bring to light the role that the Greeks gave to *lógos* and thus show that *techné* forms the idea of what needs to be done through the use of reason (Mitcham 1994: 121–122). In fact, the big difference pointed out by Mitcham between the Greek *techné* and the modern notion of technology is that while the first one involves a *lógos*, this *lógos* is not a rationalization of productive activity itself, presented without any dependency on particular knowledge of one or other technical activity. According to Mitcham, the lack of an external rationality is the main reason for the absence of a general treatise on the *techné* in the Aristotelian corpus. Mitcham also resorts to Rhetoric to show that Aristotle tried to link *techné* and *lógos* to define rhetoric as the art (*techné*) of the means of persuasion (*lógos*).

This line of thought is presented once more in the manual edited by Meijers, 14 years later (Mejers 2009). In Mitcham and Schatzberg's paper, the thoughts of Socrates, Plato and Aristotle are synthesized in the same way, stating that perhaps the limited role given to *lógos* in Greek *techné* may explain, in the work of both Plato and Aristotle, the absence of the Greek word technology, making again the same reservations as to the emergence of this term in Rhetoric (Mitcham and Schatzberg 2009:35).

Surely, it can be considered that the contributions to the reconstitution of the understanding that Aristotle elaborated on *techné* can all be extracted from many texts (as in the case of the Rhetoric as seen above), however, in Book I of Metaphysics, Aristotle points out that the difference between art, science, and the other correlated mental activities were all indicated in this text (*Metaphysics 2015*: 43–981.b25). Mitcham's perspective seems to be oriented to highlight a *lógos* that was lost or had little importance in the *techné* of Plato and Aristotle and although he himself considers that there is no background on the continued use of the word technology, the author reports that Jacob Bigellow might have adopted in English the word technology that could have been found in old dictionaries and used it to include a description of the more cultured arts, in particular those which involved the applications of science (Mitcham and Schatzberg 2009:36). It seems that, for Mitcham, the *lógos* of modern *techné* is oriented to a privileged point of observation of a universal technical process (Mitcham 1989: 129) starting from the Aristotelian *lógos* that arises in Rhetoric.

It seems to be the historical line of reasoning that explains the current meaning of the word technology, according to Mitcham. From this reasoning it is possible to infer that the need to highlight the Greek *lógos* obeys an implicit intention: sustaining the presence of *lógos* of modern science in the characterization of the term technology, in its current meaning. In fact, there is no reason to disagree on that, but certainly the Greek *lógos* is a lot more than an instrument of affirmation of modern science and, therefore, the fact that the idea that *lógos* in Greek *techné* is limited, as stated by Mitcham, cannot be considered a minor argument.

In studies on *techné* performed by Galimberti, the lack of a *lógos* in Plato's *techné* is not so evident, as there is a hierarchy of techniques in relation to the politics (the royal technique). If, on the other hand, for Plato techniques are specialized (as well as their execution, and, therefore, relatable to a social system of division of labor, i.e., adjustable), they demand a greater technique, a ratio that can measure them according to their use and not to their production. Since Greek production, according to Plato, was inferior and should have been remitted in favor of other techniques, we shall call them superior and say that it explains the primary ones, among which politics is important by being the measure (in use) of all other techniques (Galimberti 2000: 295).

The case of Aristotle is a little different, because although it doesn't have such methodological concern, it agrees with Plato in that politics is the greatest of the practical sciences. In fact, for Aristotle, philosophy (that is the totality of knowledge) involves both theoretic sciences as well as practical sciences, and it is in the latter that technique is very well characterized as a *techné* originating from *poiesis*, but sent to a lower position relative to the other form, *praxis*, both representing different fields of human action. The first is defined by an external purpose (the work) and the other being an end in itself.

From this point it can be inferred that Plato's *lógos* tends toward universal truth, to the reason and to the idea because of a primordial epistemological question: the need to explain what man (eidetic) is and not the manner by which Nietzsche explains how he – man – came into the world (genealogy). Plato's man is virtuous, and develops his virtues through just and truthful knowledge, which is the *lógos* defended by him.

The Aristotelian *lógos* is multiple, not eidetic, able to explain the sensible world, that is to say, the world of human actions, through which it is possible to build intelligible units, giving another epistemological horizon to the reality explaining *lógos*: an instrument (discourse) for straight thinking.

In *The Nicomachean Ethics*, Aristotle provides a detailed analysis of *techné*, starting by clarifying the care that this has in relation to the exactitude that so appeals to Plato, when mentioning that one cannot desire the same accuracy in all study subjects, because in some cases it is enough that something is satisfactorily established (*Nicomachean Ethics 2013*: 51-1098.b1).

In other words, in the very procedure of Aristotle, there is the possibility of inaccuracy. If a carpenter is looking for a right angle, even through a satisfactory approach, there is in this practical action a condition of realization that does not require accurate and universal measure.

There is a dimension of knowledge of the soul that, being irrational, participates in the rational principle (*Nicomachean Ethics 2013*: 64-1012.b10). This component is not madness nor anything like it, but one that cannot be explained by rationality attributed to mathematics; it has the sense of attention. In other terms, it is not the mathematical measure of proportion nor the *lógos* of theoretical or scientific rationality that Aristotle associates as a trait of *techné*, but a *lógos* that ponders, that uses a median relative to human beings, i.e., a *lógos* for practical life; the just cause of human action, which does not consider the incommensurability of the diagonal and the side of a square (*Nicomachean Ethics 2013*: 94-1112.b20) but about things that are under our control, are attainable by action and do not always produce the same final result.

Apparently, the uncertainty found in *techné* is inherent to its condition of knowledge and practical action and there is no need to translate these uncertainties into certainties. In practical life (i.e., in *techné*) there is a just ratio (*lógos*) between excess and deficiency.

Inside the soul two rational faculties coexist: one by which one can speculate on invariable principles (scientific faculties), namely, that necessarily exist, and as such, must be demonstrable and its demonstration should be convincing in its accuracy; and the other serves to speculate about things that admit variation (calculating faculties)[5] in which there is also a process of true reasoning. With this, Aristotle defines technical thought as an art that aims to bring something into existence, through a true process of reasoning (*Nicomachean Ethics 2013*:183-1140.a10).

At this point, Aristotle warns of the existence of a characteristic reasoning that is typical of technical thought, a process that is true because it exists, but not because it exists necessarily as do the objects of the scientific faculty of the soul. This *lógos*, characteristic of the Aristotelian *techné*, is, therefore, a creator *lógos*, which creates within the variable universe and operates from deliberation.

In short, the concrete fact is that both *lógos* and *techné* were very important words to the Greeks and both have an important link between them, and that within the philosophy of technology or of engineering it hasn't been observed exhaustively that the term *techné* has links of the same nature with other Greek concepts that are equally important to the characterization of the technical, for example, the *métis*, a very important word in the Greek universe.[6]

---

[5] It is important to remember that Heidegger considered calculus in his discussion about the essence of the modern technique (1953: 131).

[6] There are few exceptions in this sense. For example, the works of Mark Young (2018) and of Anagnstopoulos & Chelidoni (2008).

9.3 *Métis* and the Greek *Techné*

In Greek thought, especially for the Athenians, *techné* was valuable practical knowledge (a manifestation of *poiesis*) and several Greek terms are commonly articulated around it and cited in philosophical studies of technology,[7] but *métis* is an exception.

In this direction, Chauí (2002) draws attention to two terms that are so close to the word *techné*, that they are confused with it: *lógos*, because of the relationship established with the rational process of identification, selection, division or grouping; and *métis*, which links *techné* to an ingenious thought (practical and effective) which operates in the becoming.

According to the author, *lógos*, from the verb *legein*, is presented as a fundamental procedure which also appears in *techné* and not purely restricted to thought or speech. Chauí also points out that it was Castoriadis who established this distinction, emphasizing that the Greeks thought and spoke according to a *dissói logoi*, that is to say according to a procedure of division into contraries, a procedure that was similar to *techné* from the verb *teukkein*, operating under the same template, reuniting, joining, fabricating, on the one hand, and choosing, distinguishing, separating on the other.

Chauí (2002: 144) points out that the opposition that *legein* establishes between a being and a non-being appears in *teukkein* and *tékhne* as opposition between possible and impossible or between word and chaos.

In the case of *métis*, Chauí cites the work of Jean Pierre Vernant to explain its relationship with the word *techné*. In them, *métis* is characterized as a practical intelligence or a gift to find a way where there appears to be none (Chauí 2002: 144), being that its clear forms include ingenuity and cunning (Chauí 2002: 145).

The text cited by Chauí is extracted from the work written by Detienne and Vernant (2008: 17), where the discussion about *métis* is expanded to argue that the goddess *Métis* had been largely forgotten in the reports of Greek texts, even when she had a key role as an intelligence that exercises influence over various plans, in which the emphasis is always put on the practical effectiveness and the search for success in the field of action.

Métis deserves special attention because it is also associated with the birth of technical production through the myth of *Prometheus* (who had as one of his striking features the resourceful intelligence of métis). In the myth, the Titans cheat and steal fire from *Zeus* to pass it to men so that they should not remain at a disadvantage to the other animals. According to Greek mythology, this was the moment that men were given the fire that belonged to the gods (divine fire), which, becoming appropriate for them, also becomes technical fire, in the sense of the Greek term *techné*. So *Prometheus*, creator who creates through intellect, who thinks before acting, who bears resourcefulness itself, deems the technique as a constitutive element of

[7] *dynamis*, *physis*, *eidos* and *érgon*, for example, are cited by Mitchan when elaborating a type of pre-modern ontology about the issue (1989: 131).

men.[8] In myth, *Prometheus* is punished by *Zeus* for having stolen the divine fire, the future source of all arts for men (Aeschylus 2005: 13), and for making them inventive and resourceful, for before that they had seen and comprehended badly (Aeschylus 2005: 32–34). As for Vernant, myths are not only narratives (Vernant 1990: 15), therefore in the myth of *Prometheus* we find more than a narrative about Greek gods. In it can be placed the understanding that was developed at that time on an important part of human existence: the technique.

In Theogony, *Métis* is the goddess who knows more than anyone (Hesiod 1979: 49) and for *Prometheus* there are no secrets in the future (Aeschylus 2005: 13). *Prometheus* is the holder of a large *métis* and he, farsighted, in Greek mythology plays the role that *Hesiod* determines for the goddess *Métis*, i.e., a role of a cunning intelligence associated with, during the appearance of the human race, technical objects, i.e., objects made from a practical and cunning intelligence.

Métis appears early in an episode of *Homer*'s games when *Antilochus*, notably at a disadvantage, triumphs over *Nestor*, through procedures of a different order than strength: a cunning that is characteristic of *métis*, because it acts on a moving ground or in uncertain and ambiguous situations, typical of the experiences of *Ulysses* (the *polymetis* hero of various cunning acts) or those experienced by the octopus and the fox, two animals, according to the Greeks, who possessed a cunning higher than the rest of the animals.

Detienne and Vernant's *métis* employs all the conception the Greeks made for this type of intelligence that, instead of contemplating the immutable essences, turns directly to the practical difficulties (Detienne and Vernant 1974: 21).

This synthesis is extremely relevant, because although *lógos* can also represent an intelligence or a type of similar thought to the *episteme* in *techné*, it is no less valid than the man, when producing himself technically, has erected a *lógos* that does not arises solely in the stable context of science.

Unlike Plato, for Aristotle, *techné* is art and *episteme* is science and both terms define two different ways a man can possess the truth. *Episteme* or scientific knowledge is a kind of demonstrative knowledge by which a particular science is concerned with the causes and foundations of that object (in universal and necessary terms). More than referring to a specific method, Aristotle makes use of the relationship between the variable and invariable to differentiate *episteme* from *techné*.

The invariable can be seized, reflected and built upon a solid and rational basis, while the variable can only be understood as a unique wisdom. The elevation of this unique (although consistent) wisdom to the necessary and universal level means the movement of enhancement from art to science. But this hierarchical procedure does not exist, because for Aristotle art lies exactly in the field of production and it is not necessarily a less developed process than scientific knowledge, for art concerns itself to invent or produce something that can or cannot be, while scientific knowledge attempts to demonstrate things in a universal and necessary form. Thus, *epis-*

[8] Dinucci and Julian (2012: 38–39) refers to fire (Heraclitus's *lógos*) as the guiding principle of reality that orientates the beginning, disappearing and transformation of all things in the world.

teme – or scientific knowledge – is of an explanatory order and *techné* – technique – is of an inventive or productive order.

A good example is the understanding that Aristotle possessed on architecture as *techné*, when he describes it as a capacity to produce, that involves the right reasoning. It is understood that such reasoning reveals a purified and consistent understanding of the object to be produced, but not necessarily a demonstrative knowledge that universally exhausts its object. That is, despite the distinctions made between *techné* and *episteme*, for Aristotle, *techné* is not concerned with the universal ontological condition of a given object, but with its process of invention or production. From this statement, the presence of *métis* in the philosophy of engineering becomes extremely relevant because it allows the development of new analytical perspectives that affirm their inventive condition, establishing new relations with the speech of dominant thought.

9.4 Rethinking Engineering Through *Métis*

The inclusion of *métis* in the philosophical works of engineering does not have as a main objective to challenge the role of science in the production of contemporary technical objects. It aims at the epistemological discussion that underlies the very understanding of engineering as productive activity as though engineering has always been linked to the resolution of problems; technical inventiveness is beyond this perspective.

What is at stake is the ontological condition that sustains the dominant thought to define the engineering problems because, before being stated, a problem is merely a sensitive state that requires a shift from the present reality and, after being stated, circumscribes a problem in the field of engineering, which represents the operation of restricting their signals to those that make sense in the previously established ontological condition, and not in the problem as such.

If in dominant thought engineering is understood as problem solving, with inclusion of *métis*, it can be postulated on the basis of its own problematizing process; in the sense of inventing problems, or more broadly, to produce conditions within which the statements that promote the displacement of reality will appear (Aravena-Reyes et al. 2006; Aravena-Reyes 2014).[9] That which is invented meets a demand to update the conditions of realization for a particular mode of existence. In other words, it is in a productive existential way which brings to the present the configuration of production that one wishes to experience as a full and continuous everyday life.

In dominant thought, a problem is the rupture or the appearing of an obstacle to a pre-existing reality; however, with the inclusion of *métis*, problematization is

[9] Deluze, for example, when talking about the concept of intuition in Bergson (Deleuze 1966), defines the problematization as a criticism of the false problems and invention of true ones, making it possible to understand problematization as a mode of invention.

invention in another sense: when one invents a problem, it sets up a space for reflection that is based more on the productive dynamics than on the identification and recognition of an operationally anticipated reality, in which the inventive intellectual effort is less relevant because there is only adaptation to a set of previous statements that dictate the unquestioned form of reality, alienating the worries of life to what someone, somewhere, defined as relevant to reality.

Problematization requires the production of statements that put the problem as a starting point for the transformation of a form of existence that still doesn't exist. Thus, it cannot erect exclusively on certainty, seeking only to fulfill the logic conditions of the *a priori*; it must articulate its own technical *lógos* to the statement of the problem, in a way that reality reconfigures itself, explaining the mode of existence that integrally unfolds with its statement, in the form of a concomitant individuation of the technical object and its relationships. In light of *métis*, the problematization may be a new paradigm (Fabre 2009) or a form of emancipation of dogmatic thought of images, as Osborne interprets in the work of Deleuze (Osborne 2003).

Perhaps this kind of perspective on *métis* allows the reconfiguration of engineering in order to understand it as a set of technical acts of invention and production of new modes of existence. However, this is not its only effect: the presence of *métis* can also open other reflections on both thought and contemporary technical action.

José Ortega y Gasset (1935), for example, states that technique is not intended to meet the organic needs of man, but to the contrary: it is destined for the superfluous. It is the set of acts aimed at saving effort in order to have more time to develop the greater design of life. The word design in this context takes an unprecedented existential dimension: a *métis* that, though technical acts, enables this design. Each design of life, therefore, has its set of technical acts, whereby the engineer is the prime figure in the domain of inventive consciousness of this act, and it can certainly take the form of design of life for human collectives. In fact, despite the history of engineering still being very controversial (Picon 2004: 424), the engineer has been raised to the level of the subject best prepared for the political conduction of life in society (Layton 1962).

So, differently from what informs the dominant thought, there is indeed an emancipatory and epistemological horizon which would enable engineering to exceed the problematic established in the contemporary mode of production of technical objects, in this case, taking into account the developments that the inclusion of *métis* can promote in the philosophy of engineering.[10]

For example, unlike the dominant thought that is inspired by classical ontology, Deleuze and Guattari (1980) gathered all the prospects of being in an ontology based on multiplicity, that which allows the rhizomatic understanding of the modern technique, so that a virtuality actualizes in a becoming, as movement of desire. Thus, if the becoming responds to a complex organization articulating a desire-to-be, with the thought of these authors, engineering can be thought of, from the analysis of the articulation between technical production and the desire-to-be, not as pure

[10]Look at, for example, the recent discussion of Nolen Gertz (2018) on nihilism and technical responsibility.

psychoanalysis, but as close observation of the schizophrenic omnipresence of an economic model based on uncontrolled consumption, which requires twice as much energy as that which the planet can offer (Latour 2012: 68).

Global awareness of planetary degradation should also be directed to engineering, as operators of the concrete layers of technical production (engineers) have not yet coalesced into a critical structure in relation to its role in this geological cycle: there is a type of process of denial that has hampered any position be taken in relation to the realignment of technical production around the specific demands of the Anthropocene[11] which, obviously, are not those who order the current model of technological development.

In current engineering, rather than the existence of technical-scientific rationality, there is a methodological obedience (including education) that inhibits the possibilities of thinking about the production of technical objects from other references, about other realities or even about other modes of existence: about other desired worlds.

The engineer's compliance mechanism in the current consumerist society is detected in disempowerment of *métis* and in the exacerbation of the technical production *lógos*. Bernard Stiegler (Lemmens 2011) uses the term "proletarialization" to refer to this disempowerment of inventiveness that induces society to opt exclusively for pre-defined models of life, prejudicing its ability to invent one for itself.[12]

In this sense *métis* helps to recover the inventive condition of engineering beyond the *lógos* of modern science, also giving an opening to all kinds of knowledge that are useful in moving away from the idea of the existence of a single world generated on the support of modern science and the capitalist machine of a consumerist society. New knowledge, which may well be minor, traditional, excluded knowledge, could be accepted in academia to expand the thought, because it has the advantage of information from other worlds, creating conditions to consider other forms of existence that imply new moral, political, or cognitive forms, and consequently, techniques.

Effectively, the idea of turning towards minor knowledge doesn't mean returning to a small place, without foundations or without its own *lógos*. In them there is an epistemology, an ontological floor that can help to reconfigure technical understanding, as happens, for example, in the case of the Yanomami people, for whom the technical object is also a subject of power (Viveiros de Castro and Danowsky 2014), or being, in a universe of human-things, founded in relation of subject to subject, or being in a social relationship. Birds and hammers, for example, are potentially and essentially human, and such a perspective represents an epistemological displacement relevant to the renovation of technical understanding.

The shifting provoked by the Yanomami thought do not represents returning to a "native state's solution". It signalizes one (among several) way to think our present

[11] Term initially coined by Paul Crutzen (2002) to refer to the disruptions that man has caused in Earth's natural system.

[12] See also the Bruno Gransche's (2018) reflection on the loss of skills in a society highly assisted by technology.

that coincide with several aspect of the Simondon thought and that also could be related with the ideas of a 'technology of the spirit" and "technology of the care" developed by Bernard Stiegler.

A practical example of the epistemological shifting provoked by the Yanomami though can be extracted from that which Viveiros de Castro and Danowski call "Earthbound techniques," those which are based on minor knowledge, that prove a non-material intensification of our way of life, by the use of "workarounds" or temporary solutions to decelerate the current path, or even as cultural insurrection, that help the to dismantle the citizen-consumer logic, being sustainable techniques of agency processes, that don't pass for modern science, but are inherent to living being, and therefore, to mankind. The Anthropocene engineering seems to require an inventive form of the living being itself and not to become confined to certain harmful inventive forms, very present within particular modes of technical production.

9.5 Final Considerations

The inclusion of *métis* in the study of philosophy of engineering and technology seeks to rescue the inventive potential that exists in the production process of technical objects in order to promote new scenarios that favour new modes of existence. If the dominant thought in engineering philosophy has made science its basis, the rescue of inventive thought does not challenge this notion, just the position in time as the contemporary form of producing technical objects. From this perspective, the return of *métis* operates as an alternative philosophical thought that allows the validation of other forms of production technique as fundamental in other epistemological bases, since apparently the course that contemporary production of technical objects has taken seems to demand philosophical reflection that displaces the dominant thought.

In this context, to invent new productive methods is more than producing new methodological structures: the problem is of the philosophical order and requires the urgent inclusion of Planet Earth as the place where all of human existence is produced, and therefore, all technical objects. If there are worlds to invent, worlds less harmful to living beings and to the only place that hosts them, the first movement is to depart from the grand narrative that places engineering in a process of epistemological obedience that diminishes the potential of the production of new technical objects for these new modes of existence. This work postulates the idea that the inclusion of *métis* in studies of engineering philosophy helps to promote this displacement.

References

Aeschylus; Prometheus Bound. Brazilian edition: Esquilo. (2005). *Prometeu Acorrentado.* (J. B. de Melo e Souza, Trans.). São Paulo: Editora M. W. Jackson.

Anagnostopulos, K., & Chelidoni, S. (2008). Mêtis and the artificial. In: Paipetis, S. (Ed.), *Science and technology in homeric epics* (History of Mechanism and Machine Science Series, Vol. 6, pp. 435–442). New York: Springer.

Aravena-Reyes, J. (2014). A problematização como invenção: Fundamentos para a Educação em Engenharia. *Revista de Educação em Engenharia, 33*(2), 65–71.

Aravena-Reyes, J., Aguilar, M., & Azevedo, W. (2006). *Una propuesta de Educación para la Tecnologia.* Paper presented at the fourth international Latin American and Caribbean conference for engineering and technology, June 21–23, 2006. Mayaguez, Puerto Rico.

Aristotle. Metaphysics. Brazilian edition: Aristóteles. (2015). *Metafísica* (E. Bini, Trans.). São Paulo: Edições Profissionais Ltda.

Aristotle. Nicomachean Ethics. Brazilian edition: Aristóteles. (2013). *Ética a Nicômano* (E. Bini, Trans.). São Paulo: Edições Profissionais Ltda.

Aristotle. Rethorics. Brazilian edition: Aristóteles. (2012). *Retórica* (M. Alexandre, P. Alberto, & A. Pena, Trans.). São Paulo: Martins Fontes.

Bunge, M. (1980). *Epistemologia: Curso de Actualización.* Coyoacán: Siglo XXI Editores.

Chauí, M. (2002). *Introdução à História da Filosofia: Dos pré-socráticos a Aristóteles.* São Paulo: Companhia das Letras.

Crutzen, P. (2002). Geology of mankind. *Nature, 45*(3), 23.

Deleuze, G. (1966). *Le Bergsonisme.* Paris: Presses Universitaires de France. Brazilian Edition: Deleuze, G. (1999). *Bergsonismo* (L. Orlandi, Trans.). São Paulo: Editora 34.

Deleuze, G., & Guattari, F. (1980). *Mil Plateaux: Capitalisme et Schizophrénie.* Paris: Minuit. Spanish Edition: Deleuze, G., & Guatarri, F. (2010). *Mil mesetas: Capitalismo y esquizofrenia* (J. Pérez, Trans.). Valencia: Pre-Textos.

Detiene, M., & Vernant, J. (1974). *Les Ruses de L'intelligence. La mètis des Grecs.* Paris: Flammarion. Brazilian Edition: Detiene, M., & Vernant, J. (2008). *Métis: As astúcias da Inteligência* (F. Hirata, Trans.). São Paulo: Odysseus Editora.

Detienne, M., & Vernant, J. (2008). Les Ruses de l´intelligence: La Mètis de Grecs. Flammarion, 1974. Brazilian Edition: Detienne, M.; Vernant, J. Métis, as Astúcias da Inteligência. Trad. Filomena Hirata, Odysseus Editora Ltda., São Paulo.

Dinucci, A., & Julien, A. (2012). *O Encheirídion de Epitecto* (A. Dinucci, & A. Julien, Trans.). São Cristovão: EdiUFS.

Durbin, P. (2006). Chapter 15: Philosophy of engineering. *Techné: Research in Philosophy and Technology, 10*(2), 141–151.

Fabre, M. (2009). Qu'est-ce que problématiser? Genèses d'um paradigme. *Recherches em Education, 6*, 22–32.

Foucault, M. (1988). *Tecnologies of the self.* A seminar with Michel Foucault. Massachusettes: MIT Press. Spanish Edition: Foucault, M. (1990). *Las Tecnologias del Yo y otros textos afines* (M. A. Salazar, Trans.). Barcelona: Ediciones Paidós Ibérica S.A.

Galimberti, U. (2000). *Psiche e Techné: L'u Omo nell'età dela técnica.* Milan: Giangiacomo Feltrinelli. Brazilian Edition: Galimberti, U. (2006). *Psiche e Techné: O homem na idade da técnica* (J. Almeida, Trans.). São Paulo: Editora Paulus.

Gertz, N. (2018). Nietzsche, postphenomenology, and Nihilism-technology relations. In P. E. Vermaas, C. Didier, D. Cressman, & N. Doorn(Eds.), *The future of engineering* (pp. XX–XX). Cham: Springer

Gransche, B. (2018). Assisting ourselves to death – A philosophical reflection on lifting a finger with advanced assistive systems. In P. E. Vermaas, C. Didier, D. Cressman, & N. Doorn(Eds.), *The future of engineering* (pp. XX–XX). Cham: Springer

Heidegger, M. (1953). La Pregunta por la Técnica. In M. Heidegger (Ed.), *Filosofía, Ciencia y Técnica.* Prologue by F. Soler (1997), Santiago: Editorial Unive-rsitaria.

Hesiod. Theogony. Brazilian edition: Hesíodo. (1979). *Teogonia* (A. Cerqueira, & M. Lyra, Trans.). Rio de Janeiro: Editora da Universidade Federal Fluminense.

Jaramillo, D. (2015). Existe una Filosofia de la Ingenieria? *Universitas Philosophica, 32*(64), 313–328.

Latour, B. (2012). *Esperando a Gaia. Componer el mundo común mediante las artes y la política.*, Conference at French Institute of London (2011), Cuadernos de Otra parte. Revista de letras y artes, 26, (trans. Cucchi, S.), pp. 67–76. Argentina.

Layton, E. (1962). Frederick Haynes newel and the revolt of the engineers. *American Studies, 3*(2), 17–26.

Lemmens, P. (2011). This system does not produce pleasure anymore: An interview with Bernard Stiegler. *Krisis: Journal for Contemporary Philosophy, 2011*(1), 33–41.

Lévy, P. (1990). *Les Technologies del l'intelligence*. Paris: La Découverte. Brazilian Edition: Lévy, P. (1993). *As Tecnologias da Inteligência. O Futuro do Pensamento na Era da Informática* (C. Costa, Trans.). São Paulo: Editora 34.

Meijers, A. (Ed.). (2009). *Philosophy of technology and engineering sciences, handbook of the philosophy of science series* (Vol. 9). Oxford: Elsevier V. B.

Mitcham, C. (1989). *Qué es la Filosofía de la Tecnología?* (C. Nieto, & R. Stingl, Trans.). Barcelona: Editorial Antrophos and Servicio Editorial de la Universidad del Pais Vasco.

Mitcham, C. (1994). *Thinking through technology: The path between engineering and philosophy?* (C. Nieto, & R. Stingl, Trans.). Chicago: The University of Chicago Press.

Mitcham, C. (1998). The importance of philosophy to engineering. *Teorema, XVII, 3*, 27–47.

Mitcham, C., & Schatzberg, E. (2009). Defining technology and the engineering science. Philosophy of technology and engineering sciences. In A. Meijer (Ed.), *Handbook of the philosophy of science series* (Vol. 9, pp. 27–64). Oxford: Elsevier V. B.

Ortega y Gasset, J. (1935). *Meditación de la Técnica. Colección Austral*. Madrid: Espasa-Calpe S. A.

Osborne, T. (2003). What is a problem? *History of Human Sciences, 16*(4), 1–17.

Picon, A. (2004). Engineers and engineering history: Problems and perspectives. *History and Technology, 20*(4), 421–436.

Silva, E. (2011). *Filosofia da engenharia: O que é e porque deveria se interessar*. Master's thesis, Federal University of Rio de Janeiro, Brazil.

Simondon, G. (1958a). *L'individuation à lumiere des notions de forme et d'information*. Paris: Flammarion. Spanish edition: Simondon, G. (2007). *La Individuación a la luz de las nociones de forma e información* (M. Martinez, & P. Rodriguez, Trans.). Buenos Aires: Prometeu Libros.

Simondon, G. (1958b). *Du mode d'existence des object techniques*. Paris: Flammarion. Spanish Edition: Simondon, G. (2009). *El Modo de Existencia de los Objetos Técnicos* (P. Ires, Trans.). Buenos Aires: Editorial Cactus and La Cebra Editores.

Van de Poel, I., & Goldberg, D. (2010). Philosophy and engineering: Setting the stage. Philosophy and engineering: An emerging agenda. In V. de Poel & D. Goldberg (Eds.), *Philosophy of technology and engineering series* (Vol. 2). New York: Springer.

Vernant, J. (1990). *Mythe et Religion em Grèce Ancienne*. Paris: Seuil. Brazilian Edition: Vernant, J. (2006). *Mito e Religião na Greça Antiga* (J. Melo, Trans.). São Paulo: Martins Fontes Editora.

Viveros de Castro, E., & Danowski, D. (2014). *Há Mundo por Vir? Ensaio sobre os Medos e os Fins*. Florianópolis: Instituto SocioAmbiental and Cultura e Barbarie Editora.

Young, M (2018). Intuition and ineffability: Tacit knowledge and engineering design. In P. E. Vermaas, C. Didier, D. Cressman, & N. Doorn(Eds.), *The future of engineering* (pp. XX–XX). Cham: Springer

Chapter 10
Thorstein B. Veblen's Philosophy of Technology and Modern Capitalism

Tuna Baskoy

Abstract Thorstein Bunde Veblen, known as an American institutional economist and social critic (1857–1929), obtained his Ph.D. degree in philosophy from Yale University in 1884 and critically examined the development of economic institutions and rising internal tensions in modern Western societies throughout his life. In his analysis, modern technology plays a critical role and there is some acknowledgment of the originality of his views. Yet, their analysis is short, sporadic, and anecdotal. With the purpose of filling this void in the extant literature, this study seeks answer to the following questions: Does Veblen offer a philosophy of technology? It is maintained here that Veblen not only elaborates on the essence and properties of the modern technology, but also discusses how it interacts with its broader social economic and political environment. After reviewing the extant literature on Veblen's account of technology, the essay outlines his approach to technology, which is followed by the reverberations of technology for industrial structure, business enterprise, competition between them, and the state in his writings. The last section discusses the key findings and summarizes them as a way of conclusion.

Keywords Thorstein Veblen · Technology · Philosophy · Capitalism · Politics

10.1 Introduction

Thorstein Bunde Veblen (1857–1929) was an American institutional economist and social critic who earned his Ph.D degree in philosophy from Yale University in 1884 and studied the development of economic institutions and rising internal tensions in modern Western societies throughout his life. In his writings, the modern technology occupies a crucial place. Surprisingly, his thoughts are pretty much overlooked

T. Baskoy (✉)
Department of Politics and Public Administration, Ryerson University,
Toronto, Ontario, Canada
e-mail: tbaskoy@ryerson.ca

A. Fritzsche, S. J. Oks (eds.), *The Future of Engineering*, Philosophy of Engineering and Technology 31, https://doi.org/10.1007/978-3-319-91029-1_10

in the current discourse on technology. Only a small group institutional economists acknowledge the originality of Veblen's thoughts by mainly analyzing the economic dimension of technology. In a sense, their analysis is partial. This study asks the following question: Does Veblen offer a philosophy of technology?

It is argued in this essay that Veblen offers an encompassing view, explaining the place and role of the modern technology in its broader environment as well as its impact on the capitalist market economy by observing the changes in industry structure, business enterprise, competition between them, and the modern state. This study aims to substantiate this point by focusing on Veblen's key writings, supplemented by the secondary literature. The essay consists of the following sections. While the second section reviews the extant literature on Veblen's account of technology, the third part examines the way Veblen conceptualizes technology and its relation with its broad environment, which is followed by the implications of modern the technology for the industry, business enterprise, and competition between them, and the modern state in the fourth section. Finally, the fifth section discusses the findings before stating the major findings.

10.2 Literature Review

Recent scholars of technology have been silent about Veblen's contribution to technology. For instance, Robert C. Scharff and Val Dusek published the second edition of *Philosophy of Technology: Technological Condition, An Anthology* in 2014 which did not mention Veblen's contribution. Similarly, Donald MacKenzie and Judy Wajcman published an edited a book, entitled *The Social Shaping of Technology* in 1985 with a second edition in 1999 without mentioning Veblen at all (MacKenzie and Wacjman 1999). In his study published in 1996, Donald MacKenzie pointed out a gap between economic and sociological explanations of technology, and mentioned Karl Marx and Herbert Simon as two exceptions (MacKenzie 1996). Veblen was still missing in his work. Robin Williams and David Edge, who extensively reviewed the literature on the social shaping of technology approach (SST) between 1985 and 1996, did not touch on Veblen's contributions (Williams and Edge 1996). Major scholars of technology who claim that technology is socially constructed in their work are yet to assess Veblen's contributions (Orlikowski 2010; Feenberg 2002; Orlikowski 1992; Pinch 1996). Philosophers like Val Dusek (2006, 46–47) and Carl Mitcham (1994, 38) preferred to emphasize technocracy in Veblen's writings without introducing his broader views on technology.

As a heterodox institutional economist, Veblen is known to institutional economists whose main concern is Veblen's ideas about the capitalist market economy. They often do not see any point in exploring his understanding of technology (Horner 1989; Reisman 2012). Others are aware of his views, but short of offering an in-depth analysis (McCormick 2006; Phillips 2000, 269; Latsis 2010; Lower

1987, 1159; Dowd 1958). Finally, there are scholars who provide a good summary of Veblen's thoughts on technology and its implications for institutional economics (Samuels 2004, 297; Larson 1993, 426; Samuels 1977). Again, their analysis is far from a systematic treatment of the significance as well as the effects of modern technology on the capitalist market economy. In short, this study aims to make a contribution to the social science literature on technology by constructing Veblen's philosophy of technology and its broader economic and political implications through examining his major works and the secondary extant literature.

10.3 Modern Technology

What is philosophy of technology? A prominent American philosopher David M. Kaplan (2004, xiii) offers the following definition: "Philosophy of technology is a critical, reflective examination of the nature of technology as well as the effects and transformation of technologies in human knowledge, activities, societies, and environments." The core task of a philosophy of technology is to analyze the phenomenon of technology, evaluate its impact, and reflect on the significance of the ways it mediates and transforms our individual, social, economic, and political experience. There are different philosophies of technology. For instance, Carl Mitcham (1994) makes a distinction between engineering philosophy of technology that examines technology from within by taking its inner logic as the core of analysis, and humanities view of technology which deploys trans-technological factors in analyzing the meaning of technology. There is also emphasis on anthropological approaches to technology that explains the rationale behind the emergence and evolutions of different technologies (Gehlen 1980). Furthermore, place of technology in modern society, how human being perceive it, and the nature of interactions between humans and technology have been studied extensively (Dusek 2006). Finally, there is also a stream in the philosophy of technology analyzing the economic dimension of technology, especially Marxists interested in the role of technology in business competition (Heilbroner 1967). Within this context, Veblen's views offers another way of envisioning of the modern technology and its social, economic, political, and cultural aspects.

By using Kaplan's above definition as a guide, this section examines Veblen's definition of technology and its major characteristics. It also outlines how he conceptualizes the nature of the relationship between technology and its environment. In the modern sense, Veblen briefly defines technology as the "employment of scientific knowledge for useful ends" (Veblen 1919a, 16). Machine industry, engineering, medicine, sanitation, and economic reforms are highlighted as some of the areas where technology is deployed. To unpack this brief definition, Veblen (1919b, 328–329) offers a relatively more specific and comprehensive definition at the same time: "It [common stock of intangible, technological equipment] is a knowledge of ways and means, and is embodied in the material contrivances and processes by means of which the members of the community make their living." (Veblen 1919b, 328–9) He elaborates on the second definition throughout his works extensively and the clue is *ways and means,* a key term that he often uses to refer to technology, as the following excerpts indicates:

> Whatever passes current in this way as knowledge of facts is turned to account as far as may be, and so it is worked into a customary scheme of ways and means, a system of technology, into which new elements of information or acquaintance with the nature and use of things are incorporated, assimilated as they come. (1964[1914], 39)

In Veblen's understanding, technology refers to ways and means, practical expedients, devices, and contrivances embodying a knowledge of facts which are often modified and upgraded as new useful information becomes available to solve problems at hand related to the material means of life in an efficient, proficient, and economic way (Veblen 1964[1914], 39–40). Population growth is the main driving force behind an appreciable advance in technology to deal with the material issues of the community (Veblen 1919b, 332).

Technology, which is always plural, can be a thing, a skill or a method that embodies and shows mastery of facts which are impersonal. It epitomizes information or acquaintance with the nature and use of things. Knowledge that goes into technology is useful and applied to the control of the processes in which natural forces are engaged (Veblen 1919a, 16). Technology represents knowledge and mastery of facts about the nature, efficiency, economy, proficiency, and creative work. (Veblen 1919a, 33). Being open, dynamic, cosmopolitan, and evolving by its very nature, technology absorbs new knowledge quickly to evolve into new forms as well (Veblen 1918, 7; Veblen 1915a, 170). Finally, technology is created and held by the community at large, and advances in technology are the result of the collective and unrestrained effect of the instincts of idle curiosity, workmanship, and parental bent in the community (Knoedler 2007, 204). Nevertheless, individuals and business enterprises often claim of ownership with patents, copyrights, or other legal means sanctioned by the state (Veblen 1964 [1914], 104–105).

As part of providing a more detailed picture, Veblen (1915b, 367) describes the properties of machine technology in his works vividly: "The machine technology is a mechanical or material process, and requires the attention to be centered upon this process and the exigencies of the process." Even in its simplest form, the mechanical industry that embodies technology is based on the principle of "a reasoned procedure and a systematic knowledge of the forces employed." (Veblen 1915b, 6) The modern technology is about a sequence of mechanical action and there is no place for habitual proclivity or personal dexterity, tact, training, or routine, the rule of thumb, or the accidental surmises of the seasons (Veblen 1964[1914], 338). The ruling norm for the modern machine technology is highly impersonal and brutal, while the logic of the machine process is the logic of masses, velocities, strains and trusts (Veblen 1964[1914], 241). As a result, the modern technology is more productive, efficient, proficient, and economic which has important reverberations for the business enterprise and competition between them in the market, as elaborated in the next section (Veblen 1923, 72).

Knowledge embodied in technology can be matter-of-fact or based on convention, practical knowledge, inference and authentic opinion, or theoretical knowledge. Besides, technological knowledge may have its roots in experience, experimentation, habit, knowledge, and initiative (Veblen 1919b, 328–333; Veblen 1964[1914], 328). Technology is made up of traits of human nature (physical, intellectual, and spiritual), as well as the character of the material environment, both of which influence its efficiency (Veblen 1919b, 349). Knowledge from different sources is taken up and assimilated in a specific technological equipment to complete work at hand. As such, it reflects the predispositions and habits of a community in which it is created (Veblen 1964[1914], 41).

In modern times, science and technology have developed an intimate relationship. Idle curiosity was the engine of motivation for scientists to embark upon scientific inquiry initially (Veblen 1919a, 6–17). Scientists were not interested in technological improvements per se. This is not the case anymore because Veblen thinks that the technological situation or the state of industrial arts conditions the scope and method of learning cogently. In other words, the technological situation is characterized as a constraining circumstance imposed on higher learning (Veblen 1918, 3). The relationship between science and technology is not a one-way street anymore. Science is also used in higher education to promote technology that can be commercialized easily. There is an easy co-partnership between them that makes it feasible to play into one another's hands, creating an opportunity for commercialization of higher education eventually. This brings us to Veblen's understanding of the interactions between technology and its broader environment.

Veblen's approach to technology has its roots in evolutionary thought and he defines it as follow: "Any evolutionary science, on the other hand, is a close-knit body of theory. It is a theory of a process, of an unfolding sequence." (Veblen 1919c, 58) Both the life of individuals in the community and the evolution of social structure or institutions are subject to continuous struggle for existence and selective adaptation. "Social evolution is a process of selective adaptation of temperament and habits of thought under the stress of the circumstances of associated life." (Veblen (1934[1899], 213) In other words, culture in general and the prevailing set of social, political and economic institutions in particular change over time in response to changing circumstances and environmental pressures in broad terms (Heath 2008, 249). With its crucial role in daily life, the modern technology shapes people's habits of thought, culture and institutions at the same time (Veblen 1919a, 17).

Institutions are prevalent habits of thought with respect to particular relations and functions of the individual, the community, and the scheme of life (Veblen 1934 [1899], 190). Adaptation in the habits of thought results in the growth of institutions as well as the emergence of new ones. Changing exigencies of the environment and conditions of life bring a correlative change in human nature and hence the human material of society (Veblen 1934 [1899], 213). The problem with institutions is that they are products of past processes and adapted to past circumstances. Naturally, there is a continuous rift between them and the requirements of the present.

Progressive adaptation and adjustment takes place eventually, but it is always one step behind the changing environment of the community (Veblen 1934 [1899], 191).

Technology is a crucial factor in understanding the process of social change and adaptation by teaching regularity of sequence, mechanical precision, and impersonal and measurable cause and effect (Veblen 1915b, 309–310). The relationship between technological innovation and social change is not unidirectional, as often pictured (Person 1975, 330). New technology may disturb the existing institutional structure and be a driving force behind change. Nevertheless, the existing institutions guide the changes instigated by new technology. (McCormick 2006, 41) "Technological change enables, requires, and may be encouraged by or retarded by institutional changes." (Dowd 1958, 283). It is important to stress that institutions do not act as an obstacle for new technologies all the time. On the contrary, Veblen draws attention to decreasing gap and increasing intimacy between the institutions of European civilization and the exigencies or requirements of industry and technology after the transition to modern times in Western Europe, especially in the discipline of law and order which have become more impersonal and less coercive (Veblen 1919d, 49).

Technological knowledge is based on cumulative knowledge of the society that is passed from one generation to another by tradition, training, education, and other forms of habituation (Veblen 1964[1914], 38). Habitual conduct changes from one community to another at one point in time or in one culture over time, thereby pointing an indirect link between culture and technology which may create some complications when it is transferred from one culture to another (Veblen 1964[1914], 39). That technological knowledge is produced, held, and carried forward by the community makes it a going concern. "The novelties of today are a technologically later generation of the commonplaces of the day before yesterday." (Veblen 1923, 64). Individual members of the community continually add new elements of insight and proficiency. Yet, these new additions made by individuals are small and insignificant in comparison to the joint stock of technological knowledge (1919b, 328). In the end, all life is necessarily a group life (Veblen 1964[1914], 103–104). The nature of the relationship between the community and the individual is mutually constitutive (Milonakis and Fine 2009, 172).

Individuals may not have access to technological knowledge and resources freely. There are always limitations and restrictions they may face, even if such segregations are not that consequential for technological development in the long run (Veblen 1964[1914], 104–105). Competing technologies and access to technological knowledge and resources have significant reverberations for power and distribution of wealth. "The principles and practice of the distribution of wealth vary with the changes in technology and with the other cultural changes that are going forward." (Veblen 1919e, 360) As technology changes, so does income distribution with changing industrial structure, business enterprise and competition between them, and the state, as the following section demonstrates.

10.4 Economic and Political Reverberations of Modern Technology

As part of the broader society, technology is a factor in social change in broad terms and Veblen is aware of this role, especially in the field of economics and politics. It is clear in his writings that the modern technology has significant implications for the structure of industry, business enterprise, competition between them, and the modern state. The modern technology influences the structure of industry by creating interdependency and standardization. To start with, all industries are based on machine technology and the machine process is interdependent and interlocking by its very nature. "The whole concert of industrial operations is to be taken as a machine process, made up of interlocking detail processes, rather than as a multiplicity of mechanical appliances each doing its particular work in severalty." (Veblen 1915b, p. 7). The machine process entails increasing interdependence with other industrial processes going forward or backward or what Veblen calls concatenation. The industrial system, especially the manufacturing industry, can be characterized as a balanced system of work as well as a moving equilibrium of interlocking processes.

Efficiency of any given part is dependent on the due working of all other relevant parts. The modern industrial system as a whole is efficient and any degree of failure or restriction in rate, volume, or quality of output at any point requires a degree of limitation and inefficiency throughout the system (Veblen 1923, 248). The system at large can be characterized as a comprehensive balanced mechanical process, which makes it vulnerable to disturbances in chain reaction. Interstitial adjustment is required in other stages of production, should there be any disruption in one stage. Disturbance at any point influences all related areas at the same time and results in more or less derangement to the industrial process at large (Veblen 1915b, 16–17).

Standardization is the natural outcome of concatenation and interdependence in the modern industry (Veblen 1915b, pp. 6–12). Being a comprehensive and delicately balanced complex of sub-processes that require quantitative precision and accuracy in points of time and sequence, standardization of the magnitude of the various physical characteristics such as weight, size, density, hardness, elasticity, temperature, etc. are essential for the productivity and efficiency of the modern industry (Veblen 1915b, p. 8). Standardization of the industrial means, methods, and products enforces a close conformity in point of time, volume, and character of the product regardless of the type of the product, i.e. goods or services (Veblen 1915b, pp. 15–16). The machine process requires standardization of conduct and knowledge regarding quantitative precision as well. It teaches understanding and explaining facts in terms of material cause and effect (Veblen 1915b, p. 66).

With concatenation and standardization, the modern technology exerts a substantial influence on scale and method of production depending on the lines of the industry (Veblen 1919b, 343). Establishing a plant represents a considerable amount of invested wealth and each plant employs a large number of workers per unit of plant. Instead of working alone as independent units, plants operate in concert together as a balanced system and keep pace in common under a collective business management (Veblen 1917, 321). Industrial experts, engineers and technicians progressively take over the industrial functions of the captains of industry and leave them free to devote their attention to business alone in the modern industry (Veblen 1923, 258). As a corollary, the type of business enterprise and competition between them changes with the modern technology eventually.

Technological changes influence economic institutions immediately (Person 1975, 330). The first place to look for a concrete change is the business enterprise: "On its technological side the characteristic trait of this capitalism is that the current pursuit of industry requires a larger unit of material equipment than one individual can compass by his own labor, and larger than one person can make use of alone." (Veblen 1919b, 341–342) Family-owned small businesses would not have the necessary resources to benefit from the modern technology. The main impetus behind the emergence of the modern corporation as a dominant form of business enterprise in the second half of the nineteenth century was an advance in the scope and method of workmanship or a complex of technological changes: "The growth of business enterprise rests on the machine technology as its material foundation. The machine industry is indispensable to it; it cannot get along without the machine process." (Veblen 1915b, 375). Standardization of processes, products, and consumption required large amounts of material, which in turn, meant huge amounts of capital to benefit from technology. Technological change influenced the expansion of trade directly as well as indirectly (Veblen 1964[1914], 228–229).

Obviously, the modern corporation is a business concern, not an industrial unit, and created by a capitalization of funds which explains the rationale behind financialization of the industry. Resting on credit, it is an incorporation of absentee ownership and hence an impersonal entity in all contacts and dealings with its competitors and workers alike. As a business concern only, the modern corporation's activities are limited to business transactions of the nature of bargain and sale to make a profit (Veblen 1923, 82). If it has business in a specific industry, it still pursues a business-like outcome of net gain by employing technical experts to manage the work and oversee the production processes. Veblen describes the modern corporation as follows: "It is a pecuniary institution, not an industrial appliance. It is an incorporation of ownership to do business for private gain at the cost of any whom it may concern." (Veblen 1923, 83) Production of material wealth has a secondary place for the corporation. As a vehicle for the vested interests to reap the most benefit from salesmanship by controlling the industry and technology, the modern corporation also influences business competition.

The modern technology has three major consequences for business competition. In the first place, it makes it easy for those with legal right of ownership to control the production process (Veblen 1923, 65). The competitive system gives rival business enterprises discretion to practice interstitial adjustments such as expansion or curtailment of industrial activity and enlarged or shortened output of product (Veblen 1915b, 16–17). Veblen defines sabotage as "a sagacious limitation of productive processes to something less than the productive capacity of the means in hand." (Veblen 1917, p. 325). Systematic mutual hindrance of the management of the supply is a common practice in competitive business. Owners sabotage production and make the management of the supply difficult. Veblen's view of sabotage is broad, but business enterprises deploy such practices as obstruction, diversion, or withholding to be more profitable at the expense of less production (Sacrey et al. 2005).

Secondly, to realize efficiencies that accrue from concatenation and standardization, industrial consolidation becomes unavoidable, thereby reducing the number of effective competitors. "The competitive system has been dying at the top, and the decay has been spreading outwards and downwards as fast and as far as the other and lower branches of the industry have been sufficiently brought into line with the mechanical technology and so have become fit material for absentee ownership and control' that is to say, fit to be profitably managed on business principles." (Veblen 1923, 77–78) Capitalists tend to consolidate the ownership of all material assets to deal with competitive bidding (Veblen 1919b, 346–347).

Finally, complete monopolization is not possible because there are always areas that are not amenable to a complete monopolization of the community's technological expedients (Veblen 1919b, 346–347). One of the key characteristics of modern technology is a continued flow of new ways and means, contrivances and expedients designed to supersede the old which find their way into profitable use and displace older appliances in greater or less measure. Whenever a new machine is used in production, the price of the final product becomes cheaper, thereby pointing a new price system with the modern technology (Veblen 1919f, 218). The competitors emulate the process, which, in turn, results in the depreciation of the machine bought by the first business immediately. "Each new venture or extension goes into the competitive traffic of producing and selling any line of staple goods with a differential advantage, as against those that have gone before it, in the way of a lower scale of costs." (Veblen 1915b, 229). Machinery based on new technology underbids its older counterpart in the market, and renders it obsolete or subject to recapitalization on a lowered earning-capacity by reducing the effectual cost of products to the consumer (Veblen 1919b, 343, footnote. 8). In a way, chronic depression is the normal course of things with the higher development of the machine process based on the modern technology, despite the fact that the industrial revolution and modern technology made business competition international by nature.

In sum, Veblen is aware of the contradiction between the machine technology and business principles, i.e. free competition, and argues that they may fall into abeyance eventually (Veblen 1915b, 375). The incompatibility between the exercise of free contract and the other powers inhering in the natural right of ownership of the modern machine technology is a serious problem (Veblen 1915b, 266). Veblen characterizes the new capitalist system as a "decayed competitive system" and maintains that competition takes place between the business concerns that control production, and the consuming public in the new system (Veblen 1923, 79). The state plays a significant role in this process as well.

As an institution, the state protects the vested interests with patents, copyrights, trade secrets. It is also an avid user of technology at the same time. Technology is created and held by the community at large, but individuals and business enterprises may claim of ownership with patents, copyrights, or other legal means sanctioned by the state (Veblen 1964 [1914], 104–15). In domestic markets, business enterprises may be shielded from competition thanks to special rights and privileges they obtain through public policies. The state is the protector of the interests of the domestic business enterprises in international competition (Veblen 1915b, 292–293). Business enterprises aim to extend their markets, while trying to exclude competing traders from their domestic markets. Veblen describes the situation in the following words: "Competition may be the soul of trade, but monopoly is necessarily the aim of every trader." (Veblen 1917, 74). The state plays a critical role in the process in that it adopts policies such as subventions, tariffs, or similar privileges to shelter national business enterprises from the competition of non-domestic business enterprises wholly or partly whether at home or abroad (Veblen 1923, 73).

The state is an avid user of technology, too. Veblen recognizes the significance of the modern technology for the state: "Yet the Imperial system of dominion, statecraft and warlike enterprise necessarily rests on the modern mechanistic science and technology, for its economic foundations and its material equipment as well as for it administrative machinery and the strategy necessary to its carrying on." (Veblen 1915a, 262). Modern statecraft heavily relies on the modern technology in its economic foundation as well as in its administrative functioning. However, the modern technology also has adverse implications for the state, especially smaller ones. For instance, smaller states such as German principalities could not withstand the impact of the modern technology which was a major factor behind the emergence of larger states in Europe. "Modern warfare and modern industry alike are carried on by technological processes subject to surveillance and direction by mechanical engineers, or perhaps rather experts in engineering science of the mechanistic kind." (Veblen 1917, 303) In short, the state deploys technology and engineers not only in its daily administrative organization and functioning, but also in its ability to make wars.

In democratic countries, the state is under the heavy influence of businessmen and considerations of business expediency guide public policies, not because the state is capitalist as in the Marxist sense, but bureaucrats and legislators are chosen mainly on the grounds of their businesslike-principles and ability (Veblen 1917, 156). In a way, efficiency of the modern technology requires technocratic decision-making in government that may have adverse impact on democratic public policy making (Gehlen 1980; Dusek 2006; Mitcham 1994; Feenberg 2002). Political parties are cliques that represent business interests as well. The only time that the interests of business are ignored is in wartime. Even then, war is mainly for the protection or extension of business interests outside the state (Veblen 1917, 156). This does not mean that the state is a powerless entity after all. On the contrary, Veblen describes the state as follows: "In some potent sense, the State is a personal entity, with rights and duties superior and anterior to those of the subjects, whether these latter be taken severally or collectively, in detail or in the aggregate or average. The citizen is a subject of the State." (Veblen 1915a, 156) In a way, the state has its own powers as a sovereign entity and makes its own decisions in critical times that may go against business interests in the short run, but public policies protect them in the long term.

10.5 Discussion and Conclusion

The above analysis indicates that Veblen's philosophical framework for understanding the technology fills a major gap between sociological and economic explanations of technology. In that sense, Karl Marx and Herbert Simon are not the two exceptions, as MacKenzie (1996) claims. Veblen's approach overcomes the problem of the narrow treatment of the modern technology by neoclassical economists who often equate it with method of production, education, research or new skills broadly (Romer 1990, p. S72; Romer 1986, p. 1015; Arrow 1962, pp. 156–7; Solow 1957, 312). Technology is not something that is available to competitors whenever they want it, as the neoclassical economic theory assumes (Williams and Edge 1996, 871). Nor is there any way for the business enterprise to know the best technique of production that produces maximum profits as portrayed in the mainstream economic theory, since technology evolves constantly (MacKenzie and Wacjman 1999, 13). Compared to the Social Shaping of Technology (SST) approach, Veblen's analysis of the modern technology is more realistic in that it includes how new technology influences power relations in the market as well as outside the market, and hence distribution of wealth.

Overall, this study has demonstrated that Veblen provides a comprehensive approach to study the modern technology. Defined as ways and means, practical expedients, devices, and contrivances embodying knowledge of facts, it is created and carried out by the broader community. There are always several competing technologies which bring power, politics, and competition into play. Veblen characterizes technology as emerging and operating in a process of becoming, in opposition to being perceived as given or exogenous.

Veblen's philosophy of technology has its roots in evolutionary thought which is a theory of process and unfolding sequence to understand the distinctness of the capitalist market economy and the role of the state in it. Both individuals in the community and institutions continuously struggle for their existence and adaptation. Technology plays a significant role in the process of social change and adaptation, but the relationship between technological innovation and social change is not unidirectional or one-way. Institutions may encourage or retard technological change. After all, technological knowledge is a product of and possessed by the community at large. The broader environment influences the technological groundwork of the community, despite the fact that the individual workman is the main actor.

Technology influences the structure of industry by creating interdependence, standardization, economies of scale, and freeing of the captains of industry from their industrial functions to focus solely on finance and salesmanship. Business enterprise is the main player in the affairs of the industry thanks to technological changes. Standardization of processes, products, and consumption require large amounts of material, which, in turn, mean that big corporations with deep pockets benefit from technology. Technological change affects the expansion of trade directly as well as indirectly. Besides, technology plays a double-edged role in business competition:

limiting competition as well as creating new areas of competition. Finally, the state plays two roles as the supporter as well as the user of technology.

The scope of this study is narrowed down to Veblen's philosophy of technology. While such an endeavor created ample space to deal with abstract issues, one limitation of this strategy is the absence of recent empirical evidence to verify Veblen's approach. As a future research project, applying this theoretical framework, for instance, to the communications or auto industry will be useful to show the relevancy of his views today.

References

Arrow, K. J. (1962). The economic implications of learning by doing. *Review of Economic Studies, 29*(3), 155–173.

Dowd, D. F. (1958). Technology and social change: Japan and the soviet union. In D. F. Dowd (Ed.), *Thorstein Veblen: A critical appraisal* (pp. 283–301). Ithaca: Cornell University Press.

Dusek, V. (2006). *Philosophy of technology: An introduction*. Malden: Blackwell Publishing.

Feenberg, A. (2002). *Transforming technology: A critical theory revisited*. New York: Oxford University Press.

Gehlen, A. (1980). *Man in the age of technology*. New York: Columbia University Press.

Heath, J. (2008).Thorstein Veblen and American social criticism. In C. Misak (Ed.). The Oxford handbook of philosophy (pp. 235-253). Oxford and New York: Oxford University Press.

Heilbroner, R. L. (1967). Do machines make history? *Technology and Culture, 8*(3), 335–345.

Horner, J. (1989). The role of technology: An institutionalist debate. *Journal of Economic Issues, 23*(2), 579–586.

Kaplan, D. M. (2004). Introduction. In D. M. Kaplan (Ed.), *Readings in the philosophy of technology* (pp. 13–16). Lanham: Rowman & Littlefield Publishers.

Knoedler, J. T. (2007). Getting to the good life: Agents of change. In J. T. Knoedler, R. E. Prasch, & D. P. Champlin (Eds.), *Thorstein Veblen and the revival of free market capitalism* (pp. 202–227). Cheltenham/Northampton: Edward Elgar.

Larson, J. (1993). Elegant technology. *Journal of Economic Issues, 27*(2), 421–428.

Latsis, J. (2010). Veblen on the machine process and technological change. *Cambridge Journal of Economics, 34*(4), 601–615.

Lower, M. D. (1987). The concept of technology within the institutionalist perspective. *Journal of Economic Issues, 21*(3), 1147–1176.

MacKenzie, D. (1996). *Knowing machines: Essays on technical change*. Cambridge: The MIT Press.

MacKenzie, D., & Wacjman, J. (1999). *The social shaping of technology* (2nd ed.). Buckingham: Open University Press.

McCormick, K. (2006). *Veblen in plain english: A complete introduction to Thorstein Veblen's economics*. Youngstown: Cambria Press.

Milonakis, D., & Fine, B. (2009). *From political economy to economics: Method, the social and the historical in the evolution of economic theory*. London/New York: Routledge.

Mitcham, C. (1994). *Thinking through technology: The path between engineering and philosophy*. Chicago: Chicago University Press.

Orlikowski, W. J. (1992). The duality of technology: Rethinking the concept of Technology in Organizations. *Organization Science, 3*(3), 398–427.

Orlikowski, W. J. (2010). The sociomateriality of organizational life: Considering technology in management research. *Cambridge Journal of Economics, 34*(1), 125–141.

Person, S. (1975). *American minds: A history of ideas*. Huntington: Robert E. Krieger Publishing Company.
Phillips, R. J. (2000). Technology and institutional change from the gilded age to modern times: The impact of the telegraph and the internet. *Journal of Economic Issues, 34*(2), 266–289.
Pinch, T. (1996). The social construction of technology: A review. In R. Fox (Ed.), *Technological change: methods and themes in the history of technology* (pp. 17–36). Amsterdam: Hardwood Academic Publishers.
Reisman, D. (2012). *Social economics of Thorstein Veblen*. Cheltenham/Northampton: Edward Elgar.
Romer, P. M. (1986). Increasing returns and long-run growth. *Journal of Political Economy, 94*(5), 1002–1037.
Romer, P. M. (1990). Endogenous technological change. *Journal of Political Economy, 98*(5), 71–102.
Sackrey, C., Schneider, G., and Knoedler. J. (2005). Introduction to political economy (4th ed.). Boston, MA: Dollars and Sense.
Samuels, W. J. (1977). Technology 'vis-à-vis' institutions in the 'JEI': A suggested interpretation. *Journal of Economic Issues, 11*(4), 867–895.
Samuels, W. J. (2004). Thorstein Veblen as economic theorist. In W. J. Samuels, W. Henderson, M. Johnson, & K. D. Johnson (Eds.), *Essays on the history of economics* (pp. 271–330). London/New York: Routledge.
Scharff, R. C., & Dusek, V. (2014). *Philosophy of technology: Technological condition, an anthology* (2nd ed.). Malden: Wiley Blackwell.
Solow, R. M. (1957). Technical change and the aggregate production function. *Review of Economics and Statistics, 393*(3), 312–320.
Veblen, T. B. (1915a). *Imperial Germany and the industrial revolution*. New York/London: Macmillan.
Veblen, T. B. (1915b). *The theory of business Enterprise*. New York: Charles Scribner's Sons.
Veblen, T. B. (1917). *An inquiry into the nature of peace and the terms of its perpetuation*. New York/London: Macmillan.
Veblen, T. B. (1918). *The higher learning in America: A memorandum on the conduct of universities by businessman*. New York: B. W. Huebsch.
Veblen, T. B. (1919a). The place of science in modern civilization. In T. B. Veblen (Ed.), *The place of science in modern civilization and other essays* (pp. 1–31). New York: B. W. Huebsch.
Veblen, T. B. (1919b). On the nature of capital I. In T. B. Veblen (Ed.), *The place of science in modern civilization and other essays* (pp. 324–351). New York: B. W. Huebsch.
Veblen, T. B. (1919c). Why is economics not an evolutionary science. In T. B. Veblen (Ed.), *The place of science in modern civilization and other essays* (pp. 56–81). New York: B. W. Huebsch.
Veblen, T. B. (1919d). The evolution of the scientific point of view. In T. B. Veblen (Ed.), *The place of science in modern civilization and other essays* (pp. 32–55). New York: B. W. Huebsch.
Veblen, T. B. (1919e). On the nature of capital II. In T. B. Veblen (Ed.), *The place of science in modern civilization and other essays* (pp. 352–386). New York: B. W. Huebsch.
Veblen, T. B. (1919f). Professor Clark's economics. In T. B. Veblen (Ed.), *The place of science in modern civilization and other essays* (pp. 180–231). New York: B. W. Huebsch.
Veblen, T. B. (1923). *Absentee ownership and business enterprise in recent times: The case of America*. New York: B. W. Huebsch.
Veblen, T. B. (1934/1899). *The theory of leisure class: An economic study of institutions*. New York: Modern Library.
Veblen, T. B. (1964/1914). *The instinct of workmanship and the state of the industrial arts*. New York: Augustus M. Kelley.
Williams, R., & Edge, D. (1996). The social shaping of technology. *Research Policy, 25*(6), 865–899.

Chapter 11
Energy Technologies and Human Well-being. Using Sustainable Design for the Energy Transition

Rafaela Hillerbrand and Kathrin Goldammer

Abstract Today's electricity supply and energy access provide the foundation for health, education, economic growth, and societal development. There are manifold interconnections between individual well-being and our energy system. We are currently witnessing an energy transition, a transition towards more sustainable energy sources and uses, and this shift will also greatly impact on human well-being in multiple ways. This transition may involve more (or less) decentralized energy technologies, the transfer of funds (within a society, or internationally), and so forth. However, the impacts this transition may have on individual well-being are not well studied. The interconnection between well-being and energy systems is, in fact, rarely spelled out. Although there are various indicator sets for sustainable energy, most of these fall short of explicating the connection between (sustainable) energy and individual well-being, and most citizens, irrespective of their educational level, are not aware of the multifaceted impacts the energy system or the energy transition may have on their personal well-being.

Keywords Sustainability · Energy · Capability approach · Nussbaum · Intergenerational justice · Intragenerational justice · Ethics · Values

R. Hillerbrand (✉)
Institut für Technikfolgenabschätzung und Systemanalyse, Karlsruher Institut für Technologie, Karlsruhe, Germany
e-mail: rafaela.hillerbrand@kit.edu

K. Goldammer
Reiner Lemoine Institut, Berlin, Germany

Institut für Technikfolgenabschätzung und Systemanalyse, Karlsruher Institut für Technologie, Karlsruhe, Germany

A. Fritzsche, S. J. Oks (eds.), *The Future of Engineering*, Philosophy of Engineering and Technology 31, https://doi.org/10.1007/978-3-319-91029-1_11

11.1 Introduction

Today's electricity supply and energy access provide the foundation for health, education, economic growth, and societal development. There are manifold interconnections between individual well-being and our energy system. We are currently witnessing an energy transition, a transition towards more sustainable energy sources and uses, and this shift will also greatly impact on human well-being in multiple ways. This transition may involve more (or less) decentralized energy technologies, the transfer of funds (within a society, or internationally), and so forth. However, the impacts this transition may have on individual well-being are not well studied. The interconnection between well-being and energy systems is, in fact, rarely spelled out. Although there are various indicator sets for sustainable energy, most of these fall short of explicating the connection between (sustainable) energy and individual well-being, and most citizens, irrespective of their educational level, are not aware of the multifaceted impacts the energy system or the energy transition may have on their personal well-being.

Not only does the energy transition impact on the individual, it is individuals who create and shape the energy system in multiple ways. First, an individual can be a supporter or, even more prominently, an opponent of energy infrastructure or certain energy technologies. Local opposition, for example, made it necessary to put a portion of the high-voltage lines from northern to southern Germany underground. The new power lines are part of the advancement of the German energy system, where traditionally low-consumption areas like northern Germany are increasing their renewable energy production through wind energy and receiving new connection lines to the higher-consumption areas in the rest of the country. Another way in which the individual shapes the energy system is as a consumer, either directly by the consumption of electricity or heat, or indirectly by consuming goods that require more or less energy for their production or delivery. Consequently, a lot of research addresses the question of how to foster more sustainable consumption or sustainable behavior more generally (Reisch and Thøgersen 2015). In such research, it is commonly assumed that what constitutes sustainability in a certain context is already known; it is only a question of how the right behavior can be induced by information policies or nudging.

In this paper we challenge this approach and take as our starting point the assumption that in most contexts it is far from obvious what defines sustainable technology or behavior. The current sustainability debate has fallen short because it only argues on an aggregate level and does not connect sustainability with individual lives. In contrast, we argue that sustainability is always linked to individual well-being, but this connection may be obscured because concepts of sustainability often focus on the aggregate level only. We seek to clarify what this connection can imply by formulating what we will call 'energy capabilities.' We contend not only that sustainable energy impacts on individual well-being, but also that individual well-being is essential in defining sustainable energy.

The overarching question of this paper is: what constitutes ethical energy supply and demand? Section 11.2 argues that, despite recent criticism, the concept of sustainability seems to be the best umbrella term for addressing ethics and energy. We advocate for a fresh view on sustainability that integrates a design perspective into sustainability analysis. Section 11.3 analyses common threats to an ethically sustainable energy supply, such as egoistic preferences (or "NIMBY" and value-pluralistic societies, and outlines possible ways to avoid these threats by using the capability approach. Based on Nussbaum's version of this approach, Sect. 11.4 sketches the values we see as central to a sustainable energy system, or 'energy capabilities,' and we summarize our findings in Sect. 11.5.

11.2 Designing for Sustainability

11.2.1 Towards a Prospective and Constitutive Evaluation of Technological Futures

Current ethics of technology and engineering have outgrown their traditional image as the tardy admonisher who brings solely negative criticism, and does so only after the technology is actually already in use (cf. Hillerbrand and Roeser 2015). Far from relying on retrospective assessment, contemporary analytically-oriented ethics of engineering seeks to play a constitutive role in shaping technological progress. It aims at integrating ethical and social values into early stages of product development, when many design-related and institutional aspects of the technical artefact or system are still malleable. This integrative method appears in various design-for-value approaches (van den Hoven et al. 2015) and in forward-looking technology assessment in which ethical aspects are central (Decker 2013). Most often such assessment is performed by interdisciplinary groups that integrate the humanities and social sciences, and sometimes also include scientists and engineers.

This change in the understanding of what ethics of engineering and ethics of technology can or should do has been accompanied by corresponding societal or political changes. Expert hearing and ethics committees play an increasingly important role in discussions about future technological developments. These committees often provide representation for interdisciplinary perspectives, including engineers and natural scientists as well as ethicists or social scientists. One prominent example of such an ethics commission in the energy context is the German commission for a safe energy supply, the "*Ethikkommission für eine sichere Energieversorgung*." This commission was appointed in March 2011 by Chancellor Angela Merkel in reaction to the accidents in the nuclear power plant in Fukushima Dai-Ichi. The commission was given a mandate to develop an update of Germany's approach to energy (Bundesregierung 2010) and it eventually recommended a mid-term phase out of nuclear energy in Germany (Endbericht Ethik-Kommission 2011). This recommendation was then introduced as a legal amendment (Atomgesetz 2011) and unanimously supported by the German parliament, where it was put into legislation in June 2011. The German *Energiekonzept* now proposes that 80% of all electricity and 60% of overall energy should come from renewable resources by 2050, while there should be a 70% cut in greenhouse gas emissions by 2040 and a nuclear power phase-out by 2022. Comprised of members from various areas of academia, lobby groups, and representatives of the largest religious communities, the ethics commission served as a catalyst for gathering information. The committee conducted public hearings and interventions with various groups, including civil society groups and experts from engineering and natural sciences as well as the humanities (theology and philosophy).

11.2.2 Sustainability and the Charges Against It

(iv) In paving the way for an ethical energy supply and demand, one must first consider which values are of particular importance. Policy makers, industry, and the broader public often see affordability and profitability, ecological sustainability, and a ubiquitous, stable supply of electricity and heat as the ultimate goals for an energy system. In this context, ethical aspects are often associated with sustainability, and public or political discussion of the energy supply is likewise often cast under the broad header of sustainability.

In the academic context, however, sustainability analysis has been subjected to growing scrutiny and the term is increasingly dismissed as an outdated or at least slightly dusty concept. This is particularly apparent in research and innovation projects as well as in academic funding, where the focus has shifted to seemingly more general evaluative tools like 'responsible research and innovation' (EU), 'design for values,' or 'value-sensitive design.' These new terms seem to indicate several points. First, (i) that the focus here is on values more broadly. That is, ecological sustainability may be one of many values or one among various aspects of responsibility, alongside the social aspects of technology more generally. Second, due to a focus on general values, (ii) these appear to be more objective evaluative tools than the concept of sustainability alone. It is further argued (iii) that these concepts better accommodate possibly conflicting values (Van den Hoven et al. 2015). More generally, (iv) for many sustainability suggests personal sacrifice, sufficiency, or a general reduction of amenities. For instance, it is often associated with demands such as not to waste, not to overconsume, maybe even not to consume new things at all. An anti-progressive air clings to sustainability, and a turn towards a more forward-looking technology assessment seems hardly possible with the stuffy, dusty, outdated concept of sustainability. In the following we rebut charges (i), (iii), and (iv), showing that most of the advantages raised in favour of these new concepts are already entailed in the concept of sustainability. The question of number (ii), whether value-based approaches offer a more objective account, will be addressed in Sect. 11.3.

We begin with a quick look at the origins of the term 'sustainability.' While today many associate sustainability with technological development, the concept originates in the area of forestry, in the German term "*Nachhaltigkeit.*" Hans Carl van Carlowitz coined this term in his 1713 book *Sylvicultura oeconomica* to denote a type of forestry that does not log more trees than can grow back on a reasonable time scale.[1] Our modern use of the term is often associated with the Brundtland report, commissioned in 1983 and published in 1987. The context here was international development aid. The Brundtland report defines sustainable development as

[1] For a detailed historical account of sustainability see Grober 2013.

"development that meets the needs of the present without compromising the ability of future generations to meet their own needs" (Brundtland 1987, § 27).

Freed from its original context, "sustainability" soon became popular in the context of technological development as well. Because Brundtland's definition is rather vague, various specifications for sustainability emerged. Indicator sets such as the Energy Indicators for Sustainable Development (EISD) by the IAEA (2005), or *Energiewende-Monitoring* (BMWi 2016) were particularly introduced to measure sustainability for the field of energy supply and demand (see also Klitkou et al. 2014, Bräutigam et al. 2015). Today, the United Nations' *Sustainable Development Goals* mark a global effort to define and measure sustainability in various parts of life as wide-ranging as gender equality or access to energy (UN 2015).

Many of the indicator sets for and specifications of sustainable energy follow what is known as the three-pillar view of sustainability, which distinguishes an *economic*, an *ecological* and a *societal pillar*.[2] Classic market impacts such as issues of price and cost fall under the first pillar, while any environmental impact like water contamination or the emission of greenhouse gases belong to the ecological pillar. The societal pillar pools various effects, as diverse as child labour in coal mines, civic participation in energy projects, or simply acceptability issues within energy systems.

These three pillars can be traced back to the Brundtland report, which identified sustainable development goals such as economic and social development as well as environmental protection. The three-pillar view stresses that economic, ecological, and societal aspects are on equal footing. This central notion is utilized in many other popular concepts when considering energy supply and demand. In Germany, the so-called *Energiepolitisches Zieldreieck* (or *Energiedreieck*, 'energy triangle'), a triangle whose corners represent economic efficiency, eco-friendliness, and security of supply, is widely used by policy makers and industry alike to illustrate the goals of an energy system. Economic efficiency and eco-friendliness parallel the economic and ecological pillars, while the societal pillar has been reduced to simply the security of supply.

The goals of the *Energiedreieck* have received widespread acceptance not only in Germany, where it has even been written into law,[3] but also in the EU more generally. It inspired numerous indicator sets such as the McKinsey *Energiewende Indicator* (Vahlenkamp and Gohl 2012), the BDI *Energie Ampel* (BDI 2014) and others,[4] and thus still shapes the greater part of German energy discourse (cf. Hillerbrand and Goldammer 2017). This shows the substantial influence of the three-pillar view on the public, policy making, and industry.

---

[2] In energy systems the three pillars are often reduced to energy costs, eco-friendliness, and security of supply. We will elaborate further in the course of this section.

[3] The first paragraph of the *Energiewirtschaftsgesetz*, the law that manages electricity and gas production, transport, and supply, makes reference to the three goals: the purpose of the law is a "reliable, cost-effective, consumer-friendly, efficient and environmentally friendly (...) energy supply" (EnWG, 2017).

[4] Among other things, there is quantitative monitoring of energy parameters by the German government (the so-called *Energiewende-Monitoring*, the *Fortschrittsberichte*, etc.) and an (academic) expert commission that comments on these numbers and interprets them.

The three-pillar view of sustainability reveals that sustainability is not the short-sighted regard for environmental aspects alone that is often ascribed to it by its critics (see (i) above). Rather, it allows for a variety of different aspects that may be associated with a whole range of underlying values. Nonetheless, it must be said that in practice the focus is often on the environmental and the economic pillars. For example, the *Zieldreieck* reduces the complex and multifaceted impacts that the energy supply can have on society to one aspect only, namely, security of supply. More comprehensive evaluation tools do sometimes implicitly entail information about broader social aspects of the energy system, such as the correlation of social status to household energy costs. Most of the time, however, the social aspects are strikingly absent in existing indicator sets and standards for monitoring energy systems. The quantitative tracking that is currently set up for the UN Sustainable Development Goals might shed some additional light on the missing areas, but the scope of the SDGs is so broad and global that a particular focus on social issues in the context of energy is highly unlikely. It is also unclear exactly what a term like the "social aspects" of an energy system actually entails. This is one reason for our proposal to take a step back and view energy systems from a more human-centered perspective.

In general, the underrepresentation of societal aspects in these discussions is serious and worrying. However, this problem is not rooted in the concept of sustainability. Rather, the broader scope of the value-based approaches that, as mentioned above, are often advocated as alternatives to sustainability analysis, is already implied in sustainability analysis itself. The underrepresentation of social aspects may be due to competing political agendas or the fact that it is difficult to formulate measurable indicators for social aspects, while measurable indicators are readily available for the environmental and economic pillars. It seems most straightforward to operationalize the economic pillar as here costs and benefits are already expressed in monetary units. For the ecological pillar, instruments to assess ecological sustainability have been developed over the last several decades, measuring environmental effects such as the emission of toxic or greenhouse gases, or the use of land and water. Life cycle assessment, for example, is an elaborate and much used tool for comparing the ecological impact of a variety of products, and the method is regulated in various ISO norms (ISO 14040 and 14044). Only much more recently have similar efforts been made to specify the societal impacts of technology. Over the last decade or so, research and development of the so-called 'social life cycle assessment,' or SLCA for short, has aimed at modelling the societal pillar in a similar fashion to the ecological pillar (cf. Kloepffer 2008; Jørgensen et al. 2008). Despite these advances, many societal aspects remain much more difficult to subject to indicator sets than economic and most ecological aspects.

Alongside the narrow focus on environmental aspects, another charge against the concept of sustainability stressed by advocates for alternative approaches like design for values is its alleged inability to deal with competing or mutually reinforcing values (iii). This critique applies quite rightly to the three-pillar view and all its popular derivates like the *energiepolitische Zieldreieck*. The three pillars stand on equal footing and it is not immediately clear how to compare, for example, the ecological pillar to the economic. This is a central charge against the three-pillar view. However, it does not apply to sustainability analysis more broadly. In addition, this issue has been criticized within the sustainability community quite extensively. Though the three-pillar view is still widely used in public and political discussions, researchers advocate for alternative sustainability concepts in which the three areas – economic, social, and ecological – are seen as interconnected and the division between them is abandoned (e.g. Cato 2009, 36f.). There are a number of these integrative concepts of sustainability available for the energy context. In Germany the dominant ones are the Greifswald approach (e.g., Ott and Döring 2006), the integrative concept of the Helmholtz alliance or HGF approach (Kopfmüller et al. 2001), the normative-functional concept of sustainable development (Renn et al. 2007), or the integrative sustainability concept and sustainability strategy of the German Federal Government (Bundesregierung 2002, 2016). However, none of these address the ethical aspects of energy systems in the way that the Energy Capabilities approach does.

11.2.3 Designing for Sustainability

Two central charges against the concept of sustainability as a guiding rule for technological development were refuted in the preceding subsection. In this subsection we first outline why the concept of sustainability is particularly suited to thinking about the ethics of energy supply and demand, and will then turn to the final charge against sustainability, namely its air of sacrifice and anti-progressiveness (iv). Building on this, we advocate a fresh view of sustainability that integrates the design perspective and thus helps to overcome the charge that sustainability is only backward-looking and cannot contribute to a progressive approach to technology.

Energy technologies have a massive impact on future generations. Radioactive waste, for example, requires tens of thousands of years of safe storage in order to avoid health risks, and the use of fossil fuels (without carbon sequestration and storage) releases carbon dioxide into the atmosphere, altering global temperatures and sea levels in ways that we are only beginning to understand their effects – most of which are likely irreversible. If we accept responsibility to future generations, it is beyond dispute that intergenerational equity must be at the centre of ethical concerns about energy supply and demand. The concept of sustainability seems to be ideally suited to this task. The definition of sustainability from the Brundtland report cited in Sect. 11.2.2 above highlights that one thing is core to sustainability: intergenerational equity. Sustainability is essentially a matter of justice between generations. As a matter of justice, fulfilling the needs of present generations must not come at the expense of future generations' ability to meet their own needs. This focus on future generations may lead to a misguided overemphasis on environmental issues as the environment forms the basis of life for future generations. However, the needs of those living in the present are just as important in sustainability analysis as those of future generations. So sustainability is *per se* able to deal with a variety of ecological as well as social issues just as, for example, design for value approaches.

Summarizing our discussion thus far, we have defended sustainability against three of the four charges raised by proponents of alternative assessment concepts, and argued that, because the negative impacts of present-day energy technologies will primarily affect future generations, the focus on intergenerational justice makes sustainability ideally suited to addressing ethical issues in the energy context. However, we do not intend to pit sustainability against other assessments of technology. Quite the contrary, we hold that the rationale underlying the design-for-values approaches can provide useful guidance for sustainability analyses since they direct our attention to an important issue, namely, the fact that the technological future we want to live in is *designed*, it is deliberately created.[5] Of course, most of the time we lack complete knowledge about the impacts a design may have. The (co-)design of socio-technical systems such as the energy system is a case in point. Lack of full knowledge of the impacts of a design, however, does not change the fact that technology involves some deliberate designing in any case. This raises two important questions when we talk about a value-sensitive design of the energy transition: (a) What is being designed? And (b) who actually does the designing?

(a) As pointed out by Oosterlaken (2014), value-sensitive design has primarily been applied at the level of technical *artefacts*. The same holds for other design-for-values approaches as well. When dealing with energy supply (in Oosterlaken's context: wind energy), however, we mainly have to do with socio-technical *systems*. Applying the design perspective to energy systems raises serious problems of demarcation. It is often difficult to determine what is actually being designed. The energy transition, for example, involves an increase in intermittent renewable energy sources like wind or solar. This has an impact on the electrical grid as well as on other types of generation plants or storage systems, as the periods when electricity is not available from renewable resources need to be bridged. So when considering a sustainable design for a new wind park, it is insufficient to focus on the sustainability of the individual technical artefacts, the wind turbines or the generators, or even the wind park as a whole. Rather, one must also consider the sustainability of the changes in the grid, of the additional power plant complex, and the implications this has for individual consumers and the industry (see Kuenneke et al. 2015).

In this paper we aim at a broad scope for the ethical analysis of energy and therefore use the term "energy system." Here, an energy system is comprised of at least the following four components: (1) the provision of energy in the form of electricity, heat, and mobility; (2) the technique used for energy conversion; (3) the extraction and disposal of raw material; and (4) the distribution and storage infrastructure. In addition, we take into account that various energy systems allow for (5) different types of individual use, and may offer specific opportunities for (6) participation, e.g., financial or political participation. All these aspects will have to be taken into consideration when thinking about the ethics of the energy system that is (to be) designed.

(b) Now that the system has been demarcated in broad strokes, let us turn to the second question: Who actually designs the energy system? Here, too, we face a problem of demarcation. In a narrow understanding of the word 'design,' engineers

[5] From a practitioner's perspective, the so-called design for sustainability exploits advantages of combining the design approach with sustainability considerations, see http://www.d4s-de.org/manual/d4stotalmanual.pdf

in various industries design the forms of supply and the distribution grid when it comes to constructing, say, a power plant. However, every common citizen also shapes the energy system with her electricity usage or use of mobility technologies. Though policy makers and governing institutions use laws, regulations, incentives, and penalties to shape demand and supply within an energy market, the individual remains a central driving force – a driving force that is notoriously difficult to predict. The choice of a private petrol- or diesel-driven car over public transport is not entirely determined by fuel prices or the availability of public transport; it rather depends on many contingent features and individual preferences that are, at least in part, opaque – in some cases even to the individual herself. Taking the socio-technical nature of the energy system seriously not only requires that we consider the impact energy technologies have on society as a whole and on its various parts, it also means that we need to incorporate into our reasoning all those who presently live in the system as co-designers who shape and design the energy system day-to-day.

This brings us back to the perception that sustainability can only be achieved with personal sacrifice, sufficiency, or a general reduction of amenities. Sometimes sustainability drives itself into this corner. We may overcome this issue, however, by exploiting the link between design and sustainability and putting the focus on how we can actively realize a more sustainable future.

11.3 The Feasibility of Combining Ethics and Energy

The search for an ethical framework for evaluating energy systems with the help of the term 'sustainability' first requires further specification of Brundtland's definition of the concept (see Sect. 11.2.2). Sustainability here is understood in terms of needs, and we therefore have to delineate which needs are of primary importance. Brundtland clearly asserts that the needs of present and future generations must be given equal consideration. It is clear, however, that not all needs can be given the same weight. There seems to be an ethically relevant discrepancy between someone's need for more energy to drive an SUV in the city and the need for a reliable supply of electricity to run a hospital. In order to establish which needs are to be considered in our analysis of the energy system, it is necessary to address what is of intrinsic value. The 'needs' that correspond to what is intrinsically valuable constitute the minimum requirements for a sustainability analysis (cf. Shue 1993; Hillerbrand 2015).

In today's world, we value pluralism and heterogeneity. Can we reasonably expect to agree on what is of intrinsic value when faced with the plurality of concepts of living? In view of all the conflicting interests articulated when it comes to siting issues, e.g., the construction of a wind park or power line, it seems that no consensus can be reached. An ethical analysis of the energy system based on sustainability as introduced by Brundtland may seem unattainable. Thus before we spell out our ethical framework for energy systems in the following section, we first want to dig a bit deeper into the feasibility of ethically sustainable energy. One challenge is local opposition to new technologies that are agreed to be better in an ethical sense. These siting issues are often referred to as "NIMBY" (*N*ot *I*n *M*y *B*ack*Y*ard) – that is, the effect that one is generally supportive of a technology but not in one's own close vicinity. We will address this issue in Sect. 11.3.1. A second challenge is the aforementioned value-pluralism. In Sect. 11.3.2, following Hillerbrand and Dumke (2011), we will introduce the capability approach as an ethical framework for identifying what is of intrinsic value that avoids being paternalistic and hence may be compatible with a value-pluralistic society.

11.3.1 The Social Science of Acceptance, or: The Non-NIMBY Effect

The introduction of large-scale socio-technological systems is met with resistance even when the technology is perceived as part of a sustainable goal and people agree with this goal in general. While wind energy, for example, is endorsed by many as a valuable part of the energy system, concrete wind park locations often face significant opposition from residents and (local) environmental groups. At first glance it seems that egoistic preferences drive locals to oppose technologies that they otherwise endorse. Their own interests and needs seem to outweigh the greater good, the needs of the many future generations to come. Following this line of reasoning, for many years this and similar forms of opposition were referred to as "NIMBY" (*N*ot *I*n *M*y *B*ack*Y*ard). Over the last decade or so, however, social science research on siting issues has shown that NIMBY is not the best explanation for such opposition (cf. Haggett 2011). Rather than predominantly egoistic interests, people also raise complex, unselfish arguments against the concrete implementation of new technology. One issue that is often raised in local opposition is a (perceived) lack of procedural justice, even in cases where civic participation has been fostered. In dealing with opposition, however, the focus is often on distributive justice. Locals are offered compensation for the disadvantages they face due to the location of a particular energy infrastructure in their vicinity. These disadvantages may be an aesthetic impact on the landscape through a wind power plant and its possible effects on the local tourism industry, or simply a blocked view, as well as a greater risk for locals due to a nearby nuclear power plant. This current focus on distributive justice alone has a tendency to backfire because locals feel bribed. Instead of distributive justice, procedural justice has to be given greater weight. While most siting issues offer various possibilities for participation during the design phase of a project, locals are often given the opportunity to participate only at the end of the project timeline. Enabling more participation (and at an earlier stage) may be an effective way to deal with opposition.

To summarize, the social science literature on local opposition shows that people do not act out of egoistic motives alone. Rather than an actual NIMBY effect, we find the phenomenon of industry and policy makers alike referring to the NIMBY 'effect' when the empirical evidence shows that there is no such effect. We refer to this as the Non-NIMBY phenomenon. For the purposes of this paper it is most important to note that local opposition does not have to be a hindrance to implementing an ethical energy system.

11.3.2 Values in a Value-Pluralistic Society

The preceding subsection offered some hope that egoistic preferences need not necessarily restrict energy ethics to a mere academic exercise by showing that what is commonly referred to as the NIMBY effect does not imply that locals oppose ethical arguments. Let us now turn to the intrinsic values that will guide which needs are to be taken into account in considerations of sustainability. We argue in the following that the capability approach as devised by A. Sen and M. Nussbaum provides a suitable ethical framework for this issue. We will follow Nussbaum's version more closely in Sect. 11.4. However, Sen's understanding of the capability approach as a middle ground between a too liberal pluralistic attitude and a paternalistic approach is well-suited to our line of reasoning, and we therefore follow Sen's account in our sketch of the basic ideas of the capability approach given here.

Generally speaking, we can distinguish two very general strategies of how to determine what is important in a human life, or what is of intrinsic value: objectivism and subjectivism. Objectivism asserts that it is possible to formulate a generalized conception of what makes for a worthy and well-lived human life, a conception that is in some sense independent of what a person herself would say. The benefits of such an objective conception as a basis for sustainability analysis are obvious. It facilitates inter-individual comparison of the needs involved, and it allows the creation of general strategies for measurement. One of the most widely known approaches of this type is John Rawls' list of basic goods (Rawls 1971).

According to this conception, an energy system is only justified if it contributes to a just distribution of basic goods. However, even if it is theoretically possible to determine the distribution of basic goods, there still remains a gap between the possession of a good and the actual benefit a person can get out of it. That is, it is possible for two persons to be relatively equal in terms of basic goods but to differ vastly in terms of what is ultimately important in their lives, i.e., what they can gain from those goods. This is because human beings are irreducibly diverse and differ *inter alia* in their abilities to make use of their resources and goods. The second strategy for determining intrinsic value, subjectivism, responds to these concerns. Generally speaking, this approach assumes that what is important in human life can only be known by each individual herself. Thus every person that will be impacted by some energy decision has to be asked how their lives will be affected by this intervention by this intervention. The most common theoretical basis for this view is preference-based utilitarianism, which is commonly used in economic cost-benefit-analysis. Current versions of utilitarianism usually take utility to be the fulfillment of subjective preferences. In other words: people gain utility if they actually get what they (really) want. Consequently, in this second approach to the intrinsic good, the possibilities for measurement and inter-individual comparison, let alone inquiring about the satisfaction of future generations, are severely impeded. A common strategy for dealing with these problems is referred to as a 'revealed preference approach.' This approach assumes that whatever people choose from among a number of different alternatives reveals what they actually want. On this account, important information about people's preferences could be attained for example by analyzing their purchasing behaviour. According to the revealed preference approach, if a person buys a car, that means she really has a preference for it.

This obviously does not help in dealing with future generations, whose preferences remain unknown to us. In addition, it is highly questionable that people always choose what they would have chosen given other options or in ideal circumstances. One may buy a car, but rather prefer that the government invest in public transport or bicycle lanes.[6] Of course, ethical subjectivism might still be a valid perspective. It may well be impossible to know anything substantial about the quality of life of people whose actual preferences one does not know. Consequently, it might be that there is no sensible and reasonable account of sustainability. Nonetheless, we should not be too quick to dismiss the possibility. As Sen and Nussbaum show, there seems to be a way to overcome this problematic.

The capability approach avoids the problems of objectivism by being attentive to the inevitable diversity of human beings while at the same time avoiding the pitfalls of subjectivism by offering a general account that allows for measurement and comparison. This approach has emerged as the leading alternative to standard economic frameworks for thinking about poverty, inequality and human development generally. We suggest that Sen's capability approach can be usefully extended to questions of sustainable energy supply and demand. So what central insights does the capability approach provide? For a capability theorist, a well-lived human life is a life in which central and valuable decisions are up to the person herself. What matters is the freedom to choose and to actively realize the things one has reason to value (Sen 1992).

By assessing quality of life in terms of freedom and action, we can take into account the fact that human beings differ in their capacities to make use of goods and resources. Focusing on what people are actually able to do and to be allows for comparing elements of intrinsic value. These freedoms and activities *constitute* a well-lived life and do not merely provide the prerequisites for it. At the same time, the capability approach is not just another form of subjectivism, since it assumes that 'reasonably valuable activities' can be determined in a generalized way and can also be measured, at least in principle.

Following capability theorists, then, we replace the subjective notion of needs with the notion of capabilities. Before we specify which capabilities are central in sustainability analysis, let us note why this account is helpful in assessing the sustainability of energy systems. Firstly, the focus on capabilities brings the individual as the central agent back into considerations of sustainability. It provides an indi-

[6]This is why ethical subjectivism cannot deal with the widely held belief that sometimes human beings do not know what is best for them (cf. Nussbaum 2000).

vidualistic foundation for the concept of sustainability, which is commonly perceived as an aggregated concept. This seems to be a necessary step in the transition towards sustainable energy as it helps to illustrate the multiple connections between energy and various aspects of human life and well-being at the level of the individual. Secondly, with its central focus on capabilities this approach not only takes into account the impacts a certain arrangement of energy sources may have on various aspects of human well-being, it also allows us to consider alternative courses of action. Consider the following example: If in some near or distant future humankind should discover that wind power bears unacceptable risks, people in the present or the future may simply dismantle the wind turbines. For a hydroelectric facility like the one currently under construction in Belo Monte (Brazil), which dams the Rio Xingu on an area of more than 500 square kilometers, such a straightforward reduction of risk is impossible. This problem becomes even more severe in the case of storage for high-level radioactive waste, for which dismantling the technological facility does not eliminate the risk, even when transmutation is considered. An evaluation of energy supply must account for this loss of freedom of action as much as the concrete impact on human well-being. Current sustainability indicators do not yet adequately acknowledge this type of reversibility.

Now the question becomes: how do we select relevant capabilities for sustainability analysis? A staring point is provided by Nussbaum's "central human capabilities": life; bodily health; bodily integrity; emotions; senses, imagination, and thought; practical reason; affiliation; other species; play; and control over one's environment (Nussbaum 2006, 76–78). Influenced by Aristotelian thinking, Nussbaum puts human dignity at the center of her considerations. She justifies this list of ten central or core capabilities, derived from a Rawlsian overlapping consensus (Rawls 1993, 2001; Taebi et al. 2014), by arguing that each of these capabilities is needed in order for a human life to be "not so impoverished that it is not worthy of the dignity of a human being" (Nussbaum 2000, 72). Rather than reviewing Nussbaum's reasoning in detail,[7] we take these central capabilities as given and re-interpret them in the energy context in the following section. We aim to establish a preliminary set of 'energy capabilities,' or substantive freedoms that are predominately at stake during changes in the energy system.

11.4 A First Step Towards Energy Capabilities: Nussbaum's Core Capabilities Applied to Energy Systems

Our aim in this section is to identify those aspects of an energy system that have a significant impact on people's lives and to connect these aspects to Nussbaum's basic capabilities. In doing so we draw on preliminary work in Hillerbrand and Goldammer 2017, where the close relationship between central capabilities and various energy systems was first highlighted.

[7] For a critique of Nussbaum's list see Barclay 2003; Sen 2004; Crocker 2008; Robeyns 2016.

11.4.1 Life and Bodily Integrity

The central human capability of 'life' refers to the ability to live and not die prematurely. 'Bodily health' or 'bodily integrity' refers to the ability to enjoy good health and to eat adequately, to have shelter and to be safe from violence, as well as to move freely from one place to another. For the energy sector, we propose the following distinction: while 'life' is associated with major life-threatening accidents in the energy sector, 'bodily integrity' encompasses all further effects of energy systems on human health.

In the context of raw material extraction, the capability 'life' brings to mind the serious accidents in coalmines known from periods of early industrialization. Similar accidents still occur today, mostly in countries with lower safety regulations like China, where a portion of the hard coal used in Germany originates. Turning to electricity generation, fatal accidents may happen in construction, operation, and malfunctions of power plants. This applies not only to nuclear energy and conventional electricity generation plants, but also to renewable energies. The use of hydropower in the past few decades, for example, has caused 1.4 deaths per terawatt hour and year (TWh/year). In contrast, the number of deaths due to the civilian use of nuclear power is estimated to be only 0.01 TWh/year.[8]

Thus 'life' is impacted by (the risk of) accidents in all areas of the energy system. We relate 'bodily health' to the side effects of various parts of the energy system. This again ranges from raw material extraction to electricity production and use. These include, for example, particle emissions from diesel engines, mercury emissions from coal power plants, as well as noise and light pollution from various elements of the energy system. In contrast to the risks of an accident, these side effects have a continuous impact on people's health. Noise pollution in large cities, for example, caused by traffic or industrial plants, may lead to stress and stress-related illnesses.

Certain energy technologies impact both of these capabilities, life and bodily integrity. For example, anthropogenic greenhouse gas emissions caused by the burning of fossil fuels are associated with an increase in extreme weather events such as storms or drought. The changing climate will have direct and indirect impact on life, and will also impair bodily integrity (see IPCC 2014).

Thus far, our characterization of the relationship between energy and life or bodily integrity has only highlighted negative effects. Our focus in connection with the impacts of (energy) technologies, like that of most public debate, has been on human vulnerability. The capability approach, however, has the great advantage of not only criticizing technology, but also valuing its positive impacts on human lives. Let us therefore consider what we associate with life and bodily integrity in positive terms. Shelter, protection, and support are essential here. The human individual is in need of protection and support not just in childhood but throughout her whole life. She is not only exposed to (the presumed risks of) modern technologies, but also to

[8] These numbers take into account the accidents in Fukushima Dai-Ichi and Chernobyl, as well as the great disaster at the Banqiao dam, China, in 1975.

weather and natural forces, animals, and other people. Even in moderate climates like those found in Europe, the weather poses a great challenge to human survival and a good life. Heating in the winter plays a central role here. Heating, including heating systems with coal, oil, gas, or electricity, and the insulation of houses or thermal insulation require a large amount of energy. At the same time, life requires food, and a good life needs more than just the calories necessary for survival. Greenhouses or cold storage facilities, which provide us with vitamin-rich fruit and vegetables in the winter, consume large amounts of energy, as does the transport of food across country or continental borders. Even clean drinking water is, essentially, a matter of energy. With sufficient electricity, the global water problem could, in principle, be completely resolved: salt water could be converted into drinking water through a very energy-intensive process. The term "water-energy-nexus" points to this close connection between water and energy issues. The example of the US state of California illustrates the problem: one fifth of California's total electricity consumption goes into maintaining a clean water supply (California Energy Commission 2017). When we look beyond matters of nourishment, the list of positive effects of energy on life is practically endless: electric street lighting protects us from assaults in dark alleyways, modern electricity-based diagnostic techniques and surgical methods in hospitals save numerous lives, etc. All these technological advances depend on a stable power supply.

11.4.2 Emotions

For Martha Nussbaum, the central human capabilities grouped together under 'emotions' are the ability to avoid unnecessary mental pain or trauma, to have joyful experiences, and to feel joy. Accidents in power plants or extreme weather events due to climate change relate to this category when they provoke grief due to the loss of friends and relatives. Furthermore, people suffer emotionally from losing their habitat. Climate refugees in Africa flee from water scarcity; residents in Germany are resettled as a consequence of lignite production. Feelings of grief, loneliness, and the loss of a sense of home are often the result, as well as hatred for those who are blamed for these feelings. Perhaps future generations will hate us for not treating the ecosystem, the basis of life for future generations, more responsibly. There may well be hatred for corporations and for politicians and "elites" in general as these groups are blamed for large energy-related infrastructure measures like a high-power electrical grid, or a power plant in the immediate vicinity. Anger and hatred towards politicians may have dangerous implications if they lead to protest votes in elections. High-voltage transmission lines, which destroy the idyllic view from someone's porch, or the fear of their supposed harmful health effects lead to dissatisfaction, and all of this can be channeled in the next election.

Although the fear of radiation emitted from a nuclear power plant or fear of electrosmog from high-voltage power lines overstates the actual danger, this fear may still affect people's capabilities. This is all the more relevant because one cannot escape the (at least supposed) threats the energy system may pose: energy in all its forms is inevitable in our modern world and permeates our entire life.

However, energy technologies also have positive effects on the emotional level. A warm house not only provides protection from the cold, it also creates the feeling of security. In addition to being a highly desirable technology outcome, safety is also an important emotional state. A stable power supply helps to manage everyday life, and it gives us a sense of stability and safety. The indirect effects of energy services on security and our sense of security are manifold. Furthermore, complex emotions such as pleasure in aesthetic contemplation require the muse and the free time to develop those emotions. Technology and in particular energy services enable us to enjoy an abundance of free time that was unknown to previous generations. It is up to each and every individual how this free time is spent – whether for refinement of emotions, whether pursuing a passion or relaxing at home. The free time created by energy services thus has great potential for contributing positively to the core capability of 'emotions.'

11.4.3 Senses, Imagination, and Thought

In addition to emotion, Nussbaum lists further cognitive abilities in her catalog of central capabilities. For a dignified human life one needs to be able to use one's senses, one's imagination, and one's intellectual abilities. "The ability to use, think, and judge his senses and imagination – and to do these things in a way that is guided by appropriate education, including (but not limited to) reading and writing as well as mathematical basics and a basic scientific education" (Nussbaum 1999, 200). Nussbaum stresses that it is a political necessity not only to provide opportunities for education, but also to protect "artistic and political freedom of opinion and freedom of religion" (ibid.).

A first (negative) connection of this capability with our energy topic is where certain aspects of an energy system are placed under a taboo. Climate engineering, the conscious manipulation of the earth's climate, may be a case in point. There is striking disagreement both in academia and the broader public as to whether such interventions in climate change or research on climate engineering are immoral, unproblematic, or even morally necessary (cf. Lawrence and Crutzen 2016; Gardiner 2010). Other examples may be topics such as nuclear fission or carbon capture and storage which are considered absolutely out of the question by many in Germany. Such informal bans on thinking affect cognitive abilities: political decisions about carbon capture and storage, nuclear energy, or climate engineering make it possible to hinder or prevent both education and research on these topics.

On the positive side, we can classify various energy systems by the extent to which they constructively influence the central capabilities of "sense, imagination and thought." For example, energy systems make human life easier and thus free up time to make more use of one's cognitive abilities. Access to electricity, combined with access to education, enables the individual to continue her education. Self-learning on the computer is an option open to many today, as information on virtually any topic can be accessed on the internet from anywhere, rendering time-consuming and sometimes even impossible visits to the library obsolete.

The decision to use a particular energy technology can also increase intellectual potential as it promotes research and development in that field. The promotion of renewable energy technologies in Germany through the Renewable Energy Sources Act (EEG)[9] led to the improvement of renewable energy technologies. Not only did it produce new technologies, it also reduced the cost of these technologies, such as photovoltaic cells, to such an extent that they are now even options for the electricity supply in poorer countries. This "capability for globalization" of technologies is a politically relevant dimension of energy policies.

11.4.4 Trust

Trust refers to the capability to engage in relationships with people and things outside of ourselves. For Nussbaum, supporting this ability means supporting forms of human interaction that are demonstrably very important to human development.

When thinking about energy, we hold that the continuity of the environment is an important factor that reflects the capability of trust. Changes in the environment due to energy infrastructures or other large-scale impacts are detrimental to our ability to trust the stability of our environment. Such changes may be in the form of resettlement due to mining or migration as a result of climate change, an increasing occurrence of extreme weather events or rising sea levels. However, the continuity of the environment may be interrupted not only by conventional electricity generation, but also by the use of renewable energies. Consider the impact wind power plants or hydroelectric power stations and dams have on the landscape. Such changes to familiar surroundings affect our ability to trust that the environment will continue to exist as we have come to know it.

11.4.5 Practical Reason, or the Imagination of Goodness

Nussbaum's central capability of practical reason summarizes the capability to develop a conception of 'the good' and of 'the good life.' It also entails the capability to plan and critically reflect on one's own life. Due to the central importance that the energy supply holds for the individual and the enormous effects of energy systems on humans and their environment, this central capability includes critical reflection on personal energy use and thus reflection on the preferred energy system. This presupposes the ability to develop a concept of a good energy system and to

[9] The *Erneuerbare-Energien-Gesetz* (EEG) in its original form ensures that electricity from renewable resources can always be fed into the grid and is guaranteed a fixed price, the 'feed-in-tariff.'

critically question existing systems or political proposals. This further requires that decisions in the energy sector be sufficiently transparent and citizens be sufficiently knowledgeable in the field. Here, information about the energy system is just as essential as (at least) basic knowledge of natural sciences and social sciences, in order to form a sound opinion about energy consumption and the desired energy system.

Availability of information on the energy system is thus of central importance for this capability. The German Federal Government, for example, runs an energy monitoring process – "*Energie der Zukunft*" – and since 2012 has published energy-related data in the form of regular progress reports. In addition, four experts assess these reports independently and critically, and also publish their assessment of the energy transition's progress as a publicly accessible document. This supply of information makes public interaction, participation, and identification with the energy system more likely. It is important to note here that within the capability approach we are proposing, transparency is not itself a central capability, nor an end in itself. It is of derived, instrumental value, which has its limits. In order to protect against terrorist attacks, it may be useful to refrain from making all information publicly available – for example, details about a country's nuclear power supply.

In developing a concept of the good life, it is also important to be aware that energy technologies affect not only our environment, but also the way we perceive the world. Charlie Chaplin's "Modern Times" may come to mind, in which an assembly line worker sees wheels that need to be tightened everywhere, even in his spare time. Elements of the energy system are omnipresent in our everyday lives and thus shape our view of things. This can have positive implications, but as in Chaplin's film it can also seriously constrain our view of the world.

11.4.6 Affiliation

Affiliation refers to the capabilities of social interaction and identifying with others, the ability to "live with others and live for others," the "ability to practice justice, and make friendships" (ibid.). In the energy system, this capability hints firstly at aspects of *distributive justice*, i.e., questions regarding the effects of energy policy regulations on cost-benefit and on distribution. Who bears the costs of an energy system and who reaps the benefits? Costs and benefits here are not just economic. The costs of an energy system also include the suffering caused by, for example, losing one's home to lignite mining or the psychological effects of such a loss, or by the destruction of recreational areas. Secondly, affiliation is also about *restorative justice*. If the costs and benefits of the energy system affect different people in different ways, how can those who bear higher costs – for example because the power line or the power plant is in their immediate vicinity – be compensated? Such questions of restorative justice are not limited to a national level. It is also necessary, for example, to take into account the costs incurred in the production of raw materials beyond national borders. Thirdly, affiliation is also about questions of *trade or exchange justice*. The 'contract' between generations, for example, can be understood as an exchange in which we leave an intact environment, raw material reserves, or equivalent energy sources for future generations, and we can expect the immediate generations to care for us when we get older (cf. Höffe 1993).

This threefold classification of questions of justice in terms of distribution, restoration (or corrective justice), and exchange can be traced back to Aristotle, who is a strong influence in Nussbaum's work. In particular, she follows Aristotle's separation of 'the good life' (in our approach, capability 5) from 'the just life' (capability 6). Nevertheless, the good and the just are closely connected. Thus, in addition to its relevance for developing a concept of the good, the issue of availability of information about the energy system is also of central importance for the basic capability of affiliation. This is also reflected in discussions about the aforementioned *Energiewende Monitoring*, which has a comprehensive indicator catalogue, but often lacks social indicators (Goldammer et al. 2013).

11.4.7 Ecological Connectivity

Ecological Connectivity describes the ability to empathize//sympathize with nonhuman animals, plants, and nature in general. This capability is linked to the capability of trust (point 4), which in the context of energy we have interpreted as trust in nature. We distinguish the two central capabilities of trust (in nature) and ecological connection as follows: large-scale ecological changes, which are partly perceived as abstract, fall under the aspect of trust, while smaller, direct environmental impacts of the energy system fall under ecological connection. Let us illustrate this distinction with some examples.

Climate change and the associated risks of storms and floods, the disappearance of entire regions as a result of rising sea levels, as well as the loss of familiar surroundings, our environment, due to mining, tend to involve issues of trust: these threats destroy or impair humanity's confidence in the environment. Ecological connectivity, on the other hand, includes those effects of the energy system that directly affect people's ability to have a respectful interaction with their immediate environment: for example, when animal populations are threatened by regenerative energy systems such as hydropower and wind power. Public attention is currently focused on ecologically "sustainable" energy supplies that negatively affect the capability of ecological connectivity. We refer to these conflicts as 'green-green conflicts,' i.e., conflicts in which different ecologically meaningful goals are at odds. For example, sustainable energy generation using wind turbines or hydroelectric dams can conflict with the protection of species such as birds, fish, or bats. This issue is of such current relevance that in 2016, the German Federal Ministry for the Environment, Nature Conservation, Construction and Reactor Safety established their own "Competence Center for Nature Conservation and the *Energiewende*" in order to address green-green conflicts.[10]

[10]The homepage can be reached here: https://www.naturschutz-energiewende.de. Detailed information is currently available only in German.

Some aspects of the energy system relate to both trust in nature and ecological connectivity. For example, when large areas of the rainforest are cleared in order to cultivate energy crops, this destroys natural space familiar to human beings and makes it more difficult to trust in nature, and at the same time, animals and plants die, which violates a respectful interaction with other creatures.

11.4.8 Play

When it comes to play, it is all about the capability to laugh, to play, and to enjoy relaxing activities. When comparing different energy systems, this category raises the question of how the energy quantities or services being provided influence the central capability for relaxation. One may be quick to say "the more energy, the better," since, of course, life becomes easier with an increased energy supply and thus offers more opportunities for relaxation. However, we must also address the ways that different forms of energy conversion influence leisure activities. Consider cases in which hydroelectric power stations transform rivers and lakes such that hiking trails are affected, or offshore wind power plants restrict sailing possibilities. Thus not only the amount of energy provided, but also the lost recreational value of an unspoilt natural landscape must be taken into account.

11.4.9 Control Over One's Environment, Part A: Separateness

Separateness is the central capability to live one's own life and not someone else's. That is, separateness is about personal autonomy. In the form of electricity, energy can promote an autonomous life in a fundamental way. Once again, at first glance the appropriate response seems to be "the more, the better,", as this enables more control over one's environment. Furthermore, the reliability of energy supplies plays an important role here, because I can realize my own life and develop that life according to my own standards only if the basic prerequisites for living are always and reliably given.

Autonomy, however, has its own narrow limits. The more I depend on the availability of useful energy, the more I depend on a complex network that involves mining of raw materials, electricity and heat generation, and the less autonomous I am. For example, if I am able to navigate on my own and orient myself with the help of the sun, I am more independent than if I have to rely on the GPS function of my smartphone. In this case, more energy does not necessarily mean greater autonomy. However, Nussbaum is less concerned with this type of dependence on others here. As social beings, humans are always dependent on fellow human beings. In a globally networked society, these are not just the people in one's immediate vicinity, but also those on the other side of the globe. This dependency is also mediated via technology, in the form of hardware (e.g., satellites) or software (e.g., Google Maps, to stay with the navigation example).

Nussbaum is primarily concerned with the protection against "interventions in particularly personality-determining decisions such as marriages, births, sexual preferences, language and work." In our interpretation of Nussbaum's classification, whether it is a good idea for a national energy system to depend on, for example, a gas or oil supply from states with a completely different legal and political conception is less a question of autonomy (which in the capability approach always refers first to individuals, rather than states or other aggregated units) than a question of the good life or the just life.

11.4.10 Control Over One's Environment, Part B – Strong Separateness

While so far we have focused on autonomous decisions, the central capability of strong separateness is a matter of shaping one's environment. It is about "living his life in his own environment and his own context." That is, human beings need to be able to influence their social context, e.g., via political participation. Therefore civil rights, freedom of speech, and protection from despotism are necessary. In addition, property rights and the capability to have at one's disposal what one has created play an important role in strong separateness. However, it is worth keeping in mind that "personal property is, contrary to personal freedom, a means and not an end in itself" (ibid. 201).

The capability of strong separateness can distinguish energy systems from one another in terms of the degree of participation they allow for various components of the energy system. Here, participation is not limited to financial participation. Energy systems that allow small actors such as cooperatives and their members to design and participate fare better, as far as this capability is concerned, than energy systems with a centralized structure in the hands of a few major corporations. For example, to evaluate the German *Energiewende*, it would be necessary to examine the extent to which it enables the participation of individuals. When we look at the numbers, particularly in rural areas, participation of smaller actors has increased. In the wake of the liberalization of electricity markets and the Renewable Energy Sources Act in 2000, countless competing companies entered the electricity generation market. At the same time the number of renewable energy operators and power generation cooperatives increased. In the form of so-called *Bürgerenergie* (civilian energy) the latter enables local people, particularly in rural areas, to participate directly in the energy supply system (Engerer 2014).

Given current technological developments and the failure to promote large-scale projects such as Desertec, achieving a high percentage of renewables seems to require a change from central power plants to small, decentralized photovoltaic and wind power plants. Since the financial investment required is typically smaller in this energy generation scheme, this allows for more opportunities for participation than in centralized power plant parks.

11.5 Conclusion

To review, this paper has explored the question of what constitutes an ethical energy supply and demand system. We chose to address this question in terms of sustainability of energy systems, as we argued that the concept of sustainability has certain advantages when thinking about an ethically justifiable energy supply. A focus on design, as offered for example by value-sensitive design and other technology assessment tools that are sometimes seen as competitors of sustainability analysis, proved be advantageous. Focusing on the design aspect of technologies points to certain features of the energy system that may help in realizing a more ethically sound energy system. We therefore suggested a fresh view of sustainability that incorporates the design perspective, with attention to *what* is being designed and *who* is doing the designing. For the first question we argued that the entire socio-technical system is under construction, while for the second, relevant designers are not only policy makers and members of the energy industry, but every individual with her particular energy use profile.

We reassessed the question of what a sustainable energy supply can be and linked the concept of sustainability, which is commonly considered in aggregate terms, to various aspects of the life of the individual. In what we call 'energy capabilities,' we spelled out Martha Nussbaum's capability approach for the evaluation of a good life in such a way that it can be used to evaluate energy systems. We Although this is only the first step towards the formulation of energy capabilities, by following Nussbaum's approach this paper has shown how most public and political debate today is far removed from the concerns of the individual. From a human-centered point of view, characteristics of energy systems such as marginal increases in the price of electricity or statistics that indicate the national security of supply such as the System Average Interruption Duration Index (SAIDI)[11] are of minor or no importance. Moreover, using the capability approach we can pave the way for an ethical evaluation framework for energy systems that is broad enough to account for various individual ideologies that are compatible with a value-pluralist society. Future work will further develop these energy capabilities and demonstrate their applicability in discussions and decision-making about energy systems.

[11] The SAIDI measures the average duration of outages for an electricity customer in a given energy system in minutes or hours per year. Internationally, Germany has the lowest SAIDI with typically less than 30 min per year (Bundesnetzagentur 2017).

Bibliography

Atomgesetz. (2011). *13. Gesetz zur Änderung des Atomgesetzes*. http://www.bgbl.de/xaver/bgbl/start.xav?startbk=Bundesanzeiger_BGBl&bk=Bundesanzeiger_BGBl&start=//*%255B@attr_id=%2527bgbl111s1704.pdf%2527%255D#__bgbl__%2F%2F*%5B%40attr_id%3D%27bgbl111s1704.pdf%27%5D__1478619254328. Accessed 15 Aug 2017.

Barclay, L. (2003). What kind of liberal is Martha Nussbaum? *SATS: Nordic. Journal of Philosophy, 4*(3), 5–24.

BDI. (2014). *BDI Energiewende-Navigator, Monitoring zur Umsetzung der Energiewende*. http://bdi.eu/media/user_upload/20141201_bdi-energiewende-navigator_2014.pdf. Accessed 15 Aug 2017.

BMWi. (2016). *Die Energie der Zukunft*. https://www.bmwi.de/Redaktion/DE/Publikationen/Energie/fuenfter-monitoring-bericht-energie-der-zukunft.html. Accessed 15 Aug 2017.

Bräutigam, K. -R., Kopfmüller, J., Lichtner, P., Rilling, B., Rösch, C., Stelzer, V., & Weinberger, N. (2015). *Indicators for monitoring and assessing the German energy transition*. Poster presentation at the EST Energy, Science & Technology International Conference & Exhibition, Karlsruhe.

Brundtland. (1987). *Report of the world commission on environment and development: Our common future*. http://www.un-documents.net/our-common-future.pdf. Accessed 15 Aug 2017.

Bundesnetzagentur. (2017). *Versorgungsqualität Strom*. https://www.bundesnetzagentur.de/DE/Sachgebiete/ElektrizitaetundGas/Unternehmen_Institutionen/Versorgungssicherheit/Versorgungsunterbrechungen/Auswertung_Strom/Versorgungsunterbrech_Strom_node.html. Accessed 23 Aug 2017.

Bundesregierung. (2002). *Perspektiven für Deutschland -Unsere Strategie für eine nachhaltige Entwicklung*. https://www.nachhaltigkeitsrat.de/nachhaltigkeit/strategie/strategie-2002/. Accessed 23 Aug 2017.

Bundesregierung. (2016). *Deutsche Nachhaltigkeitsstrategie, Indikatorenbericht 2016*. https://www.nachhaltigkeitsrat.de/nachhaltigkeit/strategie/indikatorenbericht-2016/. Accessed 23 Aug 2017.

California Energy Commission. (2017). *Water-Energy-Nexus*. http://www.energy.ca.gov/research/iaw/water.html. Accessed 23 Aug 2017.

Decker, M. (2013). Technikfolgen. In A. Grunwald (Ed.), *Handbuch Technikethik*. Stuttgart: Metzler.

Crocker, D. A. (2008). *Ethics of global development: Agency, capability and deliberative democracy*. Cambridge: Cambridge University Press.

Die Bundesregierung. (2010). *Energiekonzept für eine umweltschonende, zuverlässige und bezahlbare Energieversorgung*. https://www.bundesregierung.de/ContentArchiv/DE/Archiv17/_Anlagen/2012/02/energiekonzept-final.html. Accessed 15 Aug 2017.

Engerer. (2014). *Energiegenossenschaften in der Energiewende*, DIW Roundup. https://www.diw.de/de/diw_01.c.470180.de/presse/diw_roundup/energiegenossenschaften_in_der_energiewende.html. Accessed 23 August 2017.

EnWG. (2017). *Energiewirtschaftsgesetz (Gesetz über die Elektrizitäts- und Gasversorgung)*. https://www.bmwi.de/Redaktion/DE/Gesetze/Energie/EnWG.html. Accessed 23 Aug 2017.

Ethik-Kommission. (2011). *Germany's energy transition: A collective project for the future*. Produced by the Ethics Commission for a Safe Energy Supply. https://www.bundesregierung.de/ContentArchiv/DE/Archiv17/_Anlagen/2011/05/2011-05-30-abschlussbericht-ethikkommission_en.pdf?__blob=publicationFile&v=2. Accessed 15 Aug 2017.

Gardiner, S. (2010). Is "Arming the Future" with geoengineering really the lesser evil?: Some doubts about the ethics of intentionally manipulating the climate system. In S. M. Gardiner, S. Caney, D. Jamieson, & H. Shue (Eds.), *Climate ethics. Essential readings* (pp. 284–312). Oxford: Oxford University Press.

Goldammer, K., Mans, U., & Rivera, M. (2013). *Beiträge zur sozialen Bilanzierung der Energiewende*, IASS Schriftenreihe. http://www.iass-potsdam.de/sites/default/files/files/report_beitraege_zur_sozialen_bilanzierung_der_energiewende_0.pdf. Accessed 23 Aug 2017.

Grober, U. (2013). *Die Entdeckung der Nachhaltigkeit: Kulturgeschichte eines Begriffs*. München: Kunstmann.
Haggett, C. (2011). Understanding public responses to offshore wind power. *Energy Policy, 39*, 503–510.
Hillerbrand, R. (2015). The role of nuclear energy in the future energy landscape: Energy scenarios, nuclear energy and sustainability. In B. Taebi & S. Roeser (Eds.), *The ethics of nuclear energy: Risk, justice, and democracy in the post-Fukushima era* (pp. 231–249). Cambridge: Cambridge University Press.
Hillerbrand, R., & Dumke, M. (2011). *An ethical framework on how to assess sustainability*. An Advancement of Sen's capability approach, proceedings of the 6th Dubrovnik conference on sustainable development of energy, water and environmental systems.
Hillerbrand, R., & Goldammer, K. (2017). *Energietechnik und ein gutes Leben: ein Plädoyer für einen neuen Energiediskurs*. In KWI Working Paper Special Issue Technologischer Fortschritt und gutes Leben, ISSN 2365-9092.
Hillerbrand, R., & Roeser, S. (2015). *Towards a third 'practice turn': An inclusive and empirically informed perspective on risk*. Philosophy of Technology after the Empirical Turn (pp. 145–166).
Höffe, O. (1993). *Moral als Preis der Moderne: Ein Versuch über Wissenschaft, Technik und Umwelt*. Frankfurt: Suhrkamp.
IAEA (International Atomic Energy Agency). (2005). *Energy indicators for sustainable development: Guidelines and methodologies*. Vienna: IAEA.
IPCC. (2014). *Climate change 2014: Synthesis report*. Contribution of working groups I, II and III to the fifth assessment report of the intergovernmental panel on climate change [Core Writing Team, R.K. Pachauri and L.A. Meyer (eds.)]. Geneva: IPCC.
Jørgensen, A., Le Bocq, A., Nazarkina, L., & Hausschild, M. (2008). Methodologies for social life cycle assessment. *The International Journal of Life Cycle Assessment, 13*(2), 96–103.
Klitkou, A., Hain, D. S., Munch Andersen, M., Lindgaard Christensen, J., Rennings, K., & Borup, M. (2014). *Indicators of energy innovation systems and their dynamics*. EIS Radar Report Vol. 2. Kopenhagen: Danish Ministry of Research.
Kloepffer, W. (2008). Life cycle sustainability assessment of products. *The International Journal of Life Cycle Assessment, 13*(2), 89–95.
Kopfmüller, J., Brandl, V., Jörissen, J., Paetau, M., Banse, G., Coenen, R., & Grunwald, A. (2001). *Nachhaltige Entwicklung integrativ betrachtet: Konstitutive Elemente, Regeln, Indikatoren*. Berlin: Edition Sigma.
Kuenneke, R., Mehos, D. C., Hillerbrand, R., & Hemmes, K. (2015). Understanding values embedded in offshore wind energy systems: Toward a purposeful institutional and technological design. *Environmental Science & Policy, 53*(B), 118–129.
Lawrence, M., & Crutzen, P. (2016). Was breaking the taboo on research on climate engineering via albedo modification a moral hazard, or a moral imperative? *Earth's Future, 5*, 136–143. https://doi.org/10.1002/2016EF000463.
Nussbaum, M. (1999). Women and equality: The capabilities approach. *International Labour Review, 138*, 227–245.
Nussbaum, M. (2000). *Women and human development: The capabilities approach*. Cambridge: Cambridge University Press.
Nussbaum, M. (2006). *Frontiers of justice: Disability, nationality, species membership*. Cambridge, MA: Harvard University Press.
Oosterlaken, I. (2014). Applying value-sensitive design (VSD) to wind turbines and wind parks: An exploration. *Science and Engineering Ethics* 20 (Open Access (http://link.springer.com/article/10.1007/s11948-014-9536-x)).
Ott, K., & Döring, R. (2006). Grundlinien einer Theorie "starker" Nachhaltigkeit. In K. Köchy & M. Norwig (Eds.), *Umwelt-Handeln: Zum Zusammenhang von Naturphilosophie und Umweltethik*. Verlag Karl Alber: Freiburg.
Rawls, J. (1971). *A theory of justice*. Cambrige, MA: Harvard University Press.
Rawls, J. (1993). *Political liberalism*. New York: Columbia University Press.

Rawls, J. (2001). *Justice as fairness. A restatement*. Cambridge: The Belknap Press of Harvard University Press.

Reisch, L., & Thøgersen, J. (2015). *Handbook of research on sustainable consumption*. Cheltenham: Edward Elgar Pub.

Renn, O., Deuschle, J., Jäger, A., & Weimer-Jehle, W. (2007). *Leitbild Nachhaltigkeit. Eine normativ-funktionale Konzeption und ihre Umsetzung*. Wiesbaden: VS Verlag für Sozialwissenschaften/GWV Fachverlag GmbH.

Robeyns, I. (2016). The capability approach. In E. N. Zalta (Ed.), *The Stanford encyclopedia of philosophy* (Winter 2016 Edition). https://plato.stanford.edu/archives/win2016/entries/capability-approach/. Accessed 15 Aug 2017.

Scott Cato, M. (2009). *Green economics*. London: Earthscan.

Sen, A. (1992). *Inequality reexamined*. Oxford: Oxford University Press.

Sen, A. (2004). Capabilities, lists and public reasons: Continuing the conversation. *Feminist Economics, 10*(3), 77–80.

Shue, H. (1993). Subsistence emissions and luxury emissions. *Law & Policy*, 15(1), 39–60.

Taebi, B., Correljé, A., Cuppen, E., Dignum M. & Pesch U. (2014). Responsible innovation as an endorsement of public values: The need for interdisciplinary research. *Journal of Responsible Innovation* 1(1), 118–124.

UN. (2015). *Transforming our world: The 2030 agenda for sustainable development*. http://www.un.org/ga/search/view_doc.asp?symbol=A/RES/70/1&Lang=E. Accessed 15 Aug 2017.

Vahlenkamp, T., & Gohl, M. (2012). Energiewende-index Deutschland 2020. *Energiewirtschaftliche Tagesfragen, 62*(9), 22–26.

Van den Hoven, J., Vermaas, P. E., & van de Poel, I. (Eds.). (2015). *Handbook of ethics, values, and technological design sources, theory, values and application domains*. Dordrecht: Springer.

Chapter 12
Technology, Society, and Survival

Erik W. Aslaksen

Abstract Society is presented as a complex, dynamic system, whose evolution is determined by the collective intelligence formed by the interaction of its members. It is this continuous, collective evaluation process that limits the extent of fluctuations, in the form of oppression, wars, and genocide, which can be seen as failures of the collective intelligence. As a consequence, the ability of individuals to exchange information and to exercise their intelligence by taking adaptive, goal-oriented action, becomes the central characteristic of the Good Society. Developments in information technology are increasingly influencing this ability, both positively and negatively, but the main concern is with the control of these applications of technology. The concentration of wealth and associated power in a small section of society, sometimes identified as the Transnational Capitalist Class, and its control over the flow of information, is a threat to democracy and to the stability of the evolution of society.

Keywords Evolution of society · Understanding of technology · Collective intelligence · Concentration of wealth and power · Stability of evolution

12.1 Introduction

The purpose of this chapter is to present an Australian engineer's view of how technology – in particular, information technology – is influencing the evolution of society. The approach to the subject matter reflects my long involvement with systems engineering (Aslaksen 2013) and the engineer's attitude of 'fit for purpose'; the latter sometimes viewed by philosophers as 'superficial'. My hope is that the view presented will be accepted as useful rather than necessarily true in every detail, and promote further discussions between philosophers and engineers.

E. W. Aslaksen (✉)
Gumbooya Pty Ltd, Allambie Heights, NSW, Australia
e-mail: erik@gumbooya.com

A. Fritzsche, S. J. Oks (eds.), *The Future of Engineering*, Philosophy of Engineering and Technology 31, https://doi.org/10.1007/978-3-319-91029-1_12

Following this short Introduction, the chapter is structured into two sections, each one addressing one of two distinct, but closely related issues: The meaning of "technology', and a particular view of society as a complex system. The concept of technology is central both to engineering and to philosophical investigations into many aspects of society, but the meaning given to this concept is very different in the two cases. This difference is, in my opinion, the cause of much of the difficulties experienced in the interactions between the two professions, but if properly understood, those difficulties should be eliminated. – The second section starts by defining our understanding of the concepts "society" and "system", and then develops a particular view of society as a complex system; a view in which the information exchange between the system elements – i.e. between the individual members of society – is the determining factor in the evolution of society. This then introduces the important role of information technology in mediating this information exchange, and it is argued that inherent in this role is a significant threat to the stability of the evolution, posed by the emergence of a Transnational Capitalist Class and its control of the media.

12.2 Technology in Philosophy and Engineering

The understanding of the meaning of technology within engineering was introduced at an earlier fPET conference (Aslaksen 2018) and developed further in (Aslaksen 2015). To engineers, *technology* is the resource base they use to create applications. It contains two groups of resources: A *knowledge base* comprised of textbooks, standards, published papers, operating manuals for tools and instruments, etc., and spans a continuum from advanced research results to Tables for everyday use, and a *construction base*, consisting of the millions of standard construction elements, ranging from reinforcing steel bars to microprocessors and reusable software modules, that engineers and the technical workforce can draw on in executing projects, as well as the facilities within industry for fabricating and constructing equipment and plant. Both of these bases are dynamic: new construction elements are continually being added and older elements are being phased out; new knowledge is being generated through research and experience, and what was advanced knowledge yesterday is tomorrow's accepted practice.

The identification of the construction and knowledge bases as constituting 'technology' is a deviation from the use of 'technology' by philosophers and sociologists, where it is used in a much more encompassing manner, such as 'the production and use of artefacts', derived from the Greek concept of *techné*, and often used interchangeably with *technique*. The development of a philosophical framework for technology has been described by various authors, such as (Mitcham 1994), and numerous articles and further references can be found in (van de Poel and Goldberg 2010), but it is interesting to note that some publications on the philosophy of technology make no mention of engineering at all. However, while much of what philosophers say about technology can be reflected onto engineering; it is important to keep the distinction in mind; in particular, the above definition of technology does not include any activity. Whereas philosophers see technology as an activity (or at least including activities), no engineer would speak of "doing technology". People study technology in order to become engineers, and engineering is the work performed by engineers in developing and applying technology.

A consequence of this engineering definition of technology is that what society experiences is not technology, but applications of technology, as has been pointed out by many authors, e.g. (Hughes 1994). The reason for this is that engineers are today almost completely embedded in the framework we call *industry*, which encompasses not only private industry, but also government entities involved in applying technology and educational institutions involved in developing and disseminating technology. What society experiences as technology is the product of industry. An engineer on his or her own can accomplish very little, and so what society sees is the work of the engineer through an industrial interface in which numerous people play a part, such as workers, tradespeople, marketing and sales people, business managers, financiers, etc. The position of the engineer within industry, and the relationship between engineers as employees and the industrial entities in which they are employed is not a new issue; it was raised already by Veblen (1904, 1921), Winner (1977, 1980), (Noble 1977), and Layton (1971), and more recently as part of a recent essay by Newberry (2007). In particular, he makes reference to the suggestion by Noble that industry has forcefully shaped the mechanisms for engineering education and professional socialisation in order to produce a "domesticated breed of engineers".

Here lies a main difference between science and engineering, and a source of misunderstanding between philosophers and engineers. Disregarding this difference has resulted in such unfortunate associations as "science and technology" and "technoscience", and to such a concept as "technics", which has no meaning to an engineer. The purpose of science is to gain an understanding of Nature, and the criterion is truth; the purpose of engineering is to meet needs expressed by all or part of society, and the criterion is usefulness. The work of a scientist has an intrinsic value; the work of an engineer has no intrinsic value, it only acquires a value through its realisation by industry as a successful product. There is a strong connection between engineering and science, but it is indirect, in the sense that the resource base of engineering, i.e. technology, is in part based on the results of science.

There is also a major difference to other professions, as was discussed in an earlier fPET article (Aslaksen 2017), in which medicine was given as an example of where there is a direct interface between the profession and society, as illustrated in Fig. 12.1.

The average person has a direct and reasonably good understanding of what doctors are and what they do, as evidenced by the great popularity of books, television programs and films about the medical profession; whereas the role and responsibility of engineers in the industrial process is hidden. Most people would have a very vague (if any) idea of how the "technology" they see all around them and use every day is related to the work of engineers, and, as an example, nobody would know the name of the engineer responsible for the design or production of the car they drive.

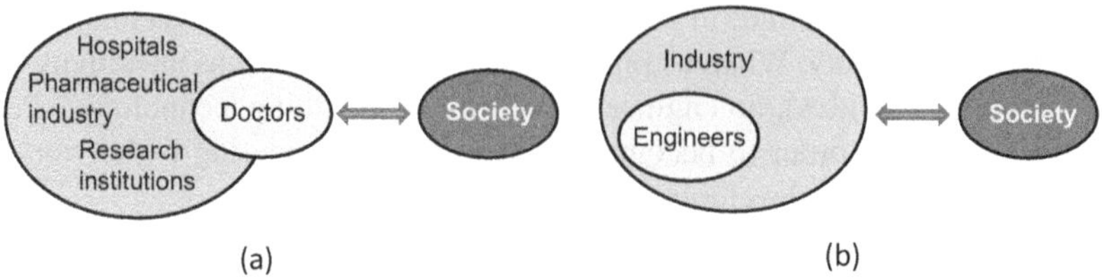

Fig. 12.1 Illustrating the indirect interaction between engineers and society, as opposed to the direct interaction between doctors and society (their patients). (From Aslaksen 2017)

It is paradoxical that as applications of technology become more and more pervasive, the relationship of engineers to these applications, as seen by society, is becoming less and less visible.

The profession has lost much of its influence and contribution to society, and these lost values are placed in a philosophical framework that provides a complementary view to that of engineering as an economic activity by someone who knew nothing about engineering. In her book, *The Human Condition* (1958), Arendt presented an analysis, and her view, of the situation in which humanity found itself in the middle of the twentieth century. The work of Arendt is taken up in more detail in another chapter in this book; here we just remark briefly on the relevance of Arendt's political philosophy for engineering. In addition to activities currently included in the process of engineering, there needs to be activities concerned with the process itself and thereby with the affairs of engineers as a major component of human affairs. Engineers must be engaged in the process of engineering, but they must not be consumed by it. When engaged in the process of engineering, engineers display their competence and their skills, and while this may involve considerable creativeness, much as a work of art does, they are basically displaying *what* they are, as opposed to *who* they are as individuals. By becoming completely absorbed in the process of engineering, in the cycle of production and consumption, engineers have become part of the species *animal laborans*, albeit as a very sophisticated version of that species. And instead of being a significant participant in shaping the evolution of society, the engineering profession is practically invisible to society, except through its products. A more detailed consideration of this issue, including the implications for engineering education, is contained in (Aslaksen 2015a).

The analysis presented by Arendt should be a cause for concern and provide a wake-up call for the engineering profession. In a vicious circle of consumption and production we have almost all become labourers, and similarly, under the demands of industry for lower cost and shorter time to market engineers have become mostly technologists and technicians. Production has become an end in itself and not primarily a means to a better life. And these two developments, in society and in engineering, are both symptoms of the same underlying cause: the role of capital as a source of income. In order to produce a return on investment, a production facility must produce more than its intended product; it must provide an additional return in the form of a profit to the capital providers. This return can be accumulated to provide more capital, which requires more return, and so on, providing a feed-back loop, and thereby introducing a time-dependence into the economic system. This can also be expressed by saying that the algebraic equations governing the money flows associated with labour have been augmented by differential equations associated with capital. A recent, and frequently cited work on the role of capital is the book *Capital in the Twenty-First Century*, by Piketty (2014).

The significance of the absence of an effective and credible input from engineers – the creators of technology – into the public discourse is illustrated by the particular view of society as a complex system presented in the next section.

12.3 Society As a Complex System

The word "society" can have a wide range of meanings. Here we shall understand it to have its most encompassing meaning and consist of the people in a defined group, such as a nation or the whole world, and their institutions, technology, and other artefacts; essentially everything these people have created. Through the interactions between these elements they form a system, and due to the number and diversity of both the elements and the interactions, this is indeed a very complex system. It is also obviously a self-organising system, unless one admits some form of divine guidance.

A system is a *mode of description*; we can choose both what we identify as elements and what we identify as their interactions (within the capabilities of the elements). There are normally several possible descriptions of an entity as a system, depending on what aspect of the entity we are interested in examining. In this manner we are hiding the complexity that is not relevant to the aspect of interest, and the system approach is a methodology for handling complexity. (For a detailed description, see Aslaksen 2013). Our view of society as a system is based on three observations:

Over the last 10,000 years or so, society has evolved from small groups of hunter-gatherers to today's world society, with its complex structure. And again, discounting any divine intervention, the driving force in this evolution must have been located within us. The world is what it is today because we made it so, and we determine what it will be tomorrow. And how did we do it? At each point in the evolution, we evaluated our situation relative to our environment, and decided on the appropriate goal-oriented adaptive action. This is basically the definition of intelligence (Legg and Hutter 2007), and so we shall represent the individual by a simple model of intelligence.

Intelligence operates on the information provided to it, and as the number and density of individuals in a society increased, this information became less a matter of observation of the environment and increasingly a matter of interaction with other members of the society. But only in the most primitive of societies is the interaction between individuals limited to direct, person-to-person interaction via speech, gestures, or physical contact. In all other societies there are additional forms of interaction that involve a medium, such as pen and paper, a telephone, a gun, etc.;

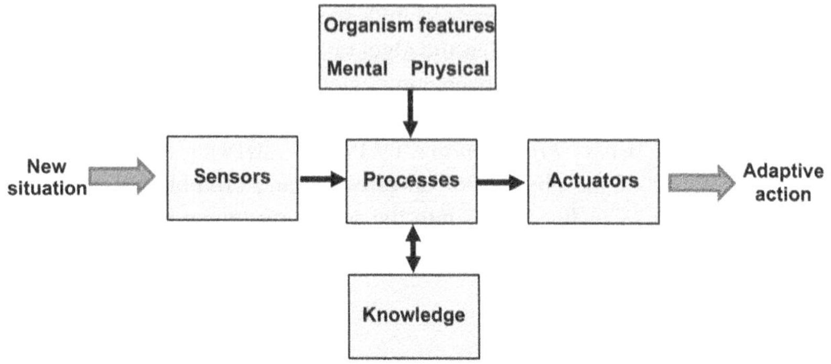

Fig. 12.2 Diagrammatic representation of the high-level model of intelligence

in short, some *application of technology*. These media become additional elements in the system, with their own descriptions of their behaviour, and contribute to the behaviour of the system, i.e. of the society.

A modern society consists of millions of individuals, each with their own characteristics. And furthermore, in all but the utopian case of a completely homogeneous society, the interactions result in a structuring of the society, i.e. the society consists of identifiable sub-societies, and depending on in which of these sub-societies an individual finds itself, it displays different properties. That is, to describe the operation of an intelligence, we would have to index it not only by individual, but by the particular *role* in which the individual finds itself, with one individual able to perform many roles (e.g. husband, father, club member, worker, etc.).

Clearly, we cannot build a model of society at this level of detail, but have to adopt a statistical approach, analogous to the way in which we form macroscopic variables in thermodynamics (or statistical mechanics), but with one very significant difference. And to explain what this difference is, we must first introduce the model of individual intelligence. The diagrammatic representation of the model in terms of its main elements and their interactions is shown in Fig. 12.2. (A detailed description of this model is contained in Aslaksen 2015b.). The organism responds to an input, in the form of the external stimuli representing a new situation, by performing an adaptive action. The action is performed by the actuators (feet, vocal chords, etc.) available to the organism, which are controlled by the processes taking place in response to the signals provided by the sensors. The processes are dependent on the knowledge accumulated by the organism through experience and education (represented by the block identified as "Knowledge"), and on inherent mental (i.e. instincts) and physical (agility, strength, reach, etc.) features of the organism. What is identified as "Mental" organism features is the same as instincts, defined as a set whose elements comprise all information, data, and beliefs transmitted to an individual through genes and to which the processes require access in formulating adaptive actions.

There are some features of this model that require further elaboration. The block identified as "Processes" represents two conceptually different groups of processes.

One contains the processes that evaluate the input on the basis of the relevant information in the knowledge base and the instincts, and determine the most appropriate action (which might be no action). The other group contains the processes that determine whether and how the knowledge base needs to be modified in order to reduce or eliminate discrepancies between the information in the input and the information in the knowledge base; it is the group that provides the above-mentioned accumulation of knowledge.

For most of our actions, such as all our repetitive daily actions – brushing teeth, catching a bus, saying "hello" to people we know, etc. – the identification as "adaptive" is trivial and unnecessary. They are actions resulting from inputs that would, at some earlier stage of our lives, have been assessed in relation to our knowledge and contributed to forming that knowledge, but they now fit into what we might call *a framework of accepted inputs*, and this type of actions is not what we normally think of as a result of intelligence. The actions we shall be primarily interested in arise from those inputs that are not only not repetitive, but where some or all of the knowledge entering into the evaluation of the inputs is of a probabilistic nature. That is, it is in the form of more or less well founded *beliefs*, and we shall further restrict this to beliefs about things that matter to the individual, the things the individual is willing to make a sacrifice for, which is clearly only a small part of our total knowledge. What an individual considers an item that matters is not determined only by the generally accepted characteristics of this item, such as family relationship, a written constitution, rules and laws, the dogmas of the Church, but by the individual's perception of the item, by additional features that are important to that individual. The relationship with each of these things that matter is peculiar to each individual, and it is something the individual establishes as it goes through its life cycle. This web of relationships is what we shall, in the present context, call the individual's *identity*, and constitutes a subset of the knowledge base. This identity determines, in effect, what the goal of the individual's "goal-oriented behaviour" is; it is the realisation of those beliefs, making them come true, and thereby it also defines what the meaning of "survival" is for that individual. Note that "truth" does not enter into this definition of identity.

From the above description of the process we note two of its important features, and they are both related to the fact that the process is a *dynamic* process. One, it is dynamic in that the process itself changes with time; the identity is influenced by the information being processed. Two, the process is dynamic in the sense that it takes time into account; time is a variable in the situations being evaluated, and the decisions and adaptive actions are based on predictions of future effects. This introduces additional variability; two persons with exactly the same moral precepts and understanding of right and wrong may take different adaptive actions because they predict different futures in which the actions will take effect.

If we now view society as a system consisting of such individual intelligences interacting through an exchange of information, it is exactly this dynamic nature of the intelligence that makes this system very different to a system of molecules, as in a gas. In the latter, the molecules are not changed by the interaction, whereas the individual intelligences are, and so, when we form a model of society as a whole in

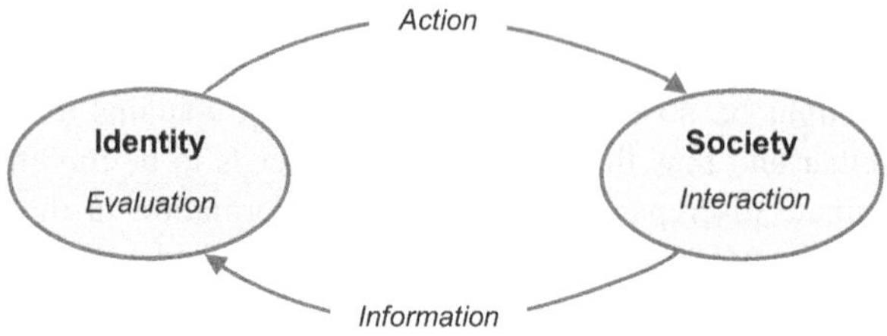

Fig. 12.3 Illustration of the mutual interaction between the individual and society. Society provides the means for realising the actions of the individual, the individual provides the means (the identity) for evaluating the performance of society

terms of what we might call a *collective intelligence* (this falls within the definition given e.g. in Wikipedia) through a process of averaging, the "macroscopic" variables that have equilibrium values are not state variables of society, but of the collective intelligence, which is a process, i.e. the evolution of society. When we speak of survival of the human race, what survives is *the continuity of the evolution of society*, where, as before, by society we understand all humans and their artefacts, the interactions between these, and the environment in which they operate. And there is no sense in asking such questions as 'Survival for what purpose, in order to reach what?'; there is no purpose, the evolution of humanity has no goal. Our evolution is not *toward* something; evolution is an eternal process of survival. It is impossible to predict or even imagine what "humanness" will be a 1000 years from now.

What we have, then, is a particular view of society as a complex, dynamic system, in which the evolution is determined by the collective operation of the intelligence of each or its individual members. The contribution of the individual to this evolution depends on the criteria developed on the basis of the information available to the individual over its history, and the ability to develop these criteria, as the individual's identity, as a reflection of the meta-criterion of survival, is the essence of life. The important corollary to this is that the goal of the "goal-seeking adaptive behaviour" can be formulated as allowing unrestricted exercise of our intelligence, and hence the importance of the quality of the information available to each individual and the ability of each individual to contribute to that information through participation in the *public discourse*, as emphasized by Kompridis (2009), Habermas (2003), Wartofsky (1992), Feenberg (2007), and Jarvie (2007). The fluctuations in the evolution, e.g. in the form of wars and persecution, are the results of restricting the ability of individuals to exercise their intelligence, mainly through the perversion of information and suppression of the public discourse.

The evolution of society is driven by the interaction between each individual and society, and it is one in which each party influences and is influenced by the other. This is illustrated in Fig. 12.3, which shows the individual receiving information by observing society through various channels and evaluating this based on the criteria making up the individual's identity. This evaluation may result in an adaptive action, and this action is carried out in the forum provided by society; that is, as an interaction

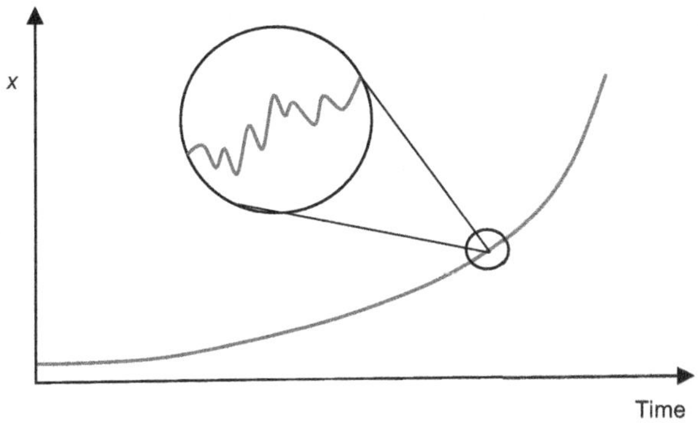

Fig. 12.4 Illustrating how the evolution of society is not a smooth progression when looked at in detail. The parameter *x* is some measure of the richness or complexity of society

with other elements of society. The new state of the society is again observed by the individual, and so on.

However, not only does society change, but the identity of the individual changes also; *identity is a social product.* Through observing its environment, the individual continuously assesses what matters for survival, and the evaluation criteria that make up the identity undergo an evolution that is related to that of society; the two are coupled. The coupling mechanism is a statistical one, in that the evolution of society is influenced, to some degree, by the actions of every individual. The degree of influence, or *power*, varies greatly from person to person, and the process by which power is developed is a highly non-linear one, in that power provides the means for its own increase through the feed-back inherent in the loop illustrated in Fig. 12.3. Let the state of society be characterised by a single parameter, such as happy life years or the richness of life, and identified as, say, x, then x will change with time as a result of the feed-back process indicated in Fig. 12.3. The dependence of x on time is shown as the smooth curve in Fig. 12.4, where x changed as humanity progressed from the cave to life in a developed society today, and at each point in time, the change in x, that is, the value of dx/dt, is what human intelligence, on the average, judged to be optimal. Any deviation from this curve is detrimental; it is not the case that the greater the value of x at any point, the better. This is, of course, the problem with different parts of the global society being on different points on this curve at the same time; what might be likened to a non-equilibrium situation in physics.

If we would look at the value of the same parameter, but averaged over only a part of society, we would find that it fluctuates around the value for the whole society, and the fluctuations are greater the smaller the part under consideration is. Consequently, if we look at the curve in Fig. 12.4 with a greater resolution, we would find that it is not a completely smooth curve, but one with fluctuations, also called "sampling fluctuations" and discussed e.g. by Cavalli-Sforza and Feldman (1981), as indicated in the figure. But because these fluctuations are detrimental, and are recognised by the rest of society as such through the evaluation process, they do not persist. However, if the averaging process is perverted, the fluctuations can become significant, and the proposition put forward here is that the fluctuations are, in the final analysis, due to failures of the collective evaluation and action process described above, or of what we might call the *social immunity system*. The nature of the failures can be characterised in terms of three causes:

(a) Interference with the flow of information as input to the evaluation, either through limiting the information, falsifying it, or by obscuring it in a mass of irrelevant data.
(b) Limiting the opportunities for taking adaptive action, either through direct repression, or by promoting a structure of society that isolates the individual from its environment.
(c) Restricting the size of the society, and thereby the averaging process, through such strategies as nationalism, sectarianism, and an unwarranted focus on local issues.

Addressing each of these three groups of causes separately, we begin with (a).

The inputs of relevance to the current subject matter are those that do not fit into what we called the framework of accepted inputs, and which therefore lead either to a change in our identity or to an adaptive action, or both. And with respect to these inputs, the current state of our society displays a couple of significant characteristics. Firstly, the amount of input presented to us on a daily basis, through a number of media, is so great that it is becoming impossible to assess and classify it properly, and as a result relevant information is lost or obscured. Secondly, it is becoming increasingly difficult to determine the veracity and accuracy of the information. Rather than being a direct result of interaction with the source of information, more and more of the information is being processed by the media delivering it, resulting in a view representing the interests and opinions of the media owners. These characteristics of the flow of information in modern society have been very well described by various authors, including Schiller (1981, 1982), Garnham (1987, 2001), and Chomsky (1980); the point to be made here is simply that these characteristics of the information industry are drivers of the fluctuations on the evolution of society.

The causes within group (b) are those that prevent adaptive actions, or that limit the range of possible actions. In order for the individual to be able to realise a desire for action, there needs to be established means of doing so. These may take the form of voting in elections or referendums, posting a comment on public media, participating in the activities of an organisation or action group, or even starting a new group, and so on. Some or all of these means can be suppressed by a dictatorship, by a legal organisation, such as a church, or by an illegal organisation, such as the Mafia or a terrorist group. But these means can also simply be rendered ineffective by structuring society so that it is not affected by them; real power rests in entities not visible or accessible to the general public.

The third group of causes that can lead to failure of the evaluation and action process, group (c), are those arising from not recognising, or taking account of, the changed environment in which the society finds itself. In particular, in not recognising that both the structure and the boundary of the society has become artificial, and that the society is holding on to an image of itself and its uniqueness that is no longer valid. Information is manipulated to fit this image. In today's dynamic world, the significant relationships are between groups of people, defined by shared beliefs and interests, and the nation should be seen primarily as a framework for managing relations between groups of people, both internally and externally. As the beliefs and interests change (not least due to increased level of education and economic development), the framework should adapt in order to manage the new relations. This is true for both intranational as well international relationships.

12.4 Social Stability and Technology

Before addressing the final issue of the stability of our present society and the role of technology in that context, it is both useful and appropriate to acknowledge that there is, within the discipline of political philosophy, a body of work related to the structure and stability of societies, and while it is not within the scope of this chapter to attempt any form of review of that body of work, the account given by one of its prominent authors, Rawls, in his book *Political Liberalism* (2005), puts forward a number of concepts that have analogies in the present work. They have to be considered analogies, as they are presented in the utopian environment of an isolated democratic society of idealised citizens, with little or no regard for the issues currently affecting real societies or for the accelerating rate at which those issues are changing, and with no mention of the role of technology, but they provide valuable insight into the issue of political stability. Briefly, Rawls defines stability in terms of an overlapping consensus of reasonable comprehensive doctrines (i.e. world views, incl. religions) on a political conception of justice. He promotes a conception of justice as fairness, but the important insight is that this conception is not the intersection of the doctrines existing within the society nor, indeed, any one of them. Its content is expressed in terms of certain fundamental, or *a priori*, ideas seen as implicit in the political culture of a democratic society and presented as freestanding and expounded apart from, or without reference to, any wider background (of comprehensive doctrines) (Part One, §2). What we have termed "collective intelligence" is close to Rawls' "public reason" in its effect. Rawls sees a sense of justice as an innate feature of being human; we see the basic feature of the human intellect as the ability to discern the moral principles that will support survival in the given circumstances. The will to survive is the basic feature of all living organisms; it is the essence of life. The human is simply the species with the (by far) greatest ability to realise this feature as a strategy for living.

The issue of the stability of society has continued to attract attention. In particular, Rawls' assertion that people with differing comprehensive doctrines would be able to affirm a common political conception of justice through the faculty of public reason has been questioned by Grose (2008), and Chow (2014) discusses issues with Rawls' argument for a congruence between the desire to act justly and a person's own good. However, none of these arguments change the importance of Rawls' work for us, it is just that we replace his sense of justice with our will to survive as the *a priori*, and insert an additional step in the process of generating the political conception of justice.

To an engineer, the significant characteristic of the three groups of causes is that they are all critically dependent on applications of technology, specifically within information technology (IT). These applications provide an unprecedented ability to generate, process, and disseminate information, and the following numbers give some feel for the magnitude of these abilities: In 2014, the amount of data generated globally was about 5 exabytes/day (i.e. 10^{18} bytes/day), the amount of data transmitted over the internet during the year was 667 exabytes, a retailer, such as Walmart, stores on the order of 5 petabytes of data, and a government agency, such as the National Security Agency, stores on the order of 2.5 exabytes (Mayer-Schönberger and Cukier 2013). If we compare this with the size of a book-sized document of alphanumeric data, which is about 1 megabyte, then storing one such document of personal data for every individual on Earth requires only about a petabyte of storage – peanuts really. So storage is not an issue; what about acquiring the data? Personal data is being collected by numerous different organisations, both government and private; some indication of this can be gained by consulting (Deibert and Rohozinski 2010) and (Microsoft 2017). Theoretically they are supposed to be separate, but in reality there is tremendous pressure on gaining access to sets of these data bases, including from national security, medical research, infrastructure planning, political promotion, and commercial applications, and besides such legal access to data, there is the illegal access through hacking.

However, the main potential for IT applications to be used to interfere with the social immunity system is not in acquisition or storage, but in processing of information and in controlling its dissemination. The basic approach and techniques have been known since Antiquity and been applied by salesmen through the ages, but technology has increasingly provided better means for applying them, both through more capable channels, such as printing, radio, television, active displays, and internet, and through selecting and targeting the information provided on the basis of information about the recipient of the information, exploiting the "cognitive advantage", as described e.g. in (Simpkins et al. 2010). And this potential is further increased by the changing distribution of wealth and the associated control of the information industry.

Wealth inequality is increasing, as is demonstrated by the data in Piketty (*op. cit.*) and in a recent Oxfam Briefing Paper (2016), a document that includes an extensive list of references to relevant data. The crucial feature of the wealth distribution data in the current context is the accelerating trend towards a concentration of wealth at the very upper end of the distribution. In the years since 2010, the share of the world's wealth owned by the richest 1% of the world's population has been steadily increasing, to the extent that it now is approximately equal to 50%. That is, this 1% of the population has as much wealth as the remaining 99%. But even more significant is the concentration within this 1%: About 4500 individuals, out of an adult population of 4772 million, each have wealth exceeding 500 million USD, and we note that most of these ultra-wealthy individuals reside in the Western part of the world, i.e. North America and Europe (Credit Suisse 2015). These individuals constitute what William I. Robinson identifies as the Transnational Capitalist Class, or TNC (Robinson 2014), and to the extent that they are able to coordinate their actions, this TNC is a force that can challenge most national governments and play a significant political role on the world stage.

The conclusion of the view of society presented in this chapter is therefore that not only is capitalism, in its current neo-liberal form, resulting in increasing economic inequality, but it is increasing the probability of a serious fluctuation of the evolution of society. By allowing an elite, instead of the collective intelligence (including the engineering profession), to control the evolution of society through the control of the information exchange, the stability of society as a complex system is put at risk.

References

Arendt, H. (1958). *The human condition*. Chicago: University Press.

Aslaksen, E. W. (2013). *The system concept and its application to engineering*. Berlin/Heidelberg: Springer.

Aslaksen, E. W. (2015). The relationship between engineers and society: Is it currently fulfilling its potential? *Journal and Proceedings of the Royal Society of New South Wales, 148*(455–456), 28–43.

Aslaksen, E. W. (2015a). The future of engineering. *Journal and Proceedings of the Royal Society of New South Wales, 148*(455–456), 159–165.

Aslaksen, E. W. (2015b). *A story about us: Evolution and the individual*. Allambie Heights: Gumbooya Pty Ltd.

Aslaksen, E. W. (2017). Engineers and the evolution of society. In D. P. Michelfelder, B. Newberry, & Q. Zhu (Eds.), *Philosophy and engineering: Exploring boundaries, expanding connections* (pp. 113–124). Cham: Springer International Publishing.

Aslaksen, E. W. (2018). An engineer's approach to the philosophy of engineering. In C. Mitcham, B. Li, B. Newberry, & B. Zhang (Eds.), *Philosophy of engineering, east and west* (pp. 85–93). Cham: Springer.

Cavalli-Sforza, L. L., & Feldman, M. (1981). *Cultural transmission and evolution*. Princeton: University Press.

Chomsky, N. (1980). *Rules and representations*. Columbia: University Press.

Chow, P. C. (2014). *Moral stability and liberal justification: An examination of the notion of stability in Rawls's theory*. London: ProQuest.

Credit Suisse. Global Wealth Report. (2015). https://www.credit-suisse.com/corporate/en/articles/news-and-expertise/global-wealth-in-2015-underlying-trends-remain-positive-201510.html. Accessed 01 Mar 2018.

Deibert, R., & Rohozinski, R. (2010). Liberation vs. control: The future of cyberspace. *Journal of Democracy, 21*(4), 43–57.

Feenberg, A. (2007). Andrew Feenberg. In J.-K. B. Olsen & E. Selinger (Eds.), *Philosophy of technology: 5 questions*. Copenhagen: Automatic Press/VIP.

Garnham, N. (1987). Concepts of culture: Public policy and the cultural industries. In A. Gray & J. McGuigan (Eds.), *Studies in culture: An introductory reader* (pp. 54–61). London: Hodder Arnold Publication.

Garnham, N. (2001, September). *The information society: Myth or reality?* Bugs, globalism and pluralism conference, Montreal, pp. 19–22.
Grose, A. W. (2008). *Democracy and the problem of stability*. Ann Arbor: ProQuest.
Habermas, J. (2003). *The future of human nature*. Cambridge: Polity Press.
Hughes, T. P. (1994). Technological momentum. In M. R. Smith & L. Marx (Eds.), *Does technology drive history? – The dilemma of technological determinism* (pp. 101–113). Cambridge: MIT Press.
Jarvie, I. C. (2007). Ian C. Jarvie. In J.-K. B. Olsen & E. Selinger (Eds.), *Philosophy of technology: 5 questions*. Copenhagen: Automatic Press/VIP.
Kompridis, N. (2009). Technology's challenge to democracy: What of the human? *Parrhesia, 8*, 20–33.
Layton, E. (1971). *Revolt of the engineers: Social responsibility and the American engineering profession*. Cleveland: Case Western Reserve University.
Legg, S., & Hutter, M. (2007). *A collection of definitions of intelligence, technical report IDSIA-07-07*. https://arxiv.org/pdf/0706.3639.pdf%20a%20collection%20of%20definitions%20of%20intelligence. Accessed 20 Dec 2016.
Mayer-Schönberger, V., & Cukier, K. (2013). *Big data a revolution that will transform how we live, work, and think*. Boston: Houghton Mifflin Harcourt.
Microsoft. (2017). *The extent of data collection and mining is indicated in Microsoft's privacy statement*. https://privacy.microsoft.com/. Accessed 20 Dec 2016.
Mitcham, C. (1994). *Thinking through technology. The path between engineering and philosophy*. Chicago: University of Chicago Press.
Newberry, B. (2007). Are engineers instrumentalists? *Technology in Society, 29*(1), 107–119.
Noble, D. F. (1977). *America by design: Science, technology, and the rise of corporate capitalism*. New York: Alfred A. Knopf, Inc..
Oxfam briefing paper. (2016). *An economy for the 1%*. https://www.oxfam.org/sites/www.oxfam.org/files/file_attachments/bp210-economy-one-percent-tax-havens-180116-en_0.pdf. Accessed 20 Dec 2016.
Piketty, T. (2014). *Capital in the twenty-first century*. Cambridge: Harvard University Press.
Rawls, J. (2005). *Political liberalism*. New York: Columbia University Press.
Robinson, W. I. (2014). *Global capitalism and the crisis of humanity*. Cambridge: Cambridge University Press.
Schiller, H. I. (1981). *Who knows: Information in the age of fortune 500*. New York: Ablex.
Schiller, H. I. (1982). *The packaged consciousness, Cyrano's journal online*. http://www.cjournal.info/CJO/packaged.html. Accessed 01 Mar 2018.
Simpkins, B., Sieck, W., Smart, P., & Mueller, S. (2010). Idea propagation in social networks: The role of 'Cognitive Advantage'. In 1st ITA Workshop on Network-Enabled Cognition: The Contribution of Social and Technological Networks to Human Cognition, Maryland, USA.
van de Poel, I., & Goldberg, D. E. (Eds.). (2010). *Philosophy and engineering. An emerging agenda*. Dordrecht: Springer.
Veblen, T. (1904). *The theory of business enterprise*. New York: Charles Scribner's Sons.
Veblen, T. (1921). *The engineers and the price system*. New York: B. W. Huebsch Inc.
Wartofsky, M. W. (1992). Technology, power, and truth: Political and epistemological reflections on the fourth revolution. In L. Winner (Ed.), *Democracy in a technological society* (pp. 15–34). Dordrecht: Kluwer Academic Publishers.
Winner, L. (1977). *Autonomous technology: Technics-out-of-control as a theme in political thought*. Cambridge: MIT Press.
Winner, L. (1980). Do artefacts have politics? In D. MacKenzie & J. Wajcman (Eds.), *The social shaping of technology* (2nd ed., pp. 121–136). Maidenhead: Open University Press.

Chapter 13
Manufacturing with a big M – The Grand Challenges of Engineering in Digital Societies from the Perspective of the Institute for Manufacturing at Cambridge University

Andy Neely, Sarah Fell, and Albrecht Fritzsche

Abstract Engineering used to be driven by a community of experts who set themselves apart from others by establishing clear boundaries of their profession. Today, however, these boundaries have become increasingly permeable, due to the increasing dynamic and complexity of technical and economic change. The manufacturing sector illustrates this process very well. Engineering is currently becoming much more deeply involved in the usage of technical artefacts and economic questions of value creation. Engineers are therefore facing the challenge of opening up their traditional domain to collaborate with other disciplines and integrate new knowledge in their theories, concepts and procedures. This contribution shows how the Institute for Manufacturing at Cambridge University copes with this challenge, expanding the scope of topics addressed in engineering and introducing new subjects in the curriculum of the students. All this seems to be a necessary prerequisite for engineers to uphold their claims of responsibility of technical development and their contribution to well-being in society.

Keywords Manufacturing · Service research · Engineering education · Well-being · Innovation

A. Neely (✉) · S. Fell
Institute for Manufacturing, University of Cambridge, Cambridge, UK
e-mail: adn1000@cam.ac.uk

A. Fritzsche
Institute of Information Systems, Friedrich-Alexander-Universität Erlangen-Nürnberg, Nürnberg, Germany

A. Fritzsche, S. J. Oks (eds.), *The Future of Engineering*, Philosophy of Engineering and Technology 31, https://doi.org/10.1007/978-3-319-91029-1_13

13.1 Introduction

When engineering students are asked why they have chosen to study engineering they give many different answers. Some say that they are fascinated by technology. Others refer to their aptitude for working with machines or their interest in practical problem solving. But this is not everything. As their teachers we frequently hear another response, too: that they want to make the world a better place. Engineering students learn to be an engineer in order to take action for the benefit of nature and mankind. They expect to acquire knowledge that allows them to leave a mark with what they do and improve the human condition. This expectation is fundamental to the understanding which engineers have of their own profession. As the British Engineering Council and the Royal Academy of Engineering puts it: "Professional engineers and technicians work to enhance the welfare, health and safety of all, whilst paying due regard to the environment and the sustainability of resources. They have made personal and professional commitments to enhance the wellbeing of society through the exploitation of knowledge and the management of creative teams (www.engc.org.uk/professional-ethics)". Engineers explicitly want to take responsibility for the future development of humanity. Furthermore, and perhaps even more importantly, they make the claim that they are the only ones who are able to do this. They state that their professional expertise enables them to change the world and direct future developments, while others can merely watch, benefit or suffer from the consequences. Wade Robison (2016) accordingly compares the engineer with the driver of a vehicle, the person who controls where the vehicle is going. Others might set the destination and suggest the best routes, but in the end, the driver is the one in charge. In a similar way, engineers consider themselves in charge of the journey to the future.

A look back into history provides much evidence to confirm the engineer's point of view. Without engineering, the Enlightenment would have hardly been experienced as an emancipatory movement which released humankind from its self-incurred immaturity. Engineering provided the material evidence as to how humans could make use of reason to purposefully alter the physical conditions under which they lived and to subject them to their own intentions and interests. They put the harness on nature which allowed modern societies to grow and expand in new directions which were not predetermined by the physical conditions of their environment, with cities, factories, water and energy supply and transport networks all across the planet. Furthermore, they made sure that this harness remained intact over time by maintaining the technology they created, such made it appear like a second nature into which we are born.

Like so many others, the profession of engineering can be said to have gradually evolved from philosophy and early modern science in the reflection of practical problem solving and craftsmanship. Over time, the domain of engineering was rather clearly set apart from other professions in the static structures of early industrialized societies. Politics, law, business and philosophy were separated from engineering not only by the output they created or the knowledge that was formed, but also by the communities of people who were involved in them and the traditions they handed down from one generation to the next. The University of Cambridge played its own role in this process with its strong emphasis on applied mathematics and mathematical physics from the seventeenth century onwards, which contributed significantly to the foundations of modern engineering knowledge.

The more clearly labour is divided in a society, the easier it is to connect certain responsibilities with certain professions. For engineering as a closed community of specialists, claims of being in charge of future development were easily made and emphasized by the formation of influential associations and societies. To this day, they play a major role in public decision making about technical development and are actively involved in the setting of general standards and safety requirements for the operation of technical devices and installations. At the same time, however, the community of engineers has gone through significant changes which put their claims about being in charge of future development into question. Social structures which have previously determined careers and relationships have become more permeable. In the course of the twentieth century, immense efforts have been undertaken to ensure that career paths are no longer determined by birth, race or gender. Education systems are designed to give equal opportunities to everyone and provide the foundations young people need in order to enter whichever profession they choose in the coming years. At the same time, increasing market dynamics shorten economic cycles and lead to the emergence of new industrial sectors, such as microelectronics, IT-consulting, online retail, personal logistics, computer gaming and many more. Company structures become more fluid, hierarchies change and the creation of new departments put new people into managerial positions. Technical change entails frequent revisions of the body of knowledge needed for a profession and creates new, hybrid qualification profiles in practice.

Like any other group of professional experts, engineers today face a world of huge opportunities, but also threats resulting from the openness of contemporary socioeconomic structures. Making claims of being in charge of something becomes increasingly difficult. Where engineers used to be solely responsible for their own domain of work, with their own schools, professional associations, rules and laws which described their rights and responsibilities, they are now involved with other experts: designers, computer scientists, economists, accountants, managers, sociologists and many more who would not consider themselves to be engineers, although the work they do is so closely related to the work of engineers that it is hard to draw lines distinguishing one from another. The grand challenges of engineering as a profession in the years to come can accordingly be considered to result from the fact that there is increasing confusion about who is in charge, due to other stakeholders who have found their way into the core of the engineer's domain (Fritzsche 2017). What is challenged by them is the ability to act as an engineer itself, to take responsibility and contribute to making the wold a better place.

Some areas of engineering are protected from too much change by rigid regulatory structures. For manufacturing, this is not the case. Manufacturing is therefore particularly affected by the grand challenges of engineering and the questions that entails: how is it possible to maintain a professional identity for engineers? How can research and education provide orientation and direction for the future development of the field? And how can students be supported in their wish to take responsibility for making the world a better place? These questions have been extensively discussed at the Institute for Manufacturing (IfM) at the University of Cambridge. The following pages arise from the insights emerging from our discussions.

13.2 Manufacturing with a big M

Manufacturing matters. After a steady decline of employment numbers and the share contributed by manufacturing to the GDP of most industrialised countries, the recent economic crisis has reawakened interest in it and increasing consideration is being given to rebalancing the economy in its favour. Manufacturing, however, is not the same as it was decades ago. It has turned away from the simple concept of a market with a single point of exchange between production and consumption. Technical artefacts are not designed and constructed any more as products which are then handed over to consumers who are left alone with the decision how to use them. Today, engineers are more involved in the usage of technology, due to the changing nature of technical architectures and the strong interrelatedness of single operations.

With a high grade of complexity in technical artefacts and short intervals for maintenance and updates, users need extensive expertise to ensure that tools and machines can be operated. While this leads on the one hand to increasing technical skills among consumers, especially in handling information technology, producers also take over more responsibility for the operation of the tools and machines they provide. This process is often described as servitisation, as it turns the attention away from the transfer of material goods and towards the activities which can be carried out using those goods (Lightfoot et al. 2013). A philosophically satisfying explanation of the contributions of technology providers to user activities is hard to give, since there are various different patterns to consider. Providers can offer maintenance services, they can rent out products and grant their functional integrity over time, or they can get actively involved in the usage of the products by being on location and working together with the users in a team. The latter happens, for example, in a military context, where ships and aircraft are not only rented out, but the maintenance crews on location are employees of the technology providers and not military personnel themselves, with interesting implications for the understanding of governmental responsibility for national security.

The Internet of Things also facilitates other ways in which engineering can engage with user activity. In manufacturing products as cyber-physical systems which use sensors and actuators to establish a direct correspondence between operations in the physical world and information systems, it becomes possible to observe, control and regulate the application of technical devices from a distance (Gölzer and Fritzsche 2017). Engineers can thus track in detail how the tools and machines they provide are actually used, gain a better understanding of practice to improve their designs in the future, or recommend different usage patterns. An early application case of such recommendations is the management of traffic at construction sites. Companies such as Caterpillar have equipped the heavy machinery and vehicles they sell with multiple sensors which collect extensive information about vehicle movement. This information can be collected by the company to create a comprehensive picture of the whole construction site, which makes it possible to identify further potential to optimize routes, cargo, usage times etc.

In short, engineers in manufacturing expand their perspective to encompass not only the design of certain artefacts or devices for certain purposes, but also a deeper understanding of why the purposes or outcomes sought by the user are framed in the way they are, and if they could be framed more effectively. Where they used to be concerned with the output of industrial activities, single transactions, output volumes, supply routes and single technical devices, they are now concerned with their outcomes in a larger sense: holistic solutions, relationships between contributions, partners in networks and ecosystems of value creation. This transition is emphasized by the phrase 'manufacturing with a big M', meaning that it covers the full cycle from understanding markets and technology through product and process design to operations, distribution and related services. The old distinction proposed by Weber (1978) between technical problems as problems of designing and applying means opposed to economic problems as problems of defining and pursuing ends becomes obsolete. Engineering has to consider both at the same time, not only questioning the ends to which technology is used, but anticipating them and shaping them from the perspective of those who make them accessible.

Research and education in engineering accordingly have to cover a much larger horizon. In particular, they have to include content from business and management studies about innovation, value creation, strategy and the organization of industrial activity, in addition to traditional engineering topics like design, material science or operations management, which have to be adjusted to the larger body of knowledge which now defines the discipline.

13.3 The Engineer's Place in Innovation Ecosystems

Engineers, of course, would claim that innovation has always been at the heart of their profession. From the perspective of Manufacturing with a big M, however, it is necessary to ask questions about innovation which have so far mostly been out of scope for engineers, because they go far beyond the performance of systematic research and development activities in industry and address the general conditions under which they take place. Such conditions are set by the national and international policies to support innovation by a suitable infrastructure in terms of money and knowledge as well as legal regulations about ownership, rights of use and accountability for failures.

During the last years, for example, many countries have started huge initiatives to support the digital transformation of industry. This includes programs like

Germany's Industry 4.0, Japan's New Robot Strategy/Monodzukuri, Catapult in the United Kingdom, and the parts of EU's Horizon 2020 program related to digitalisation (see e.g. Oks et al. 2017). Such programs have a huge impact on the direction which industry takes in the respective countries. Traditionally, engineers might have been first and foremost interested in the enablement for their own work resulting from the programs. From a policy perspective, it is more interesting to look at the overall results which have been achieved by these initiatives. The question is what makes some national innovation systems more effective than others at translating science and engineering research knowledge into new technologies, industries and economic wealth. The answer to this question will not only inform politicians how to set up new innovation programs, but it will also tell engineers how their own work is related to the efforts of other stakeholders in technology development and clarify the terms of their mutual engagement. These terms are well known to be quite different in different countries, up to the general understanding of engineering as a discipline (Kornwachs 2017).

Another question related to innovation concerns the topics on which engineers spend their time. In a globally connected world where solutions are designed and implemented across many organisational boundaries, it becomes fundamentally important for engineers to decide where to put their own efforts and where they leave it to others to become active. During the development of new solutions, engineers can make or buy, and the optimal choice between the two depends on various factors which need to be understood in detail. This goes along with a better awareness of the fact that engineers are not alone in their efforts to make the world a better place. Quite the contrary, this is a highly collaborative exercise. The greatest impact is likely to come from a coalition of specialists. There is much more to be achieved if they divide their efforts such that they complement each other, even if this means that some have to give up topics which they hold very dear. The more engineers expand their own horizon, the more they have to think about the alignment of their own activities with those of others, and focus their efforts on those which are most related to their core contribution to the overall system, the basic definition of their business and the competitive advantage they have before others.

Choices between making and buying become particularly important in the context of new additive manufacturing techniques like 3-D printing, which weakens the connection between the material properties of a product and the elaboration of the environment in which it is produced. Under the right conditions, additive manufacturing techniques can make many complex factory installations and expert knowledge which has evolved over hundreds of years obsolete. It is therefore important to distinguish between the context of value creation, the operative procedures to capture value and the use cases which determine the actual value in the application of technology. Again, this forces engineers to take a more reflective stance towards their own work in the view of its contribution to the overall result which is achieved.

13.4 Re-organizing Manufacturing

By putting engineering in a wider context, Manufacturing with a big M also forces engineers to review their own repertoire of views, concepts and methods in daily practice. Once engineers start thinking about the broader outcomes of their work, it is no longer sufficient to design functional artefacts in large volumes, which are easy to use and custom-made to fulfil certain tasks. Doing so, they only address the core benefits of their work, but not the wider implications they have in the lives of their users. Engineers look at the performance of the artefacts they produce, their reliability, build quality and durability. They also consider their producibility in terms of modular designs and component manufacture, assembly and test processes, and platform strategies. They are also well aware of other factors which influence the adoption of their products, such as the different facets of usability and desirability. All this is embedded into a larger system of service operations which go along with the usage of a product and augment its value for the users, including support, maintenance, financing, warranty, delivery and installation. All of them can, again, be reflected on a higher level with respect to their consistency with each other, the regulative measures necessary to ensure their function, their development over time and the strategies and visions related to it. In this sense, 'manufacturing with a big M' talks about different layers of design, concerning the core product, the augmented product together with the service offerings around it, and the meta-level on which all this is planned, enabled and controlled.

The different layers of design and their various interdependencies add enormous complexity to the work of engineers in manufacturing. Finding ways to manage this complexity is one of the major tasks in research and education. During the past years, design research has developed a large variety of new tools, methods and laboratory environments (Chakrabarti and Blessing 2014). They make design practice more efficient and avoid unnecessary overload by supplementary activities which do not usefully contribute to the design process. Furthermore, modern information technology has proven to be very helpful in rearranging the boundaries between different fields of expertise in engineering to encapsulate similarly structured activities and exclude the others, which make the management of complexity significantly easier.

Another approach to re-organizing engineering topics in manufacturing draws more deeply on the roots of engineering in driving innovation. The idea is to identify key technologies which will determine future socio-economic development and to work systematically on core products, augmented products and "meta-products" based on these technologies. As they are expected to set the pace and direction in which industry and society are going to change, they also seem to be the right points of reference to organize resources around them, define research agendas and teaching curricula. Lasers, additive manufacturing, inkjet and other fluid-based manufacturing processes and nanotechnologies are some of the fields in which engineering is evolving rapidly. IfM is researching both these technologies themselves and the broader business and policy contexts in which they sit: in other words, 'manufacturing with a big M'.

1. An example of how work is becoming highly interdisciplinary is being led by the IfM's Fluids in Advanced Manufacturing research group. This project has brought together researchers from across the University of Cambridge together with colleagues from the University of Ghana with expertise in diagnostics, synthetic biology, microbiology and parasitology, colloidal science and manufacturing engineering. Collectively, they are developing innovative biotechnology and manufacturing technologies to develop low-cost diagnostic tests for infectious diseases for use at the point of care.
2. Another area in which manufacturing research is seeking to have a significant and lasting impact is industrial sustainability. In order to address this we need an understanding of the whole industrial system, as the optimisation of any one part is ultimately constrained by the others. IfM research is currently focused on resource efficiency (eco-efficiency), manufacturing systems transformation (eco-factories) and sustainable value exchange (sustainable business models).
3. Food safety is another topic on which 'manufacturing with a big M' is seeking to make an impact, in this instance through the study of supply chains and their resilience. This involves mapping existing supply chains to identify potential points of vulnerability and how best practice can help mitigate risk. The research agenda is expanding to bring in other research domains to integrate the use of sensors and big data to develop a digital food supply chain framework.
4. Understanding intellectual property, another expanding area of IfM research, is critical for bringing new products and services successfully to market. In a commercial environment, protecting IP is a key aspect of business strategy, but IP can also be used strategically to support the development of more sustainable products and processes.
5. At the IfM, 'manufacturing with a big M' also encompasses methods for bringing together the diverse specialists who need to collaborate in order to address societal 'grand challenges'. Challenge-led research can be very different to traditional academic research. For example, addressing manufacturing related grand challenges may require significant advances in engineering tools and infrastructure – requirements for success which may not be evident from the outset. It generally requires the bringing together of academic disciplines, practitioners and stakeholders – many of whom are not used to collaborating with each other. Grand challenge research also needs ongoing input from fundamental science, technology development, engineering demonstration and scale-up efforts.

The scale, scope and complexity of these endeavours means that in most cases they need new sets of partnerships and completely different cultures of collaboration. The IfM is also exploring the implications of this complexity for those government R&D agencies charged with supporting challenge-led innovation. How can funding agencies strategise, cooperate and coordinate around grand challenge research endeavours? How can they design effective programmes for supporting initiatives which can cope with all this complexity? How can they ensure research activities are informed by the right mix of disciplines and communities – not only researchers and industrialists, but users and stakeholder groups? How can they identify appropriate challenge 'stretch goals', assess the potential for impact of particular proposals, and measure the success of such complex endeavours?

13.5 Conclusion

In the course of the past centuries, engineering has changed the world in numerous ways, and there is no reason to assume that it will stop doing so in the future. On the contrary, the increasing possibilities of addressing physical objects through sensors and actuators within large information systems rather suggests that engineering will have an even stronger impact on the world in the future, as it makes many items and structures which have so far dominated the fabric of our society replaceable or dispensable. At the same time, however, the nature of engineering is changing as well. Boundaries to other disciplines dissolve and new field of work emerge which do not fit to the traditional canon of engineering. The field of manufacturing illustrates this phenomenon in numerous different ways.

A new view of 'manufacturing with a big M' to express the larger horizon which it covers, can help us find better ways to cope with the challenges related to this change. To summarize the overall objective, one can say that engineering needs to find a way to reinvent itself as a discipline which can claim responsibility for the future development of economy and society through the introduction of new technical solutions, and to shape it in the best way possible. Making the world a better place is a strong motivation for engineers to do their work, and its contribution to a functioning society can hardly be underestimated. What we are currently learning to understand better is the need to align the engineers' efforts with those of others. Without engineers, the world might not become a better place in the future, but engineers can also not make it happen without a deep and continuous engagement with the rest of society. This paper has outlined various different ways how this engagement can be fostered by institutions for research and education, and it is our job as academics to keep a leading role in the discourse on these issues, so that engineering will continue to be ready to face the grand challenges ahead of us.

References

Chakrabarti, A., & Blessing, L. T. M. (2014). Theories and models of design: A summary of findings. In A. Chakrabarti & L. Blessing (Eds.), *An anthology of theories and models of design* (pp. 1–45). London: Springer.

Fritzsche, A. (2017). *Open innovation and the core of the engineer's domain. Philosophy and engineering: Exploring boundaries, expanding connections*. Dodrecht: Springer.

Gölzer, P., & Fritzsche, A. (2017). Data-driven operations management: Organizational implications of the digital transformation in industrial practice. *Production Planning and Control, 28*, 1332.

Kornwachs, K. (2017). Is technology a science? Recent developments in German philosophy of technology and engineering. In D. P. Michelfelder, B. Newberry, & Q. Zhu (Eds.), *Philosophy and engineering: Exploring boundaries, expanding connections* (pp. 61–72). Dordrecht: Springer.

Lightfoot, H., Baines, T., & Smart, P. (2013). The servitization of manufacturing – A systematic literature review of interdependent trends. *International Journal of Operations & Production Management, 33*(11/12), 1408–1434.

Oks, S. J., Fritzsche, A., & Möslein, K. M. (2017). An application map for industrial cyber-physical systems. In H. Song, S. Jeschke, C. Brecher, & D. B. Rawat (Eds.), *Industrial internet of things: Cybermanufacturing systems* (pp. 21–46). Cham: Springer.

Robison, W. L. (2016). *Ethics within engineering: An introduction*. London: Bloomsbury Publishing.

Weber, M. (1978). *Economy and society: An outline of interpretive sociology*. Berkeley: University of California Press.

Chapter 14
Fostering Subjectivity in Engineering Education: Philosophical Framework and Pedagogical Strategies

Cecilia Moloney, Cecile Badenhorst, and Janna Rosales

Abstract This paper argues that fostering subjectivity in engineering education will aid engineers in understanding the connections between their own life values and motivations and their career choice and development. By fostering subjectivity in engineering education, we mean linking the person who studies with what they are studying, a definition that can be situated within the philosophy and methodology of Bernard Lonergan. This paper also presents evidence for pedagogical strategies to foster subjectivity based on our definition of subjectivity in engineering education. We analyze data collected during a pilot offering of a co-curricular course for engineering graduate students (the *Lead by Design* Institute) to determine to what extent the *Lead by Design* pedagogy fostered subjectivity. The paper concludes with reflections on implications for future engineering education, from the philosophical framework of Lonergan's model of the human subject, and from our analysis of the *Lead by Design* pedagogy.

Keywords Engineering education · Integrative pedagogy · Subjectivity · Generalized empirical method · Bernard Lonergan's model of the human subject

14.1 Introduction

This paper develops an argument that fostering subjectivity in engineering education will benefit individual engineers in their self-understanding of themselves as engineers, notably in the connections between their life values and motivations and their career choice and development. We define the fostering of subjectivity in an educationally relevant manner, as the explicit linking of the person who studies or

C. Moloney (✉) · J. Rosales
Faculty of Engineering and Applied Science, Memorial University of Newfoundland, St. John's, NL, Canada
e-mail: cmoloney@mun.ca

C. Badenhorst
Faculty of Education, Memorial University of Newfoundland, St. John's, NL, Canada

A. Fritzsche, S. J. Oks (eds.), *The Future of Engineering*, Philosophy of Engineering and Technology 31, https://doi.org/10.1007/978-3-319-91029-1_14

works (i.e. their subjective "who" and "why") with what they are studying or working on (i.e. the objective "what" and "how" of technical content and design). We situate our definition within a philosophical framework grounded in the philosophy and methodology of Bernard Lonergan (1992).

In addition to a theoretical argument in favour of fostering subjectivity, this paper also suggests pedagogical strategies that may aid in fostering students' subjectivity. We analyze research findings from a pilot offering of a co-curricular course called the "*Lead by Design* Institute on Leadership, Diversity and Dialogue for Engineering Graduate Students" that was developed and offered at Memorial University of Newfoundland.

14.1.1 The Context of Twenty-First Century Engineering

Engineering now works within a world of rapid change, not only in technology but also in the social, economic, environmental, and political contexts, at scales from local to global; indeed, it is increasingly the case that engineering problems are situated within global contexts, and of concern to the whole human community (e.g. Smith 2010). For these reasons, engineers are working on problems that are becoming ever more challenging, both technically as well as in their contextual breadth that extends beyond engineering to other domains of human expertise. For example, the US National Academy of Engineering has identified 14 "grand challenges" for engineering in the twenty-first century (NAE 2008); these challenges include providing access to clean water, securing cyberspace, and managing the nitrogen cycle, amongst others.

In addition, there is an increased call for diversity within the engineering workforce, linked to the complex nature of many twenty-first century engineering problems and the consequent need for engineers to work on large multidisciplinary teams. Diversity can have many dimensions, including gender, ethnicity, expertise, personality, ethical stance, and motivation to choose to study engineering, amongst others. A wider diversity in engineering will tend to increase perspectives on problems and their solutions (Engineers Canada 2016a, b; McLean 2003; Michelfelder and Jones 2016). At the same time, the benefits to be derived from a diversity of perspectives and backgrounds must be balanced by the additional care and effort, as well as time, needed to manage collaboration within a diverse group. The care and time may be even greater when collaboration involves engineers and other professions or communities, as the motivations and approaches to problems may vary. Nonetheless, since the benefits of diversity can be significant, we can focus on the expected outcomes, and then seek the methods needed to enable the finding of common ground for successful collaboration. For example, such methods derive from dialogue (Bohm 1996) and dialogue education (Gunnlaugson and Moore 2009), as well as from functional collaboration (Lonergan 1973). Thus, while increased diversity in engineering will tend in the first instance to increase the complexity of working together on problems, ultimately the goal of engaging diverse perspectives is to enable engineers to enhance their abilities to forge important connections between the profession of engineering and other professions, and to dialogue between engineering and other human communities (Catalano 2011; Cech 2013; Goldberg et al. 2014).

14.1.2 The Challenges for Engineering Education

Engineering educators and key educational institutions are responding to both the increasing contextual complexity of engineering and the goal of attracting more diversity into engineering. The new engineers need to understand engineering principles at fundamental levels and to develop creative design and practical skills; they also need process skills to enable them to integrate a wide range of relevant factors into innovative engineering solutions, to work on interdisciplinary teams, and to provide leadership to themselves as well as others. In addition, the new pedagogies to achieve the educational goals of twenty-first century engineering should also address the criteria to attract and retain diversity within engineering.

Much work has already been done towards understanding, proposing and implementing new approaches to educating the twenty-first century engineer (e.g. Goldberg et al. 2014; Grasso and Burkins 2010; Montano 2008; NAE 2005; Sheppard et al. 2008). However, questions remain about the most effective pedagogies to achieve these educational goals.

Our work in this paper is motivated broadly by these questions. Within this broader context, this paper focuses on two specific questions: first, why is subjectivity important in engineering?; and second, does the *Lead by Design* pedagogy developed and implemented by the authors tend towards the goal of fostering subjectivity within engineering education?[1]

14.2 Subjectivity and Engineering

Subjectivity, as it is commonly understood in everyday life, pertains to an individual's perspective, thought, opinion, or perception. This common view of subjectivity stems from its binary placement opposite to objectivity that is commonly viewed as impartial and thereby truthful (Hall 2004). These understandings of subjectivity and objectivity derive from René Descartes who noted the limited and often error-prone perception of human beings (Hall 2004). By extension, subjectivity is sometimes understood to connote that which is based on feeling or desire rather than grounded in evidence (Morelli 2016).

[1] The *Lead by Design* Institute is part of a larger project, the MetaKettle Project, that aims to contribute to the development of integrative pedagogies for engineering (Moloney and Rosales 2011), situated within the ongoing transformation of engineering education (e.g. Goldberg et al. 2014).

Thus, subjectivity is often viewed negatively against the desired objectivity of the Western scientific worldview that we have inherited as a cultural legacy from Descartes. Moreover, from Francis Bacon we have inherited a privileging of an objective and empirical approach to knowledge; as such, logical positivism continues to persist as a perspective held by some in engineering and the natural and other quantitative sciences (Keller 1985; Dusek 2006).

When subjectivity is explicitly mentioned in engineering or technical papers, it is usually as a problem or to be avoided (e.g. Deo and Strong 2003). In general, indications of subjectivity are minimized, in favour of a deliberate objectivity of thought and expression, manifest as the impersonal discourse of science and engineering (Montgomery 1996). One notable exception is that subjectivity is acknowledged in professional development contexts, when it is generally considered to be useful to know one's self in order to further one's life and career development (e.g. Cohen and Cohen 2012).

On the other hand, some scholars and scientists consider subjectivity as positive, even necessary, and to be included explicitly in the work of engineering and science. In her classic work *Reflections on Gender and Science*, Evelyn Fox Keller develops the notion of dynamic objectivity, from which the knower views the external world as an independent reality but also acknowledges and relies upon her connectivity with that world (Keller 1985). Keller includes a quote from Piaget that is pertinent to her model of dynamic objectivity: "So long as thought has not become conscious of self, it is a prey to perpetual confusions between objective and subjective, between the real and the ostensible" (Keller 1985, 117).

Another positive perspective on subjectivity is that of quantum physicist and educator Arthur Zajonc who argues for the inclusion of a contemplative component in educational programs to nurture the whole person. The whole person clearly includes both their subjectivity and objectivity. Zajonc notes the importance of repeated attentive engagement as essential for an integrated and embodied education (Zajonc 2016). In a similar vein, the MetaKettle Project was started at Memorial University of Newfoundland from the realization that a key to attracting a wider diversity into engineering is an explicit focus on why engineering matters in people's lives. Within the educational context of a university, this translates into pedagogies that can acknowledge and work with the broad set of motivations and talents that students bring to engineering, thus enabling the whole person to be engaged in engineering (Moloney and Rosales 2011).

More precisely in philosophical terms, the subjective, while still often viewed as opposed to the objective, refers to an existence in the mind of the subject about a possible object that may be apart from the subject.[2] That is, the subjective is connected with an object as known in the mind, as distinct from the thing itself. While

[2] Space precludes a full review of the various philosophical perspectives on subjectivity and objectivity, and their evolution over time. Here we start with a generic philosophical definition of subjectivity (Hall 2004; Dusek 2006), then refine the definition from within the philosophical framework of Bernard Lonergan's cognitional theory and model of the human subject (Lonergan 1992).

this gives a more precise meaning to subjectivity than the everyday understanding noted above, the well-known fallibility of human perception and of human knowing have tended to reinforce the everyday suspicions of subjectivity.

The negative perspective on subjectivity is amplified by modern science that has sought to overcome human subjectivity by calling for reliable external measuring devices by which we may tend towards knowing nature objectively (Hall 2004; Dusek 2006). That is, the modern sciences seek knowledge about nature that is true (even if that goal can be reached only at the limit of an indeterminate number of refining approximations). By contrast, engineering differs from science in that its ultimate goal is to produce actions in the world. These intended actions mean that the engineer must ultimately be concerned for human beings, and hence for their combined subjectivity and objectivity. Thus, engineering needs to overtly acknowledge and work with human subjectivity.

As a response, within the philosophy of engineering there are calls to understand the role of subjectivity in engineering, and to consider how best to work with it. Korte notes in a paper examining a philosophical grounding of engineering work: "Engineering work is necessarily entangled with non-engineering human systems in all their complexity, subjectivity and unpredictability" (2015, 6).[3] In examining the need for a philosophy of engineering, Goldman (2004) contrasts two clusters of cognate concepts, the principles of sufficient reason (seen to be associated with scientific reasoning) and the principles of insufficient reason (seen to be part of engineering reasoning); in the latter cluster he lists subjectivity, and as such, associates subjectivity with the contingent reasoning of engineering. In their examination of the philosophy of Bernard Lonergan applied to engineering, Frezza and Nordquest ask about the postmodern relativistic challenges to truth and value; they find that these challenges are addressed, not by epistemological theories, but "by exploring the data, the experience of human knowing; knowing in the context of subjectivity…" (Frezza and Nordquest 2015, 1).

The importance of subjectivity in engineering is heightened by the profession's increasing recognition of the complexity of the problems of twenty-first century engineering; for example, the grand challenges of the NAE (2008) and many other problems may qualify for the label of "wicked" (Rittel and Webber 1973; Conklin 2005; Lönngren and Svanström 2016). These problems may not admit a precise definition, or their definition may become focused only when potential, and often many-dimensional, solutions are envisaged. Therefore, scientific rationality cannot necessarily lead to their optimal solution; rather, solutions for complex problems in engineering and in other fields are found as "good enough" rather than as globally optimal (Simon 1981, Lönngren and Svanström 2016). In cases such as these, as Lonergan notes (1973), we need to employ methods based on heuristics, rather than

[3]With personal digital technologies such as smartphones, tablets and wearables, the entangled relationship of the engineer's work with the complexity and subjectivity of human beings (the engineer included) is paralleled in a new way by the relationship of the user's subjectivity and personal experience with her smart communicating device (Fritzsche 2018).

strict logic, and the methodical heuristics must explicitly and intentionally include human subjectivity.

In this paper, we are interested in the question of subjectivity in connection with engineering. Since engineering intends actions and products for human living, our focus on subjectivity also points to a wider set of questions about how to include human need, and human intentionality and creativity more fully in our collective seeking of the basic human goal of successful living. While, as noted above, subjectivity is frequently considered in our culture as a negative quality, and is often posed as over against the objectivity of science and our cultural scientific worldview, we argue that subjectivity is not only an unavoidable human characteristic, but one that we ignore to our detriment. Moreover, subjectivity has a deeper existential aspect, and a more important imperative, one written about extensively by the Canadian philosopher Bernard Lonergan who argued that authentic subjectivity is necessary for genuine objectivity (Lonergan 1973, 292).

Thus, we claim that a foundational philosophical framework for subjectivity in engineering (as in other fields of human life and endeavour) can be found in Bernard Lonergan's cognitional theory (Lonergan 1992). We also claim that the "authentic subjectivity" that Lonergan finds at the core of his cognitional theory brings the subject and object into one viewpoint, thus balancing and refining previous understandings of subjectivity and objectivity.

14.3 Lonergan's Model of the Human Subject

In his key work, *Insight: A Study of Human Understanding*, Lonergan (1992) built from the Archimedean "eureka" experience of insight to a model of human cognition. Lonergan's model of the human subject as a knower and actor has four levels: experience, intelligence, reasonableness, responsibility. Each level has its key operations and their corresponding inputs and outputs; movement from level to level occurs in response to spontaneous questions that arise. For example, from the level of experience with its operation of attention to the data of experience, the question "What is it?" sparks the operations of intelligence that may lead to an insight into the data, or to an understanding, concept, formulation, etc. that explains the data (Benton et al. 2005). Since an insight can be incorrect or insufficient, the further question "Is it so?" moves the subject to a higher level where the operation of judging or verifying leads to a reflective insight, or affirmation. This is Lonergan's cognitional model, leading from experience to knowledge.

Further spontaneous questions lead the subject to the next levels of deliberation and action, based on the questions in response to values, "What to do?" and "Is it worthwhile to do?" (Benton et al. 2005). Portions of the cycle of four levels, or the entire cycle (from attention to data, through knowledge, to action) recur until the subject has grounds for satisfaction in the knowledge as true and their actions as worthwhile in the light of their values. Several (or an indeterminate number) of repetitions may be needed, as knowledge always remains to some extent uncertain or contingent, and values-based decisions seldom clear-cut. Nonetheless, since the process is methodical as opposed to purely logical, its recurrence can lead to "cumulative and progressive results" (Lonergan 1973, 4). Moreover, this "method of the human mind" (Lonergan 1973, 4) applies to the data and contents of all human experiences.

Central to Lonergan's model of the human subject is the notion of self-appropriation, the coming to an understanding of what one is doing when one is knowing (Lonergan 1992, 343). Elsewhere, Lonergan expressed the same notion in what he termed the transcendental precepts: Be attentive; Be intelligent; Be reasonable; Be responsible (Lonergan 1973, 20). These four precepts match the four levels of Lonergan's model of the human subject.

In self-appropriation, the subject attends to the data of the object (i.e. as in the above description of the cognitional model); these data are paralleled by the data that is the subject. By self-presence while operating, the subject comes to affirm herself as a knower and actor. While the process of knowing and doing are always operative, in response to the internal norms that arise from the spontaneous questions, the self-attention needed for self-appropriation can be difficult (Morelli 2016). Self-attention is most easily reached via puzzles or problems in mathematics, where the operations are clearer and the content more stripped down; Lonergan starts *Insight* with the simple question "what is a circle" and invites the reader to come to knowledge of what a circle is while attending to the levels involved in the process of knowing (Lonergan 1992, 31).

Thus, self-appropriation is an advertence to one's own subjectivity. Lonergan famously wrote that genuine objectivity is the fruit of authentic subjectivity (Lonergan 1973, 292), calling this nexus of subjectivity and objectivity generalized empirical method— the paying of attention to the data that is the subject as well as the data of the object (Lonergan 1985, 141).

Empirical or scientific method is one of the great achievements of human thought from the previous millennium; scientific method entails systematic attention to an object (even when that object is the human person herself). In order to reach objective knowledge about the world, scientists strove to exclude the subject and her subjectivity. In doing so, past scientists have advanced and refined our understanding of method; however, they did not include all the relevant data. To extend empirical method to allow for a methodical study of the "problems of everyday life in this messy world" (Korte 2015, 1) —which, per Korte, includes engineering— we need to add subjectivity back in, but do so methodically. Lonergan's notion of generalized empirical method involves the paying of attention to the data that is the subject (i.e. consciousness, self-presence) as well as the data of the object (1985, 141). Hence, in educational contexts, the key point in fostering generalized empirical method is to encourage students [or myself] to puzzle over their [or my] own experience of insights in engineering, in conversations, or in puzzling or other activities, so as to understand what they [or I] are [am] doing.

Thus, Lonergan's subjectivity applies both to the "who" and "what" of the person, as well as the "how" and "why" of the content of the objects of engineering. The authentic subjectivity of the self-presence of oneself with one's operations of knowing and doing leads to genuine objectivity, whether that objectivity is about a mathematical derivation, a scientific observation, an engineering design, or about what one should choose to do in one's own life based on the data of one's experience (including feelings), and with the operations explicitly informed by personally affirmed values. The overall result is self-knowledge: highly sought, although not easy to achieve, as it is the result only of a long and deliberate engagement with self and the world (Lonergan 1992; Morelli 2016).

Influenced by the ideas of Bernard Lonergan and other philosophers of consciousness (Webb 1988), as well as by the theory of reflective practice (Bolton 2010) and by feminist and other theories of liberation (Freire 1970), each of the co-authors of this paper is oriented to fostering subjectivity, in various ways, in our professional practice and research. In the *Lead by Design* study of the MetaKettle Project (discussed further below in Section 14.4), we are seeking to introduce subjectivity into engineering. While we are not yet explicitly promoting Lonergan's authentic subjectivity (since that is extremely challenging), we have started with the looser and more easily understood and explained definition of subjectivity of the MetaKettle Project. That is, by fostering subjectivity in engineering education, we mean introducing students to the link between the subject they are who studies or works (i.e. their "who" and "why") and the objects they are studying or working on (i.e. the objective "what" and "how").

Our working hypothesis is that doing so will further the education and professional sustainability of engineers themselves and will tend to attract and also retain an increased diversity of persons and expertise within the engineering profession.

14.4 Pedagogical Strategies: The *Lead by Design* Institute

14.4.1 Description and Methodology

We developed a 5-day intense workshop-style co-curricular course, and facilitated a pilot offering in April 2014 of the "*Lead by Design* Institute on Leadership, Diversity and Dialogue for Engineering Graduate Students". The pedagogy of the *Lead by Design* Institute was developed around three specific threads of narrative reflection, dialogue, and self-leadership.

Over a one-week period, 14 participants engaged daily in a variety of activities that included dialogues and reflections, facilitated discussions, skills building exercises, case studies on leadership and diversity, reflections on objects that inspired them to study engineering, etc. Through discussion and free-writing, they explored questions such as "What is engineering?", "What attracts you to engineering?", "What does empathy have to do with engineering?", etc. The curriculum plan wove the three threads of reflective practice, dialogue, and leadership over the 5 days around a thematic arc of "Understanding how we got where we are, designing (engineering) a new future," with a focal point on Day 4 of "Re-engineering the foundations of my career." While the pedagogy emphasized the motivations and identity of engineering students, it also included elements of engineering content and problem solving. In particular, participants engaged in a team-based engineering challenge for which they produced a rapid technical solution that they presented at a public event at a local art gallery.

In addition to facilitating the Institute, we also studied the *Lead by Design* Institute via a basic qualitative study (Merriam 2009), with the purpose of collecting data to show to what extent the pedagogy fostered participants' deeper understanding of their own motivations and experiences as students and engineers.[4]

14.4.2 *Lead by Design*: Results and Observations

To help participants to spontaneously access their subjectivity, we devised a series of free-writing exercises within the reflective writing thread of the Institute.[5] The free-writes in the *Lead by Design* Institute employed a technique based on Badenhorst (2007) and grounded in reflective writing theory (Bolton 2010): participants were invited to write in landscape mode, using coloured paper and fine-tipped coloured markers, as these aesthetic differences from academic writing tend to separate free-writing from the often anxiety-laden task of academic writing (Badenhorst et al. 2016; Moloney et al. 2016a). It is worth noting that free-writing is meant to be a spontaneous recording of immediate thoughts, that, when expressed in written form, may become a source for critical reflection and insight. As such, free-writing tends to consist of existentially connected sentences or phrases, in contrast to the more structured argument of academic writing. Despite participants' initial reluctance to free-write on the spot, we observed an increasing ease with free-writing over the 5 days of the Institute. By the end, participants wrote effortlessly, even eagerly, on the suggested free-write topics.

To examine the fostering of subjectivity in the *Lead by Design* Institute, we use as research data the free-writings by participants in response to specific questions, and our written observations as participant-observers (per Merriam 2009). Below are several groupings of data from the Institute; each grouping consists of quotes from participant free-writes followed by our analysis comments that reference Lonergan's framework. Although we did not introduce Lonergan's notion of self-appropriation to the participants, nor explain the structure of his model of the human subject, we found it more effective and foundational to analyze the data through the

[4] More information on the *Lead by Design* Institute and research study can be found in Moloney et al. (2016a, b).

[5] By the theory of reflective practice (Bolton 2010), spontaneous writing over a short period of time is a process that can elicit genuine insights. Thus, subjectivity is an element of reflective practice, one of the three threads of the Institute. The second thread was dialogue; subjectivity can be a key element of dialogue, when it is properly conducted (Bohm 1996). The third thread was leadership, defined to include self-awareness and self-leadership (Cohen and Cohen 2012). Thus, subjectivity was a potential element in all three threads of the Institute.

lens of Lonergan's model of cognition; that is, it is easier to pick up traces of the "method of the human mind" (Lonergan 1973, 4) than to detect evidence for the more loosely phrased (albeit more easily understood) definition of the "who" and "why" connected to the "what" and "how". Also, since we didn't formally introduce Lonergan's model to the participants, we didn't seed their responses with phrases that they could mirror back to us in their responses, as could be the case with other material we presented to them on reflective practice, personality types, self-leadership, etc.

Day 1: In response to a free-write question "Are there boundaries in engineering? If so, how are they manifested?"

- "Boundaries: I think there [are] no boundaries… everyone can be an engineer of himself. Planning our own life, is also kind of engineering." [Charlotte][6]

Charlotte's phrase "planning our own life" aligns with Lonergan's cognitional level of "responsibility". Moreover, it is interesting that she saw "planning our own life" as "also a kind of engineering", as this shows that she understands that the method of her own mind can apply equally to the existential dimension of her life as well as the more overtly "objective" technical dimensions of engineering.

Day 2: In response to a free-write on "What attracts you to engineering?"

- "It was something developed throughout my life experiences." [Lilly]
- "When I was a child electricity was the most curious topic for me. I wonder[ed] how a simple press in a switch can rotate a fan or turn on a light. It was beyond my mind. Then I set up my mind that I need to know this mystery. Which stimulates me to become an engineer… Because my parents told me if you become an engineer you will understand everything what is happening." [Colt Tropper]
- "When I was a kid I was always curious about how things work. How clock works? After I know how it works I wanted to make some products in my mind." [Stephen]

Many of the responses to this question referenced pivotal experiences in their childhoods. The selected responses echo words or phrases from Lonergan's cognitional theory, including: "experiences", "set up my mind", "I need to know", "understand", "products in my mind". As well, note the sequencing in Colt Tropper's response in which "beyond my mind" yields to "set up my mind" which yields to "you will understand everything".

Day 3: In response to a free-write on "What is something significant that you've learned this week? [i.e. Days 1-3]"

- "What impressed me most is the way of finding what is engineering and the ways to know more about myself." [Charlotte]

[6]All participant names are anonymized.

- "... the most important thing was self-awareness how I can look inside myself and discover my interest, my talents, my abilities to communicate with other members of group and how we can respect others and at the same time how we can react when we are facing diversity in our team work, how these aspect[s] can affect our behaviours and to be successful what are the main points of becoming a good engineer. It was really good experience for me." [Melisa]

Charlotte is perceptive and tuned into her subjectivity, as she was impressed by "the ways to know more about myself." As on Day 1, it is telling that Charlotte puts the two aspects "finding what is engineering" and "ways to know more about myself" in the same sentence —indicating, perhaps, that she was aware of or coming to understand that the two are outcomes of the same method of the human mind. The quote from Melisa demonstrates her enthusiasm for the Institute and the value she sees from attending it. She uses the words "experience" and "self-awareness," and alludes to the level of responsibility in her phrase "affect our behaviour". However, as researchers we need to be wary of reading too much into the free-writes. Students can feed back a facilitator's points and suggestions; such could be the case even with mature graduate students, a tendency that can be attributed to traditional engineering education.

Day 4: In response to a free-write "Reflect on the narratives – what do you want to know?"

- "I found the more I write (free writing) the more comfortable I am ... describ[ing] my situation." [Sheri]
- "The answer sometime is very simple, right there in your hand [writing]." [Kelly]
- "I think the free writes help to get your mind in the zone and generate ideas for the narrative. ... I'm actually planning to use these techniques to organize my thoughts for my thesis writing. ...Can narratives be more on other things than self reflection?" [Amy]
- "I don't know whether I'm fooling myself or making up the narratives sometimes." [Charlotte]

The first two quotes from Sheri and Kelly point to increasing ease with the process of spontaneous free-writing, as well as the simplicity of this method of finding out about themselves. Kelly's comment that "the answer ... is right there in your hand [writing]" echoes Lonergan's idea of self-presence leading to self-knowledge. These two comments point to the power of free-writing, as well as the welcome that many people will give to themselves spontaneously as they start to use methods that allow them access to their own method of knowing and doing.

The quote by Amy reveals a broadening of her insights and perspectives on the process of free-writing. She spontaneously starts to make a connection between the method applied to knowing about herself and the method for finding out what she thinks in other contexts, including engineering and writing her thesis.

The quote from Charlotte reveals honest doubt, and as such, is evidence of the Lonergan level of judgment and reasonableness, in response to the question "Is it so?". By expressing her doubt, she reveals possibly the best evidence of all the quotes for movement towards self-appropriation and generalized empirical method.

Day 4: In response to a free-write on "Re-engineering the foundation of your career."

- "… I can do something to make people's life much easier. Meanwhile I am also a person full of curiosity. … I really enjoyed the exploring process, it includes both work and myself. … I can't wait to see how my academic skills fit into the real world." [Kelly]

This quote demonstrates enthusiasm and hope. Moreover, by noting that she is "full of curiosity" she is referencing the spontaneous movement of human beings towards knowing and subsequent doing.

Day 4: In response to a free-write on "How will you contribute to re-designing engineering?"

- "Redesigning engineering, maybe in the way [we] work together in a project. We are [from] different backgrounds, cultures, languages, levels. If we work effective in a group, the quality of results will be better." [T. Smith]

This quote about diversity and working together indicates insights into how to improve the outputs of engineering based on understanding and working with the subjectivities of the group. As such, this quote indicates inter-subjectivity, and the insight that, to be effective, we must pay attention, in order to understand group dynamics, then feedback and refine initial understandings into reasonable judgments, before deciding and acting as a group in a responsible way.

14.5 Discussion

In Sects. 14.2 and 14.3 we have made an argument towards the first goal of this paper which was to answer the question, why is subjectivity important in engineering? The argument is developed in Sect. 14.2 from the literature in the philosophies of science and technology and the emerging philosophy of engineering, and then is made more precise in Sect. 14.3 via the philosophy and methodology of Bernard Lonergan.

The second goal of this paper was to examine the pedagogy that was developed for the *Lead by Design* Institute, to determine if and to what extent the pedagogy tended to foster subjectivity for engineering students. Section14.4 presents evidence, albeit preliminary and possibly fragile, that bolster the claim that the *Lead by Design* pedagogy can foster subjectivity, or at least can start to help participants to become more present to themselves. From our observations in the *Lead by Design* Institute (and elsewhere), we find that students are eager to undertake exercises that reveal themselves to themselves. However, we note several caveats. One week is very short to change habits and orientations or to show sustained development of self-attention. Also our research findings are from a small group of self-selected graduate students. As such, the best we can say is that the *Lead by Design* Institute provides a "proof of concept" and preliminary evidence that the pedagogical approach and its strategies may foster subjectivity.

As a more significant caveat, we must also acknowledge the challenges of designing and realizing pedagogies to foster subjectivity in engineering education. First, it is challenging to design pedagogies to work effectively with a broad range of student interests, motivations and experiences. Second, there is also a prior, and potentially greater challenge, in that the self-presence of Lonerganian subjectivity is difficult to sustain, in large part due to our tendency to focus our attention preferentially outwardly and on the experience of our senses rather than on the non-sensible experience of self-presence (Morelli 2016).

The *Lead by Design* research project was designed to study students' self-understanding of the connections between their life values and motivations and their career choice and development; the evidence for fostering subjectivity presented in Sect. 14.4 only addresses these connections. However, the argument for subjectivity made in Sects. 14.2 and 14.3 points to a broader potential of Lonerganian subjectivity and self-presence to aid engineers in their self-understanding of themselves as engineers, e.g. when working on technical problems, or in collaboration with others. With respect to collaboration, Lonergan scholar Morelli notes that in as much as objectivity is achieved for one person by their authentic subjectivity, so objectivity is also what is achieved by authentic subjectivity in authentic community (2016). We leave this broader view of subjectivity in engineering for future theoretical and experimental study.

At the present time, there are many calls for change in engineering coming from the profession itself as well as from wider stakeholders and communities (e.g. Catalano 2011; Cech 2013; Engineers Canada 2016a, b; Goldberg et al. 2014; NAE 2005). Canadian professional engineers wear an "iron" ring on their writing hand to remind them of their ethical obligations to work for the greater good of society. But when the contexts are rapidly changing, it can be difficult to determine what exactly constitutes the greater good of society. Engineers may need more refined tools to help them determine what to do. In light of the Deepwater Horizon tragedy in the Gulf of Mexico, Catalano suggests that we need to develop a new engineering ethic, one which "focuses more on the results as evidence by its adoption across a myriad of disciplines rather than the purity or elegance of the argument" (Catalano 2011, 3). Catalano thus seeks the dynamism of method that can change in response to the evidence of the data of experience. But change in method occurs because of the prior "method of the human mind" (Lonergan 1973, 4) operative in the subjects who are working to change the method; hence, in order for change to yield progress, all of the data need to be considered, including the data of subjectivity. We posit this as the foundation of future engineering.

While this is a distant vision, the key point is to orient efforts towards it. Inviting students to advert to their subjectivity in order to benefit their own lives and careers is good in its own right; but doing so will tend to open their minds to other data about themselves and their work. This will also tend to reveal some deeply embedded assumptions, e.g. about the objectivity of engineering and the permanence of the ways things are (Cech 2013), thus adding to individual growth and, through collaboration, to the whole enterprise of engineering and the quality of its solutions. Engineers would more truly become, as they wish to be, "enablers of dreams" (Canadian Engineering Leadership Forum 2009).

14.6 Conclusions

This paper developed an argument for the potential positive role of subjectivity in engineering, framed within Bernard Lonergan's model of the human subject; then the paper turned to the specific question: how well did the *Lead by Design* pedagogy foster subjectivity? Analysis of the research data allowed us to extract some positive findings about the pedagogy. We acknowledge, however, that this is one step in what will be a long and collective journey towards a possible future for engineering in which generalized empirical method is operative. We hypothesize that, in the long run, the benefits of fostering subjectivity in the individual will tend to benefit the quality of engineering contributions to the solutions of problems facing the global community. This is clearly a longer-term project that will require further scholarly examination as well as evidence gained in a methodical and cumulatively progressive way.

Acknowledgments This research was supported by a grant from the Hebron Diversity Fund 2013-2014 and by an award in 2014 from the Quick Start Fund for Public Engagement (Memorial University). We acknowledge with thanks the participation of the Engineering graduate students in the *Lead by Design* Institute in April 2014, as well as the contributions of guest speakers and other supporters. Ethical approval of this research was granted by the Interdisciplinary Committee on Ethics in Human Research at Memorial University.

References

Badenhorst, C. M. (2007). *Research writing: Breaking the barriers*. Pretoria: Van Schaik.

Badenhorst, C. M., Moloney, C., & Rosales, J. (2016). *New literacies for engineering students: critical reflective writing practice*. Poster presented at EARLI SIG writing conference, Liverpool Hope University, Liverpool, UK, 4–5 July 2016.

Benton, J., Drage, A., & McShane, P. (2005). *Introducing critical thinking*. Vancouver: Axial Press.

Bohm, D. (1996). *On dialogue*. New York/London: Routledge.

Bolton, G. (2010). *Reflective practice: Writing and professional development* (3rd ed.). London: Sage Publications.

Canadian Engineering Leadership Forum. (2009). *Leading a canadian future: The new engineer in society*. Montreal Declaration, National Engineering Summit, Montreal, QC 21 May 2009. http://www.engineerscanada.ca/. Accessed 15 Jan 2017.

Catalano, G. (2011). *Tragedy in the Gulf: A call for a new engineering ethic* (Synthesis Lectures on Engineers, Technology and Society). San Rafael: Morgan and Claypool Publishers.
Cech, E. A. (2013). The (mis)framing of social justice: Why ideologies of depoliticization and meritocracy hinder engineers' ability to think about social justices. In J. Lucena (Ed.), *Engineering education for social justice: Critical explorations and opportunities* (Philosophy of Engineering and Technology series, pp. 67–84). New York: Springer.
Cohen, C. M., & Cohen, S. L. (2012). *Lab dynamics: Management and leadership skills for scientists* (2nd ed.). Cold Spring Harbor: Cold Spring Harbor Press.
Conklin, J. (2005). *Wicked problems and social complexity*. CogNexus. http://www.cognexus.org. Accessed 15 Jan 2017. Also Chapter 1 in Conklin (2005). *Dialogue mapping: Building shared understanding of wicked problems*. Hoboken: Wiley.
Deo, B., & Strong, D. (2003). Fixing the problem of subjectivity in the concept of 'activity' in activity based costing (ABC) – An engineering perspective. In *Proceedings of the IIE Annual Conference*.
Dusek, V. (2006). *Philosophy of technology: An introduction*. Oxford: Blackwell Publishing.
Engineers Canada. (2016a). *Reaching 30 by 30: Promising practices for increasing diversity and inclusion in engineering*. http://www.engineerscanada.ca/. Accessed 10 Oct 2016.
Engineers Canada. (2016b). *Welcoming workplaces: Diversity in the engineering profession*. http://www.engineerscanada.ca/. Accessed 10 Oct 2016.
Freire, P. (1970). *Pedagogy of the oppressed*. New York: Herder and Herder.
Frezza, S. T., & Nordquest, D. A. (2015). Engineering insight: The philosophy of Bernard Lonergan applied to engineering. In R. Korte, M. Mina, I. Omidvar, S. T. Frezza, D. A. Nordquest, & A. Cheville (Eds.), *Philosophical and educational perspectives on engineering and technological literacy, II*. Wicklow: American Society of Engineering Education. http://lib.dr.iastate.edu/ece_books/2. Accessed 10 Oct 2016.
Fritzsche, A. (2018). Dancing the device: A translational approach to technology. In J.C. Pitt & A. Shew (Eds.), *Spaces for the future: A companion to philosophy of technology*, pp. 216-223. New York: Routledge.
Goldberg, D. E., Somerville, M., & Whitney, C. (2014). *A whole new engineer: The coming revolution in engineering education*. Douglas: Three Joy Associates.
Goldman, S. L. (2004). Why we need a philosophy of engineering: A work in progress. *Interdisciplinary Science Reviews, 29*(2), 163–176. https://doi.org/10.1179/030801804225012572.
Grasso, D., & Burkins, M. B. (Eds.). (2010). *Holistic engineering education: Beyond technology*. New York: Springer.
Gunnlaugson, O., & Moore, J. (2009). Dialogue education in the post-secondary classroom: Reflecting on dialogue processes from two higher education settings in North America. *Journal of Further and Higher Education, 33*(2), 171–181.
Hall, D. E. (2004). *Subjectivity*. New York: Routledge.
Keller, E. F. (1985). *Reflections on gender and science*. New Haven: Yale University Press.
Korte, R. (2015). Pragmatism, practice and engineering. In R. Korte, M. Mina, I. Omidvar, S. T. Frezza, D. A. Nordquest, & A. Cheville (Eds.), *Philosophical and educational perspectives on engineering and technological literacy, II*. Wicklow: American Society of Engineering Education. http://lib.dr.iastate.edu/ece_books/2. Accessed 10 Oct 2016.
Lonergan, B. J. F. (1973). *Method in theology* (2nd ed.). London: Dartman Longman and Todd.
Lonergan, B. J. F. (1985). *A third collection*. New York: Paulist Press.
Lonergan, B. J. F. (1992). In Collected Works of Bernard Lonergan, v.3, R. M. Doran & F. E. Crowe (Eds.), *Insight: A study of human understanding*. Toronto: University of Toronto Press.
Lönngren, J., & Svanström, M. (2016). Systems thinking for dealing with wicked sustainability problems: Beyond functionalist approaches. In W. Leal Filho & E. Nesbit (Eds.), *New developments in engineering education for sustainable development* (World Sustainability Series, pp. 199–212). Berlin: Springer. http://www.springer.com/us/book/9783319329321

McLean, D. (2003). Workplaces that work: Creating a workplace culture that attracts, retains and promotes women, Canadian federal/provincial/territorial ministers responsible for the status of women.

Merriam, S. B. (2009). *Qualitative research: A guide to design and implementation*. San Francisco: Jossey-Bass.

Michelfelder, D. P., & Jones, S. A. (2016). From caring about sustainability to developing care-ful engineers. In W. Leal Filho & E. Nesbit (Eds.), *New developments in engineering education for sustainable development* (World Sustainability Series, pp. 173–184). Berlin: Springer. http://www.springer.com/us/book/9783319329321

Moloney, C., & Rosales, J. (2011). The MetaKettle project: A journey to the heart of higher education. In *Proceedings from 15th international conference of women in engineering and science*, Adelaide, Australia, July 19–22. www.mun.ca/springinstitute/publications/MoloneyRosalesICWES15FinalMay27.pdf. Accessed 10 Oct 2016.

Moloney, C., Rosales, J., Badenhorst, C. & Roberts, J. (2016a). Fostering reflective practice for sustainable professional development: *Lead by Design*, a pedagogical initiative. In W. Leal Filho & E. Nesbit (Eds.), *New developments in engineering education for sustainable development* (World Sustainability Series, pp. 199–212). Berlin: Springer. http://www.springer.com/us/book/9783319329321.

Moloney, C., Rosales, J., & Badenhorst, C. (2016b). A methodological evaluation of an integrative pedagogy for engineering education. In *Proceedings of IEEE Frontiers in Education (FIE) 2016*, Erie, PA, October 12–15, 2016.

Montano, G. (2008). *A quantitative analysis of first-year engineering student persistence and interest in civic engagement at a Canadian university*. M.Ed. thesis, Memorial University, 2008.

Montgomery, S. L. (1996). *The scientific voice*. New York: The Guildford Press.

Morelli, M. (2016). *Self-possession: Being at home in conscious performance*. Los Angeles: Encanto Editions.

National Academy of Engineering (NAE). (2005). *Educating the engineer of 2020: Adapting engineering education to the new century*. Washington, DC: National Academies Press.

National Academy of Engineering (NAE). (2008). *NAE grand challenges of engineering*. http://www.engineeringchallenges.org/cms/challenges.aspx. Accessed 10 Oct 2016.

Rittel, H. W. J., & Webber, M. M. (1973). Dilemmas in a general theory of planning. *Policy Sciences, 4*(2), 155–169.

Sheppard, S. D., Pellegrino, J. W., & Olds, B. M. (2008). On becoming a 21st century engineer (Guest editor's forward). *Journal of Engineering Education*, July, Special Issue on Educating Future Engineers: Who, What, and How, *97*(3), 231–234.

Simon, H. A. (1981). *The sciences of the artificial* (2nd ed.). Cambridge, MA: MIT Press.

Smith, L. C. (2010). *The world in 2050: Four forces shaping civilization's northern future*. New York: Dutton.

Webb, E. (1988). *Philosophers of consciousness*. Seattle: University of Washington Press.

Zajonc, A. (2016). Contemplation in education. In K. A. Schonert-Reichl & R. W. Roeser (Eds.), *Handbook of mindfulness in education* (pp. 17–28). New York: Springer.

Chapter 15
Managing the State of the Art of Engineering: Learning from Medicine

Édison Renato Silva, Roberto Bartholo, and Domício Proença Jr

Abstract This chapter briefly presents the management of the state of the art (sota) in Medicine as a possible learning opportunity for the future of Engineering. Engineering and Medicine are sibling disciplines for intervention in reality, "sciences of the artificial" according to Herbert A. Simon. They seek to enlarge and disseminate their state of the art (sota) for greater scope and effectiveness. Both seek to convert scientific knowledge or technological possibilities into answers and procedures in tune with practical needs. In different ways, each seeks to improve the quality of the data it considers and the rigor of the methods it employs. Medicine has arrived at one striking, unique arrangement to support individual practitioners: a system that collects, classifies and qualifies medical knowledge comprehensively, and culminates with access through Patient-Intervention-Comparison-Outcome (PICO). PICO protocol allows a medical practitioner to access an up to date configuration of the whole of medical knowledge, being available as readily as in a smartphone. The chapter argues for the opportunity, propriety and desirability of translating the PICO experience to Engineering.

Keywords Engineering knowledge management · Engineering heuristics · Evidence-based medicine · Evidence-based engineering · Sciences of the artificial

15.1 Introduction

The aim of this chapter is to offer elements of Evidence-Based Medicine as a promising experience for the future of Engineering. Engineering and Medicine are what Herbert A. Simon called "sciences of the artificial", disciplines for intervention in reality. They operate change through diagnosis, prognosis and action on situated, individual cases of variable complexity. Neither can be reduced to being "applied

É. R. Silva (✉) · R. Bartholo · D. Proença Jr
Management & Innovation Area, Production Engineering Program, Universidade Federal do Rio de Janeiro, Rio de Janeiro, Brazil
e-mail: edison@ufrj.br

A. Fritzsche, S. J. Oks (eds.), *The Future of Engineering*, Philosophy of Engineering and Technology 31, https://doi.org/10.1007/978-3-319-91029-1_15

science" (Freidson 1988; Bunge 2013; Mitcham 1994; Koen 2003). Rather, like all disciplines for intervention, Engineering and Medicine rely on heuristics, a concept Polya (1957) popularized and which Billy V. Koen reconceptualized as "anything that promises to aid in the solution of a problem" (Koen 2003:28), the same posture and practice of Medicine when dealing with situations for which there is only "sub-optimal evidence" (Sur and Dahm 2011).

Each discipline for intervention possesses an ever-evolving state of the art (sota) that admits expression in the discipline-specific dynamics among the sota of individual practitioners, the sum of the sota of all practitioners, and the shared sota of a given discipline. Each sota admits enlargement by new heuristics. Heuristics, by definition, may come from anywhere, including Science, so long as they promise to be useful (Silva and Proença 2015). Hence, Engineering and Medicine are not insulated, self-contained. Rather, they are open to any source that might provide heuristics – including each other's states of the art (sotas).

For H.A. Simon, the key to understanding choice of courses of action and heuristics for a given application is bounded rationality (Simon 1956). Such concept that has been revisited and further detailed using the notion of "adaptive toolboxes" (Gigerenzer and Selten 2002), which emphasizes that in those disciplines which Simon conceptualized as "Sciences of the Artificial" (Simon 1969), attempts towards an optimal solution take place within constraints. In such cases, however, the strategies of the "adaptive toolbox" of such disciplines rarely seek to optimize, to compute the maximum of some function. Gigerenzer and Selten (2002) explains that "optimization is feasible in only a restricted set of problems, typically on the basis of simplifying assumptions" (Gigerenzer and Selten 2002:40). He concludes, then, that the aim of an adaptive toolbox [of heuristics in a state of the art, using Koen (2003)], is to "achieve proximal goals and include learning mechanisms that allow an adjustment of the tools when environments change" (Gigerenzer and Selten 2002:40).

Engineering and Medicine have developed different adaptive toolboxes to achieve such ambition. Engineering and Medicine have distinct processes through which they manage – identify, test, establish and disseminate heuristics to enlarge their states of the art (sotas). As preface for what follows, it is opportune to recall that this is a tale of two disciplines, Engineering and Medicine, each seeking to improve practice, each desirous of acting on the best available information, each seeking to enlarge, to perfect, its sota and the sota of individual practitioners, each faced with the need to keep practitioners up to date.

Engineering courses of action admit pairing problems with sector-specific "engineering toolboxes", based on the demand for solution to a given, specific kind of problem. In this pursuit, Engineering established standards, benchmarks or best practices, subjected practice to codes, cumulated experience, and offered heuristics in consensus reports, differentiating approaches that best deal with different kinds of projects. This stems from the fact that the choice of an applicable "engineering toolbox" can be made, *prima faciae,* beforehand. For instance: building a bridge would call for a bridge-building toolbox, manufacturing airplanes for a different one, and so on.

Medical courses of action, in contrast, must deal with each patient's "unique biology, values and circumstances" (Straus et al. 2011: 3). The pairing of the needs of a particular patient with possible approaches that might offer solutions can only take place once this patient's biology, values and circumstances have been established. So as to avoid biases and thus errors in this process – to support prevention, diagnosis, prognosis, therapy or causation – Medicine developed clinical trials and Evidence-Based Medicine. These seek to provide individual practitioners with the most reliable, up to date and applicable information about the effectiveness of existing heuristics. And this is something Engineering might profit to learn from Medicine.

A note of caution, ethical caution, must precede what follows. To take the position that unrestricted access to knowledge is a good in itself is to presume objectivity, ethical neutrality or the strict value-free instrumentality of knowledge. The case has been argued, with well-known concerns, for Science (e.g., Daston and Galison 2010). But that cannot stand in the case of Engineering or Medicine's heuristics. They allow capabilities for which the current ethical attitude is to restrict access to knowledge – the manufacture of weapons of mass destruction, forcible methods of manipulation, e.g..

Further, the issue cannot stand at all when one considers the consequences of the heuristic fundament of disciplines of intervention. This is a matter of the ethical consequences of heuristical epistemics. Heuristics are situated and personal in application. The presumption of neutrality cannot stand given the situated, personal, nature of the interference that must take place between practitioner and reality. Were Medicine or Engineering sciences, descriptions and explanations about reality with no consequence but understanding… But they are not: they are interventions that seek to effect a particular change in reality. This consideration must apply to Engineering, Medicine, and to any of Simon's "sciences of the artificial". As a result, heuristics cannot be considered ethically neutral in any case whatever (Mitcham 1994; Petroski 2011).

15.2 Learning from Medicine

The bulk of the present chapter delineates, in broad lines, how Medicine approaches the management of its state of the art (sota). It presents a mosaic that reconstructs some aspects of Evidence-based Medicine's practice, with the cautionary remark that successful as this has been, it should not be mistaken for an universal, complete, definitive framework (Howick 2011). The presentation seeks to illuminate the most salient aspects of Evidence-based Medicine that might provide for the case of its translation to Engineering in what concerns sota management. It offers a deliberate gradient that considers practices that admit close parallel to those of Engineering (Translational Medicine), to a divergence of emphasis (Hierarchy of Evidence), and finally to practices that are very different from those of Engineering (the system that culminates in PICO – Patient/Population – Intervention/Indicator – Comparison/

Control – Outcome). Its purpose is to support the proposition that an Engineering analogue of PICO (that expresses a whole systemic framework that combines content *and* function) might prove an opportune addition to engineering practice.

15.2.1 Translational Medicine

Largely in common with practices and concerns of Engineering – which might be argued to be evidence-based in its totality (Van Aken and Romme 2012) –, Evidence-based Medicine seeks "to have practice as its focus" (Heneghan and Badenoch 2006). This means choosing priorities and directing processes to the enlargement of Medicine's state of the art (sota) so as to answer and to provide for the needs of individual practitioners.

Translational Medicine expresses Evidence-based Medicine's procedures that seek to connect the results of basic research in science or technology with the needs of individuals and populations, to connect epidemiology with clinical practice (Mittra and Milne 2013: 6; Wehling 2015:3). This is deliberately understood as a two-way bridge, with the needs of practice guiding research that seeks to further and ease practical application. It is subject to both internal and external controls that seek to validate results and avoid or correct errors.

Efforts at translational medicine admit variety, and comprise various phases that must follow ever-evolving procedures and detail. In general lines, it goes from basic research, to health application discovery and design, to pre-clinical development, to various phases of clinical trials, to certification, and then to monitored application to a given population (Thornicroft et al. 2011: 2016–2018). This, in broad terms, is bringing the benefits of art and science into medical practice, translating knowledge into medical heuristics. This is much the same as in other disciplines, including Engineering, although it is not usually described as "translational engineering", rather as a broader movement of "dissemination and implementation research" (Colditz 2012:8; Fixsen et al. 2005:vi).

15.2.2 Hierarchy of Evidence

Diverging to some extent from the practices and concerns of Engineering (which admits different sota in individual practitioners and proprietary knowledge), Evidence-based Medicine seeks to learn with practice/practitioners (Wyer and Silva 2009; Claridge and Fabian 2005). This means bringing the shared sota of Medicine closer to the sum of the sota of all individual practitioners and making it widely available to medical practice's stakeholders.

Evidence-based Medicine developed a Hierarchy of Evidence (Fig. 15.1), a set of rules that expresses the relative importance of various sources of heuristics by appreciating their quality in terms of eliminating biases (in data, methods, and so

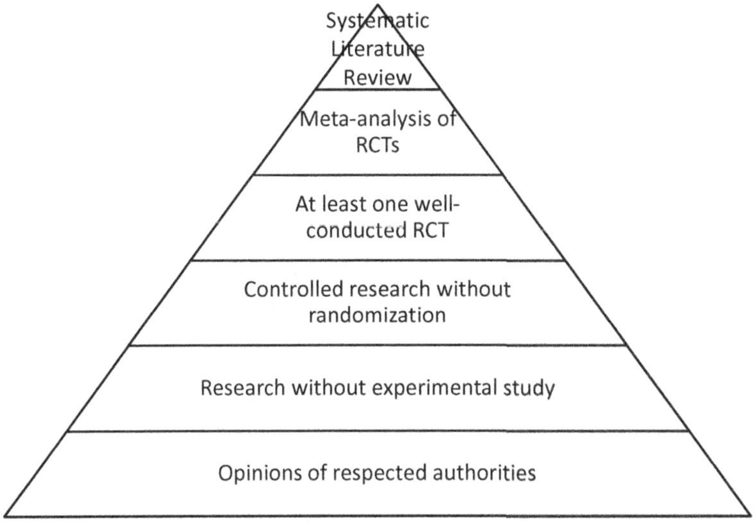

Fig. 15.1 Hierarchy of evidence. (Source: Adapted from Mantzoukas (Mantzoukas 2008:217))

on) and considering the largest possible samples (Evans 2003). Succinctly, the Hierarchy of Evidence (which, as Evidence-based Medicine itself, is not settled and final) admits methodological and data concerns (Bigby 2001; Bluhm 2005; Borgerson 2009). It ranges from expert opinion – one practitioner's experience (even with one single patient), through case studies with one or more cases, to various types of observational studies, to randomized trials and, finally, systematic literature reviews and meta analysis (that appraise and synthesize the whole of existing knowledge, weighing relative methodological robustness) (Mantzoukas 2008). Cochrane Collaboration, an independent, international volunteer organization of recognized medical professionals, is devoted to offer comprehensive systematic reviews of evidence. Their Handbook for Systematic Reviews of Interventions (Higgins and Green 2011) is one of the most important frameworks for conducting systematic reviews. Higgins and Green (2011) recognize that there are different evidence hierarchies for studies (e.g. Eccles et al. 1996; National Health and Medical Research Council 1999; Oxford Centre for Evidence-based Medicine 2001) but do not adopt any of them. In their view, "[d]ifferent designs are susceptible to different biases, and it is often unclear which biases have the greatest impact and how they vary between clinical situations". Hence, each study should determine beforehand which evidence hierarchy to use.

This admits some parallel to Engineering in many ways. For instance, aeronautical engineering deals with possible biases of a local optimum with experiments that assess the performance of given aerodynamical arrangements under controlled circumstances (Vincenti 1993), but usually falls short of the full requirements of clinical trials. Broader collaborative efforts may lead to Engineering benchmarks or consensus reports (National Academies 2016), but these usually lack the comprehensiveness of Medicine's systematic literature reviews and meta-analysis, or the scope and reach of Cochrane Collaboration's Handbook.

15.2.3 PICO

The sota management practice of Evidence-based Medicine that finds less correspondence in Engineering, and hence offers the most promising learning opportunity, is the system that culminates with PICO (Patient/Population – Intervention/Indicator – Comparison/Control – Outcome). Unique to Medicine, PICO expresses the concern with near-real-time access to the whole of medical knowledge in such a way that it answers a medically relevant question, supporting better-informed decision-making by practitioners. This means allowing the sota of individual practitioners to benefit from access to the sota of Medicine. The system that supports such an ambitious achievement is composed of four main components.

The first component is MEDLINE. MEDLINE is the contemporary face of the longwinded ambition of the nineteenth century *Index Medicus*. MEDLINE aspires to collect and index the meta-data of everything that might be medically relevant (O'Leary 1997; Wilson 1997; Lindsey and Olin 2013; Kurata et al. 2013). While it serves researchers in keeping up with research, it aims at much more than academia, being attentive and appreciative of the needs of the many audiences and stakeholders involved with Medicine and their different contexts, concerns, interests and language.

The second is MEDLINE's robust and universal controlled vocabulary – Medical Subject Headings, MeSH. MeSH structures a unified taxonomy that allows multiple denominations to converge on precise denotations, connecting different perspectives and approaches with phenomenological rigor. It accommodates natural language as well as customary and specific medical terminology by a unified nomenclature based on scientific criteria. Its goal is to provide a "medical ontology" that would prevent stakeholders from being uninformed (Lipscomb 2000; Richter and Austin 2012).

The third is PubMed Central, the online access to MEDLINE through MeSH that aims at a universal, free on-line access to all medical knowledge (Gordon 1999; Caelleigh 2000). Nowadays, it faces limitations due to publication embargoes or commercial restrictions, ameliorated by PubMed's licensing free-access to commercial products funded by the USA's *National Library of Medicine.* PubMed Central offers free-access to MEDLINE-indexed content, allowing retrieval worldwide. Its intended user community is envisioned as exhaustive: researchers, practitioners, patients, the public, and policy makers.

Finally, the fourth is PICO, a ubiquitous protocol that divides clinical or research questions into its four elements (Patient/Population – Intervention/Indicator – Comparison/Control – Outcome). PICO allows questions to return results filtered by practice-relevant variables in such a way as to allow the consideration of expected outcomes from existing alternatives (Heneghan and Badenoch 2006).

15.3 Conclusion

The transposition of heuristics for state of the art (sota) management from Medicine to Engineering, as mentioned in the Introduction, seems both admissible and possible and, given the above presentation of the components of Medicine's management of its state of the art (sota), might prove to be opportune. It admits, as vision and provocation, the idea, the symbol, of "engineers with PICO". To put it differently: as final end goal for Engineering learning from Medicine, an E-PICO with all the systemic arrangements it requires, and the systemic dynamics it induces, that would provide engineers with what PICO provides medical practitioners. The purpose of this conclusion is to share some cautions as to what is meant by such a vision, with some conclusive remarks that argue that this might, indeed, be *desirable*.

To argue for the opportunity and desirability of seeking, as vision, an E-PICO does not mean relinquishing the *a priori* formulation of the "case before the practitioner" of Engineering for the *a posteriori* dynamics of Medicine mentioned in the Introduction. Rather, it admits, requires, actual translation of the original medical PICO system (and all it requires and induces) to engineering PICO, E-PICO.

This admits the acknowledgement of dynamics that cannot be anticipated completely. It admits, in principle, two nested possibilities, at least at start. The use of E-PICO to access the whole of Engineering knowledge (the Engineering analogue of MEDLINE, e.g.) and the use of E-PICO to access the sector-specific (arguably, over time, project- or problem-specific) Engineering knowledge. The broader E-PICO accesses Engineering ENGLINE in full, and relies on a comprehensive Engineering Subject Headings (ESH) and Engineering Hierarchy of Evidence. For simplicity, sector-specific E-PICO might be more specialized, accessing ENGLINE in part, at least as a start and perhaps as a practical necessity. This is something to be found out, not defined, as it might incorporate undesirable bias. Conversely, there might be heuristics germane to one engineering toolbox alone, but that does not compromise the comprehensive foundations of the elements of an ENGLINE, an ESH or an Engineering Hierarchy of Evidence.

To take an example of a possible application domain for an ESH and an E-PICO: the field of software engineering has been developing Evidence-based Software Engineering (EBSE) since Kitchenham et al. (2002) was published. This field gained momentum after the publication of two methodological guidelines for EBSE research (Kitchenham 2004; Kitchenham and Charters 2007). The guidelines offered suggestions and insights on how to adapt Evidence-based Medicine's practices of hierarchy of evidence, systematic literature reviews and meta-analysis to software engineering. As a result, the number of academic publications using EBSE as their methodological framework dramatically increased. Da Silva et al. (2011), Kitchenham et al. (2009) and Kitchenham et al. (2010) identified 68 software engineering systematic mapping studies published between 2004 and 2009. A Mapping study is a prerequisite for systematic literature reviews that aims to classify the literature on a specific topic (Kitchenham et al. 2011). Also, Kitchenham and Brereton

(2013) counted 68 systematic literature reviews in software engineering published between 2005 and mid-2012. Recent publications in EBSE have been discussing research instrumentalities to support its practice, such as evaluating softwares that would support EBSE. However, their evaluation criteria have focused on whether such tools are economically viable, easy to introduce, supportive of all phases of a systematic review and whether they allow for appropriate process control and management (Marshall et al. 2014, 2015). Hence, the need for a (software) Engineering PICO and a corresponding (Software) Engineering Subject Headings persists: with more EBSE publications there are more variables under study, and one must be able to cope with research biases that cannot be eliminated by systematic review softwares. Software Engineers have even conducted experiments that showed that structured abstracts have better readability than unstructured ones (Budgen et al. 2008; Kitchenham et al. 2008), but it has not been set as a standard and would only be a palliative measure.

Software engineering can be seen as a rather successful, yet recent and therefore still incomplete, case of emulation of evidence-based medicine to the engineering domain. Among the possible points for improvement of the field's sota management are the use of an (Software) Engineering Subject Headings and a correspondent (Software) Engineering PICO.

That being said, what the vision glimpses and this chapter proposes is the translation of the system, the heuristics of sota management of Medicine by Engineering. The proof of the pudding is such an E-PICO that can answer engineering relevant inquiries, providing up to date, near real time comprehensive access to the sota of Engineering in support of the individual practitioner.

It is important to conclude by stating clearly that this is not just a matter of an improved information system for Engineering *qua* information system, but rather, and emphatically, the translation of the system of sota management of Medicine. This comprises much more than an improvement of access. It seeks enhancement of the management of Engineering sota, with an eye on Medicine's. It requires comprehensiveness, comparison, elimination of bias, hierarchy of evidence on informational and methodological terms, and further the virtuous circle of research guided by, and for, the practical needs and ease of use by the individual practitioner.

A medical practitioner has at hand the whole of the sota of Medicine when faced with a patient – and medical practitioners, medical practice, have become better, more effective, as a result. The vision of E-PICO as a foundation for the future of engineering aspires the same for the engineer – to have the whole of the sota of Engineering at hand – becoming a better engineer as a result.

References

Bigby, M. (2001). Challenges to the hierarchy of evidence. Does the emperor have no clothes? *Archives of Dermatology, 137*(3), 345–346.

Bluhm, R. (2005). From hierarchy to network – A richer view of evidence for evidence-based medicine. *Perspectives in Biology and Medicine, 48*(4), 535–547. https://doi.org/10.1353/pbm.2005.0082.

Budgen, D., Kitchenham, B. A., Charters, S. M., Turner, M., Brereton, P., & Linkman, S. G. (2008). Presenting software engineering results using structured abstracts: A randomised experiment. *Empirical Software Engineering, 13*(4), 435–468. https://doi.org/10.1007/s10664-008-9075-7.

Bunge, M. (2013). *Medical philosophy: Conceptual issues in medicine*. London: World Scientific Publishing.

Borgerson, K. (2009). Valuing evidence: Bias and the evidence hierarchy of evidence-based medicine. *Perspectives in Biology and Medicine, 52*(2), 218–233. https://doi.org/10.1353/pbm.0.0086

Caelleigh, A. S. (2000). PubMed central and the new publishing landscape: Shifts and tradeoffs. *Academic Medicine: Journal of the Association of American Medical Colleges, 75*, 4–10.

Claridge, J. A., & Fabian, T. C. (2005). History and development of evidence-based medicine. *World Journal of Surgery, 29*(5), 547–553. https://doi.org/10.1007/s00268-005-7910-1.

Colditz, G. A. (2012). Dissemination and implementation research. In R. C. Brownson, G. A. Colditz, & E. K. Proctor (Eds.), *Dissemination and implementation research in health: Translating science to practice* (1st ed.). New York: Oxford University Press.

Da Silva, F. Q. B., Santos, A. L. M., Soares, S., Franca, A. C. C., Monteiro, C. V. F., & Maciel, F. F. (2011). Six years of systematic literature reviews in software engineering: An updated tertiary study. *Information and Software Technology,* 53(9, SI), 899–913.

Daston, L., & Galison, P. (2010). *Objectivity*. Zone Books.

Eccles, M., Clapp, Z., Grimshaw, J., Adams, P. C., Higgins, B., Purves, I., & Russel, I. (1996). North of England evidence based guidelines development project: Methods of guideline development. *British Medical Journal, 312*, 760–762.

Evans, D. (2003). Hierarchy of evidence: A framework for ranking evidence evaluating healthcare interventions. *Journal of Clinical Nursing, 12*, 77–84. https://doi.org/10.1046/j.1365-2702.2003.00662.

Fixsen, D. L., Naoom, S. F., Blase, K. A., Friedman, R. M., & Wallace, F. (2005). *Implementation research: A synthesis of the literature.* Tampa: University of South Florida, Louis de la Parte Florida Mental Health Institute, The National Implementation Research Network (FMHI Publication #231).

Freidson, E. (1988). *Profession of medicine: A study of the sociology of applied knowledge.* Chicago: University of Chicago Press.

Gigerenzer, G. & Selten, R. (Eds.). (2002). Bounded rationality: The adaptive toolbox (Reprint edition). Cambridge, MA.: The MIT Press.

Gordon, D. (1999). PubMed central: A boon to information access, a bane to quality control, or neither? *Gastroenterology, 117*, 1040.

Heneghan, C., & Badenoch, D. (2006). Evidence-based medicine toolkit. Malden/Oxford: BMJ Books/Blackwell Pub.

Higgins, J. P., & Green, S. (2011). *Cochrane handbook for systematic reviews of interventions* (ver. 5.1.0). http://handbook.cochrane.org

Howick, J. (2011). The philosophy of evidence-based medicine. Chichester/West Sussex/Hoboken: Wiley-Blackwell, BMJ Books.

Kitchenham, B. A. (2004). *Procedures for undertaking systematic reviews*. Joint Technical Report, Computer Science Department, 2004, Keele University (TR/SE-0401) and National ICT Australia Ltd (0400011T1).

Kitchenham, B. A., & Brereton, P. (2013). A systematic review of systematic review process research in software engineering. *Information and Software Technology, 55*(12), 2049–2075. https://doi.org/10.1016/j.infsof.2013.07.010.

Kitchenham, B. A., & Charters, S. (2007). *Guidelines for performing systematic literature reviews in software engineering.* Technical Report EBSE-2007-01, School of Computer Science and Mathematics, Keele University.

Kitchenham, B. A., Pfleeger, S. L., Pickard, L. M., Jones, P. W., Hoaglin, D. C., & El Emam, K. (2002). Preliminary guidelines for empirical research in software engineering. *IEEE Transactions in Software Engineering, 28*(8), 721–734.

Kitchenham, B. A., Brereton, O. P., Owen, S., Butcher, J., & Jefferies, C. (2008). Length and readability of structured software engineering abstracts. *IET Software, 2*(1), 37–45. https://doi.org/10.1049/iet-sen:20070044.

Kitchenham, B., Brereton, O. P., Budgen, D., Turner, M., Bailey, J., & Linkman, S. (2009). Systematic literature reviews in software engineering – A systematic literature review. *Information and Software Technology, 51*(1), 7–15.

Kitchenham, B., Pretorius, R., Budgen, D., Brereton, P., Turner, M., & Niazi, M. (2010). Systematic literature reviews in software engineering – A tertiary study. *Information and Software Technology, 52*(8), 792–805.

Kitchenham, B. A., Budgen, D., & Brereton, O. P. (2011). Using mapping studies as the basis for further research – A participant-observer case study. *Information and Software Technology*, 53(6, SI), 638–651. https://doi.org/10.1016/j.infsof.2010.12.011.

Koen, B. V. (2003). *Discussion of the method: Conducting the engineer's approach to problem solving*. New York: Oxford University Press.

Kurata, K., Morioka, T., Yokoi, K., & Matsubayashi, M. (2013). Remarkable growth of open access in the biomedical field: Analysis of PubMed articles from 2006 to 2010. *PLoS One, 8*, e60925.

Lindsey, W. T., & Olin, B. R. (2013). PubMed searches: Overview and strategies for clinicians. *Nutrition in Clinical Practice: Official Publication of the American Society for Parenteral and Enteral Nutrition, 28*, 165–176.

Lipscomb, C. E. (2000). Medical subject headings (MeSH). *Bulletin of the Medical Library Association, 88*, 265–266.

Mantzoukas, S. (2008). A review of evidence-based practice, nursing research and reflection: Levelling the hierarchy. *Journal of Clinical Nursing, 17*(2), 214–223. https://doi.org/10.1111/j.1365-2702.2006.01912.x.

Marshall, C., Brereton, P., & Kitchenham, B. A. (2014). Tools to support systematic reviews in software engineering: A feature analysis. In: *Proceedings of the 18th international conference on evaluation and assessment in software engineering* (p. 13). ACM.

Marshall, C., Brereton, P., & Kitchenham, B. A. (2015). Tools to support systematic reviews in software engineering: A cross-domain survey using semi-structured interviews. In: *Proceedings of the 19th international conference on evaluation and assessment in software engineering* (p. 26). ACM.

Mitcham, C. (1994). *Thinking through technology: The path between engineering and philosophy* (1st ed.). Chicago: University Of Chicago Press.

Mittra, J., & Milne, C.-P. (2013). Introduction and definitions. In J. Mittra & C.-P. Milne (Eds.), *Translational medicine: The future of therapy?* Hoboken: Pan Stanford.

National Academies. (2016). *Guidelines for the review of reports of the National Academies of Sciences, Engineering, and Medicine*. http://www.nationalacademies.org/nasem/na_067075.html

National Health and Medical Research Council (1999). *A guide to the development, implementation and evaluation of clinical practice guidelines* [Endorsed 16 November 1998]. Canberra: Commonwealth of Australia.

O'Leary, M. (1997). PubMed initiates new MEDLINE era. *Database, 20*, 70–72.

Oxford Centre for Evidence-based Medicine (2001). *Levels of evidence*. http://www.cebm.net/oxford-centre-evidence-based-medicine-levels-evidence-march-2009/. Accessed 14 Jan 2016.

Petroski, H. (2011). *The essential engineer: Why science alone will not solve our global problems* (Reprint ed.). New York: Vintage.

Polya, G. (1957). *How to solve it: A new aspect of mathematical method* (2nd ed.). Princeton: Princeton University Press.

Richter, R. R., & Austin, T. M. (2012). Using MeSH (Medical Subject Headings) to enhance PubMed search strategies for evidence-based practice in physical therapy. *Physical Therapy, 92*, 124–132.

Silva, E. R., & Proença, D., Jr. (2015). Não ser não é não ter: Engenharia não é Ciência (nem mesmo ciência aplicada). In A. Proença et al. (Eds.), *Gestão da Inovação e Competitividade no Brasil* (pp. 197–218). Bookman: Porto Alegre.

Simon, H. A. (1956). Rational choice and the structure of the environment. *Psychological Review, 63*, 129–138.

Simon, H. A. (1969). *The sciences of the artificial*. Cambridge, MA: The MIT Press.

Straus, S. E., Glasziou, P., Richardson, W. S., & Haynes, R. B. (2011). *Evidence-based medicine: How to practice and teach it*. Edinburgh: Elsevier/Churchill Livingstone.

Sur, R. L., & Dahm, P. (2011). History of evidence-based medicine. *Indian Journal of Urology: IJU: Journal of the Urological Society of India, 27*(4), 487–489. https://doi.org/10.4103/0970-1591.91438.

Thornicroft, G., Lempp, H., & Tansella, M. (2011). The place of implementation science in the translational medicine continuum. *Psychological Medicine, 41*(10), 2015–2021. https://doi.org/10.1017/S0033291711000109.

Van Aken, J. E., & Romme, G. (2012). A design science approach to evidence-based management. In D. M. Rousseau (Ed.), *The Oxford handbook of evidence-based management* (pp. 43–57). New York: Oxford University Press.

Vincenti, W. G. (1993). *What engineers know and how they know it: Analytical studies from aeronautical history* (Revised ed.). Baltimore: Johns Hopkins Univ Pr.

Wehling, M. (2015). Introduction to translational medicine. In M. Wehling (Ed.), *Principles of translational science in medicine, second edition: From bench to bedside*. Amsterdam: Academic Press.

Wilson, J. W. (1997). PubMed: A winner on the web for the National Library of Medicine. *Online, 21*, 60–61.

Wyer, P. C., & Silva, S. A. (2009). Where is the wisdom? I – A conceptual history of evidence-based medicine. *Journal of Evaluation in Clinical Practice, 15*(6), 891–898. https://doi.org/10.1111/j.1365-2753.2009.01323.x.

Chapter 16
What Ethics Owes Engineering

Glen Miller

Abstract The relationship between ethics and engineering has largely been depicted as one-directional in which ethics enlightens engineers and their practices. The opposite contribution, from engineering to ethics, which has received far less attention, can be organized in three main categories. First, engineering leads to a wider separation between intention and ends, which are often unclear and sometimes not obtained, and a multiplication of means and mediation, which can increase uncertainty in ethical assessments. Second, engineering reduces the amount of time spent satisfying basic needs, which increases the amount of time and energy directed toward voluntary goals while amplifying human power and providing the material and social conditions that have been associated with philosophical activity since the time of the ancient Greeks. Third, engineering successes have also brought into relief an innovative understanding of desire and its social implications, catalyzed a more expansive scope of moral reasoning and universal imperatives, and illuminated the interrelated nature of existence between humans and the non-human world. Taken together, such insights have rejuvenated ethical inquiry and so have led to better understandings of the "good life" and authentic development. Engineering can be thought of as a "gadfly" that can shake the dogmatic ethical "horse" into action.

Keywords Engineering and technological mediation · Engineering ethics · Engineering, needs, and desire · Engineering, technology, and freedom · Environmental and communal dimensions of engineering work

The relationship between ethics and engineering is usually considered unidirectional: ethics is seen as a source of enlightenment for engineers about their practices. In the realm of descriptive ethics, anthropologists, psychologists, and sociologists study and evaluate the actions of engineers, their professional societies, and their

G. Miller (✉)
Department of Philosophy, Texas A&M University, College Station, TX, USA
e-mail: glenmiller@tamu.edu

A. Fritzsche, S. J. Oks (eds.), *The Future of Engineering*, Philosophy of Engineering and Technology 31, https://doi.org/10.1007/978-3-319-91029-1_16

work in organizations, usually without considering how engineering affects their own fields. Similarly, in the normative space, illumination flows from ethics to engineering. This is true whether the approach taken is a fruits-of-theory approach that admits the relevance of multiple theories, the application of mid-level principles or general norms such as autonomy or justice to certain situations, as is common in medical ethics, or the "engineering approach," by which a particular theory is applied to problems faced by practitioners and patients. In all of these approaches, engineering is the patient and ethics the agent.

A nuanced line of reasoning about the relationship of subordination of the engineering to the humanities has been argued in Carl Mitcham's influential *Thinking through Technology*, which has been translated into many different languages (Mitcham 1994). His concern is broader—the relationship of engineering philosophy of technology to a humanist philosophy of technology—but related. Mitcham rejects the adequacy of an engineering philosophy of technology, noting that even if an argument for the historical primacy of engineering over humanities were successful, it does not imply logical primacy. Yet Mitcham immediately acknowledges that if there is any continuity between premodern and contemporary technology, "a philosophy of technology that takes its bearings solely from the humanities rather than from technology must be deficient." On account of this recognition, he concludes that a "humanities philosophy of technology must in principle remain open to the engineering perspective" (Mitcham 1994, pp. 138–140).

In this paper, I move beyond a principled openness by offering some specific ways that the field of ethics and those who practice it owe a debt to engineers and engineering. Engineering practices, successes, and failures have been catalysts to engineering ethics, and more broadly, professional ethics and ethics in general, yet these contributions have not received their just deserts, perhaps in part because they have not been explained in an organized manner. The practical problems that arise with the creation, adoption, use, and disposal of new technologies have spurred ethicists beyond their disciplinary strictures, where they were prone to intellectual circularity that led to the *odium theologicum* that festered between theologians of the late medieval period. As social epistemologist Steve Fuller notes, engineering ethics, as part of professional ethics, is one of the rare "growth areas" in philosophy (Fuller 2009).

A structure for situating the contributions that engineering has made to ethics includes three components.

- Engineering emphasizes the separation of intention and ends. In the engineered world, means and mediation are multiplied. The mediating nature of technology in human interactions tends to obscure the connection between act and consequence, especially in the complex webs of action found in the globalized economy, where developed usually means intensively engineered. Such concerns are exacerbated by the lack of *telos* inherent in engineering itself.
- Two kinds of second-order implications become obvious in the engineered world. One, the proportion of energy spent satisfying needs is shifted to pursuit of desires. Conveniences provided by engineering have reduced the proportion

of time needed to satisfy basic needs while at the same time increasing the number of possible means and ends available. Two, engineering has challenged the dominant trend toward freedom and autonomy in ethics by highlighting the communal and ecological dimensions of human actions. Similar to the advantage that Socrates' defense of justice found by considering the *polis* and not just the individual, the reciprocal effects of individual actions on other humans and on the non-human world become clearer when the technological effects are *writ large*.

- The nature and secondary consequences of engineering have resulted in material and social conditions markedly different than what situated humanity in earlier eras. In short, engineering serves as a gadfly, stinging the ethics horse to reflect on its habits, and, if nothing more, to reframe and reformulate the points of entry into the "Brave New World" that most of us inhabit. The third section enumerates several important arguments on desire, moral reasoning, the "good life," and authentic development that have arisen in engineered environments.

For the purposes of this paper, including the title, by "ethics" I mean those who study free human behavior, i.e., that which does not occur by necessity, and the body of the work they generate on the topic. I similarly use "engineering" in a dual sense that encompasses practitioners and the products of their efforts.

16.1 Re-Engineering Means and Ends

Engineering work produces technologies that mediate and amplify human action. Mediation complicates the analysis of human agency while separating and obscuring the relationships between agents, actions, and patients. Mediating technologies also often constrain human actions and favor some behaviors over others. In this section, I look at some of the conceptual and ethical insights and challenges that result from mediation, especially as it exists and is produced in the profession of engineering.

The ethics of technically-mediated action, in decision-making, analysis, and assessment, is more complex than that of direct action, where there is either no intermediary or a small one between act and end. These difficulties are exacerbated in the convoluted, interwoven sociotechnical systems and supersystems in which contemporary acts occur, where an individual agent affects and is affected by a myriad of other actors, known and unknown, some intentionally and others unintentionally. These three characteristics, separation in time and space, uncertainty, and complexity, make assessment of praise and blame problematic, even in hindsight. One response to them is resignation, which supports nihilistic philosophical positions that have come to age in post-modernity; another is to focus simply on perfecting means while letting consideration of ends recede into the distance. These characteristics also exacerbate the long-standing tension between expected and actual consequences found in utilitarian reasoning, illuminating an inconsistency many ethical theorists find problematic.

The general challenges raised by mediation and the associated temporal and spatial separation found in sociotechnical collective agency are of specific interest in ethics for engineers and designers, who have varying degrees of control and influence over the objects that they create. While engineers and designs impart a script that favors certain actions on their products, the products themselves can be repurposed in the hands of the public, i.e., the products possess multistability, to use terminology from post-phenomenology (Verbeek 2011). Historically, commonly cited examples of multistability were often limited to using a product as it was intended but in order to reach a different goal. One well-known example of such reappropriation is Andrew Feenberg's frequently cited historical research on the French *Minitel*, a network designed to foster democratic deliberation that became a gossip and dating site (Feenberg 1995).

The idea of multistability was developed in a period when the function of a thing was largely assumed to be fixed—in part because of the tangible and mechanical characteristics of engineering products—and the purpose for which it was deployed was all that was altered. The trend at the present, enabled by the proliferation of cheap microprocessors and digital data, is to alter the product or service, to hack it, which, by altering the function, structure, or configuration of a thing, permits a more radical redirection of its purpose. Hacking vastly increases the scope of possible actions and ends that can be achieved. It diminishes the control and responsibility that designers and engineers have over their products and shifts them toward users, who now possess additional creative agency.

The shifts in agency and control, in means and ends, take place in the shadow of an ambiguous *telos* of engineering. Mitcham, when reflecting on the development of engineers, argues that engineering should be considered a weak profession (Mitcham 2009). It lacks a determined end compared to strong professions such as medicine, whose end is health, and law, whose end is justice. For law and medicine, the end is clear, regardless of how imperfectly it is realized, and it is an essential part of the training of a lawyer or doctor. By contrast, "engineering as a profession is analogous to what medicine might be if physicians had no expert knowledge of health or to law if attorneys knew nothing special about justice" (Mitcham 2009, p. 339). This indeterminacy is somewhat cloaked in obscurity in such contemporary sources such as the Preamble to the U.S. National Society for Professional Engineers Code of Ethics, where the "direct and vital impact" of engineering is on "quality of life." Perhaps in less developed economies, quality of life is an adequate clarification for the aim of engineering, but in contemporary, pluralistic societies—precisely the kind of economies that engineering helps to make possible—the phrase is the topic of conceptual dispute. Moreover, it seems clear that engineering work produces some goods, those that can harness the efficiencies of the engineering method, more easily than others, favoring some "qualities of life" over others, even if engineering and technology appear neutral at first glance. Neutrality and goal indeterminacy are both captured clearly in Thomas Tredgold's classical definition of engineering, "the art of directing the great sources of power in nature for the use and convenience of man" (Institution of Civil Engineers 1828).[1]

Mitcham connects "use and convenience" to David Hume's proto-utilitarianism and argues that "use and convenience is simply assumed to be synonymous with the advancement of commercial and industrial interests" (Mitcham 2009, p. 345). As Mitcham notes in a later paper with Adam Briggle, "use and convenience" falls under Hume's heading of utility, where it serves as an unproblematic foundation for morals (Mitcham and Briggle 2012). Mitcham's critique highlights the instrumental nature of the benefits of engineering rather than their intrinsic or final goals. In Tredgold's enumerated list, engineering should create a transportation infrastructure, yet the purpose of motility is not mentioned, and allow the drainage of cities and towns, but there is no mention of the goods of creating more habitable land or preventing disease.

Mitcham's line of thought emphasizes the first term, "use," which signifies the character of instrumentality and implies the mediation of actions and goals or ends discussed earlier, whereas convenience suggests a "coming together," based on its Latin origins, of *desiderans* and *desiderata*. Convenience arises from the amplification of human power that engineering makes possible. Engineering technology harnesses the "great sources of power of nature" to work over larger temporal and spatial scales and to reorganize nature at a more fundamental level, e.g., at atomic or nano scales. Such power allows humans to redesign their world, to materially reconfigure it, and alter the principle that guides the function and end of a being, sometimes forever. The College of Engineering at Texas A&M University is currently using the slogan "Engineers change the world," to which should be added, "often for good, and sometimes for the better."[2]

These amplifications, which increase the number and proportion of means and ends, have derivative effects that affect free human action. Given the separation of intention and action from ends, the uncertainty, and the indeterminate goals of engineering work, a reflective inquiry on intentions, practices, decisions, and consequences is appropriate.

16.2 Second Order Implications of Re-Engineering Means and Ends

The alteration of means and ends through engineering and technology has several secondary effects on free human action. Fundamentally, the convenience promised in Tredgold's definition has largely been fulfilled. Humans today spend far less time

[1] As Mitcham notes, this definition has largely been accepted to the present. The same definition holds in McGraw-Hill Concise Encyclopedia of Science and Technology, s.v. "Engineering," accessed December 01, 2012, https://libproxy.library.unt.edu/login?qurl=http%3A%2F%2Fwww.credoreference.com/entry/conscitech/engineering.

[2] I owe the inspiration for this clever comment to C. E. (Ed) Harris.

satisfying basic needs and biological demands than at any previous time in human history. By increasing time and energy directed toward voluntary goals, a sense of liberation is gained and freedom can be said to expand. An increased space for voluntary action expands the space where ethical thought is required for a growing number of people. The amplification of power that results in these conveniences also suggests the need for a new ethical attentiveness.

Critics of engineering and technology note that much of the freedom from these efforts has been redirected toward satisfying what John Stuart Mill called lower pleasures and another part toward conspicuous consumption. Yet it should be noted that the same prosperous material conditions have been connected to the rise of Western philosophy by the Greeks, even in light of Aristotle's critique that good life as pleasure held by *hoi polloi* is suitable for fatted cattle, not humans (Aristotle 2002, 1095b10). In his expansive *History of Greek Philosophy* series, W. K. C. Guthrie notes that "the environment of the Milesian philosophers, then, provided both the leisure and the stimulus for disinterested intellectual inquiry, and the dictum of Aristotle and Plato, that the source and spring of philosophy is wonder or curiosity, finds its justification" (Guthrie 1962, p. 30). Today, the potential for a Golden Age is no longer restricted to a small geographical area: the conditions that accompanied the birth of Western philosophy hold for the majority of individuals in developed countries. Additionally, regardless of how widespread the search for pleasure or consumption purchased by engineering becomes, the same freedoms make it possible to experience personal events, such as new life, loves, fights, and death, and political events, from contemplation of justice in one's own society to geopolitical collisions, more reflectively. In contrast to mundane tasks that yield comfort and convenience, these kinds of events stand out as more intense and more real.

In an ironic twist, at the same time that engineers have contributed to unseen prosperity, the value of their efforts has become imperiled by their success. Engineers often justify the value of their projects by appealing to improvements in the quality of life that will result, i.e., they will satisfy a need or desire, they connect *desiderans* and *desiderata*. The benefits of projects that address basic human needs are obvious. In developed countries, though, most basic needs are already adequately met, and engineering projects address desires. In these conditions, the assumption that engineering projects are valuable, as long as they do not result in illness or environmental damage, must be reconsidered. Living in the engineered world gives us reason to question the assumption that the more desires are satisfied, the better, and it causes us to rethink the nature of desire. The concept of desire is usually thought to be direct, one where the object of desire is sought by the desirer, either through inclination or after rational deliberation. If this were the case, the contemporary abundance of objects of desire that engineering makes possible should provide proportionally improved quality of life, yet this does not seem to be the case. These facts suggest that our concept of desire demands more nuance, an inquiry pursued in more detail in the next section.

It is certainly not the case that an engineered world only leads to increased freedom. Many contemporary technically-mediated actions never escape the instrumental space, i.e., they never escape the creation and maintenance of technological society and infrastructure. Jean-Pierre Dupuy, following the lead of Ivan Illich, captures the ensnarement lucidly:

> When one steps back in order to leap forward, one must keep one's eyes fixed on the obstacle to be surmounted. If one steps back while looking in the opposite direction, one runs the risk of forgetting one's objective and, seeing one's regression as progress, of taking the means for ends. Then rationality turns into counterproductivity, it takes the form of the torture of Tantalus. (Dupuy 2002, p. 194)

As one point of evidence suggesting this risk exists outside of Gulliver's imagination, Dupuy cites his investigation into the French automobile culture in the 1970s, repeated again in the 1990s. It showed that the average rate of locomotion that cars provide is about four miles per hour, if the distance traveled were divided by the total of the time spent traveling and the costs of purchasing and owning the vehicles, converted using the inverse of the monetary value of time commonly used in economic analysis (Dupuy 2002, p. 195). A tremendous amount of effort must be allocated toward building and maintaining systems that are valued for their use with little reflection on how well they contribute to particular ends or the opportunity costs as they preclude many other ends. To put it another way, use can be unconsciously valued for use's sake. System maintenance also serves as a constraint for the actors, for sociotechnical systems do not decompose into readily reusable components like plants do when they die. In such situations, the value of engineering and the human affairs that surround it deserves greater scrutiny.

Another consequence of engineering is the prompting of a reconsideration of ends that are sought, which are now amplified by modern technologies. The instrumental view of artifacts held by earlier societies ignores the transformed nature of our actions, which deserve further attention because of their amplification, precision, extension, and durability, as discussed earlier. This recognition leads to the realization that more concern should be given to other humans and non-humans, those alive today and future generations. To put it another way, it demands the development of an environmental ethic and emphasizes the political dimension of human action. Such ethical considerations call into question the primacy given to autonomy—hailed as the "ground of all moral principles" by Immanuel Kant—in recent Western philosophical tradition, and they check the trend toward subjectivity.[3] These limitations become clearest in what Jacques Ellul called "the technological society," where devices and artifacts produced by engineering are ubiquitous and efficiency, closely linked to convenience and use, is one of the fundamental values, if not the fundamental value (Ellul 1964).

The second order consequences of engineering have changed the conditions of ethical reflection, which in turn has brought rise to a number of innovative or clearer ethical insights that I develop in the third section.

[3] In fairness to Kant, the individualist aspect of this thought should be balanced by his efforts to move between the subjective and universal, which undergirds his *First Critique*.

16.3 Philosophical Insights Revealed Through Engineering

The ethical implications of engineering and its second-order effects have been touched on briefly in the previous section through reference to Aristotle, Kant, and Mill and more recently and more specifically in the works of Anders, Dupuy, and Ellul. In this section, I catalog ethical insights on desire, moral reasoning, and our interconnected existences, which, taken together, elicit new views on the "good life" and authentic development.

Desire is commonly understood to be structured linearly between *desiderans* and *desiderata*. A person desires something, be it material or immaterial, primarily because of the thing's properties or capabilities. On this line of thinking, I desire a certain fast car, respect from others, or even a particular sense of self-image on account of their intrinsic or instrumental value, i.e., what they are or will do for me, understood rather simply. Desire is simple.

René Girard has argued that such a view of desire is inadequate for objects that are not necessities. While the connection between *desiderans* and *desiderata* exists, it is secondary. Desire is complex, mediated by what he called the triangulation of desire (Girard 1966). My desire of a certain car is determined largely by what a respected mediator desires. This phenomenon is seen in the use of spokespeople, who are often exemplars in some sense, be it athletics, fashion, or politics, in advertising. I buy Gatorade not because it is the best recovery drink, but because I want to be like Mike, and I want Beats headphones because Dr. Dre is shown, in advertisements and outside of them, using them. Desire, as Girard argues, is largely mimetic, and absent a mediator, many of these desires would not arise.

Girard's definition of desire applies to objects that do not satisfy basic human needs, which aligns with the shift in engineering work toward satisfying a broad range of voluntary desires rather than basic biological human needs. As argued in the second section, engineering has yielded socioeconomic conditions in which actions based on desires are widespread. While these background conditions should not be considered necessary conditions for Girard to come up with his theory from a rather limited analysis, which depends on a literary analysis of just a few historical novels, it seems that they likely facilitated its generation and the widespread interest shown in it.[4] Similarly, one factor that seems more than coincidental in the reception of Girard's ideas is that the mediation central to mimetic desire reflects the mediated structure of much of human agency today.

A decrease in the energy spent satisfying basic needs means that more energy is directed toward non-essential desires. When such desires are mimetically guided, they result in tension between individuals who share the same mediator and the same goods and between individuals and their mediators, a relationship that always retains distance, which Girard considers primarily social or intellectual. This tension

[4]A wealth of social science data that supports Girard's claims has also been produced after the publication of *Deceit, Desire, and the Novel* (1961 in French, 1966 in English), such as Richard Easterlin's "paradox" that the correlation between happiness and wealth weakens after basic needs are satisfied.

manifests itself as envy and jealousy through unconscious and somewhat elaborate psychological maneuvers. Envy and jealousy accumulate and lead to violence unless they are discharged through some kind of scapegoating process, understood in its original sense. The scapegoat, which can be a person, a select group, or animal, must be thought to be either the cause of discord or suffering experienced by the community or a suitable target on which the responsibility for discord or suffering can be projected. It can then be ostracized or destroyed to offer relief from tensions, discord, and suffering. (The current understanding of the term, in which the scapegoat is innocent, renders the act absurd.) (Girard 1986) If Girard is right, to the extent that engineering achievements lead to the cultivation of desires and promote mediation, they deserve some responsibility for social disintegration, even if they are usually considered to be contributions to "quality of life," and to the unconscious creation of the scapegoats of the twenty-first century.

The reconsideration of desire based on engineering is a specific form of a more general expansion of the nature of moral reasoning, where ethicists have responded with innovative ideas. Considering the amplification of human power that technology provides led Hans Jonas to craft a new ethical imperative, one that is universal like Kant's categorical imperatives but that is informed by attentiveness to material concerns and consideration of long-term and cumulative consequences. What is needed, Jonas says, is "an imperative responding to the new type of human action and addressed to the new type of agency that operates it might run thus: 'Act so that the effects of your action are compatible with the permanence of genuine human life'; or expressed negatively: 'Act so that the effects of your action are not destructive of the future possibility of such life.'" (Jonas 1984, p. 11) Note how ethical concerns have significantly changed: in the engineered age, having too much success in our technological acts is a legitimate concern, perhaps one that overshadows concerns about failures.

When reflecting on the changed nature of human action, Jonas concludes that "the changed nature of human action calls for a change in ethics as well: this is not merely in the sense that new objects of action have added to the case material on which received rules of conduct are to be applied, but in the more radical sense that the quantitatively novel nature of certain of our actions has opened up a whole new dimension of ethical relevance for which there is no precedent in the standards and canons of traditional ethics." (Jonas 1984, p. 1) If Jonas is right, ethicists are derelict if they only retread historical arguments. To put it another way, ethicists have work to do in this engineered age.

Another call for a new approach to ethics arises in Günther Anders's reflection on the atomic age. He also offers a "commandment" that has much in common with Kant's categorical imperative but responds to the interrelationship between our things and ourselves: "Have and use only those things, the inherent maxims of which could become your own maxims and thus the maxims of a general law." (Anders 1961, p. 134) In a world where devices are ubiquitous, the world engineering has provided, human interactions are almost always mediated by devices, which often structure these interactions according to their non-human properties: in this world, Anders's commandment makes sense and matters. Anders cites two new tasks, incumbent on everyone but especially appropriate for philosophers, namely, to develop a more expansive moral imagination and an extended sense of time than has developed to date in the history of philosophy. Seen in this light, technologies developed by engineers have animated ethics, necessitating new developments and exposing the limitations of dogmatic acceptance of inherited ethical claims.

In addition to prompting a reconsideration of desires and their satisfaction and expanding the space of ethics from an investigation into transcendental conditions or consciousness, modern engineering has brought into focus the political and ecological aspects of life. These aspects were often neglected as ethics became increasingly centered on the individual in the modern period. While these trends were not universal, take the signal contributions of Immanuel Kant and G. W. F. Hegel. Kant had accepted the expansive early Greek understanding of ethics as the study of free human behavior, distinct from physics, the study of objects that behave according to necessity, and logic, the study of relationships that do not depend on any objects. The political is always present yet always in the background of his ethical reflections, and his emphasis on human rationality excluded non-humans as moral subjects. Hegel's inward turn, toward reflexive consciousness, starts with the individual subject in a radical departure from the Aristotelian and Scholastic sense of the human as a political animal, one meant for the *polis*.[5]

The engineered world and its accompanying modifications of the non-human world serve as a correction to the narrowing scope of ethics because their language is, at it were, the environment that surrounds us, which, unlike a book on ethical theory, need not be picked up and cannot be put away. Just as Thomas Hobbes's mechanical body politic in *Leviathan* corresponded to the technological state of the art at the time, it should be no surprise that Bruno Latour's development of actor-network theory arose as radio, television, and telephone networks became increasingly common (Latour 1987). As Latour argued, the view of individuals as atomistic, independent agents, long the focus of ethicists, overlooks the connections and responses that meet can diminish or amplify one's agency. Thanks to social networking technologies such as Facebook and Twitter becoming a dominant cultural paradigm, his arguments have increased purchase. Along similar lines, Albert Borgmann's influential critique of devices, which disrupt social, spatial, and temporal continuities and result in a diminished existence where use and convenience is elevated over excellence, is both easier to imagine and more persuasive when devices are widely present (Borgmann 1984, 2006). Likewise, Langdon Winner's argument that Robert Moses's racist intentions were impressed into the New York City transportation infrastructure, regardless of its accuracy, is more believable when artifacts and systems are as developed as they were (Winner 1980).

The work of Latour, Borgmann, and Winner highlights the shortcoming of ethics as myopic individualism by arguing that free will and autonomy should be complemented

[5] Whatever its other shortcomings, it is worth noting that utilitarianism addresses these problematic tendencies, at least to a degree.

by a study of the decisions and actions of others, who have left us or provide our tools, habits, and devices. Jonas and Latour, among others, have also paid considerable attention to the non-human, non-artifactual environment. Jonas argues that amplified and refined nature of human action requires breaking through the "anthropocentric confinement" of previous ethics, one that alters the conception of nature and forces a reconsideration of "the doctrine of being, that is, metaphysics, in which all ethics must ultimately be grounded" (Jonas 1984, p. 8). In Latour's theory, actors can include other humans, artifacts, and non-human organisms. An ethics of human action, then, must include complementary political and environmental dimensions.

That the academic specialization of environmental ethics has developed over the same time period as philosophy of technology is no coincidence. The growing encroachment of human development, propelled by engineering, and its consequences, including pollution, has drawn attention to the value of nature, which is now no longer a threat to humans that must be overcome, but something that should be protected, and, according to many, something that possesses some intrinsic value.

The reconsideration of the relationship between humans and non-human organisms can be seen not only in the development of environmental ethics but also in ecology, the science most closely connected to it. Some of the most significant advances in the field over the last 50 years were made possible by the development of ecosystem sciences, whose origin can be traced to same origin that animated Anders's ethical reflections: the International Biological Program, which was funded in part by the U.S. Atomic Energy Commission (McIntosh 1985, pp. 215–217). In this program, the ecological and technological were tightly coupled. Ecological knowledge multiplied as movements of energy and matter through ecosystems could finally be traced using engineering technology, which allowed scientists to quantitatively measure these movements. The reintegration of humans as part of the environment, rather than as an anthropocentric exception, is also becoming more common, as is shown in a Mark J. McDonnell and Steward T.A. Pickett's *Humans as Components of Ecosystems* (McDonnell and Pickett 1993), Luisa Maffi's *On Biocultural Diversity: Linking Language, Knowledge, and the Environment* (Maffi 2001), and the shift from long-term ecological research to long-term socioecological research (Haberl et al. 2006).

Taken together, these new understandings of means and ends, of desire, of politics and environmental ethics necessitate a new understanding of human flourishing. Kate Soper has developed an updated understanding of the "good life" as "alternative hedonism" (Soper 2009). The good life, she argues, must acknowledge the unintended consequences of consumption and affluence. Time and energy spent in acquisitive tasks comes at the expense of personal relationships. In addition to its environmental impacts, a jet-setting mentality and energy-intensive lifestyle also weakens communal bonds. Increasing inequality results in community strife. Soper's conception of the good life is a necessary improvement to the Aristotelian understanding of *eudaimonia*.

Sustainability concerns prompted Edward Spence (2012) to develop an innovative understanding of epicureanism that is aligned with sustainability by harnessing

Epicurus' original trifold taxonomy of desires, which has similarities to Girard's division between needs and desires. Epicurus separated desires that are natural and necessary, which in Girard's taxonomy would be needs, from those that are natural but unnecessary and those that are unnatural and unnecessary. The second and third kinds of desires, Spence argues, should only be pursued when they are sustainable, even for hedonists. Spence has limited faith in the masses to be elevated epicureans, and he argues that political institutions will need to nudge or coerce their members to focus on desires in the first category and limit the pursuit of those that fall in the second or third.

A similar challenge to the mindset of development has arisen in the work on the capabilities approach, where the work of economist Amartya Sen is intimately related with Martha Nussbaum's ideas (Johnstone 2012). While Sen's work has forced a reconsideration of how development should be evaluated, Nussbaum's philosophical approach develops an expanded understanding of human abilities and commonly shared potential, and it carefully considers the characteristics of an environment in which these abilities can be exercised and potential realized.

16.4 Conclusion

The preceding inquiry has organized and developed insights that depend, at least in part, on engineering and the engineered world. These insights arise from reflection on the nature of engineering and technological action, its social and economic consequences, and philosophical advances made in the wake of these developments. It has sought to articulate the debt that ethics owes to engineering based on what the latter has contributed to the former.

The word "owes" can also be interpreted as an obligation to give someone or something what it rightly deserves.[6] Following this line of reasoning reveals other aspects of the relationship between ethics and engineering. Ethics owes engineering respect as a field of inquiry whose active mode differs in nature from it. In philosophical efforts, a contemplative mode of reading and thinking is naturally paired with an active mode of dissemination through teaching and publishing. There seems to be a centripetal force that narrows the courage of philosophers to adopt a critical attitude that revolves around prohibitions. The active expression of engineering, while determined to a degree by negative "prohibitions" found in the laws of nature and inherited engineering practice, is essentially constructive and substantive, even in situations of risk. Ethics should respect the difference in the kind of action that engineering is, and it would do well to imitate engineering in this sense. Moreover, if it is to have optimal influence in engineering, it may need to.

Ethics also "owes" engineering continued attention for two reasons. One, the status quo does not hold for engineering systems: entropy is always at work, and they require periodic, if not continual, maintenance. To the degree that ethics aims

[6] I am indebted to Diane Michelfelder for this insight.

to increase the space of human freedom, such concerns require attention. Two, the preceding reflections have been restricted to engineering as applied to the non-living world. Advances in genetic engineering have the potential to dramatically change our environmental conditions and concerns, and such advances are at the initial stages of being applied to humans. While there are a number of "bioconservatives" working against such progress, there is also a vocal and well-funded group of cheerleaders, including Max More and Nick Bostrom, advocating an ambitious effort under the rubric of transhumanism.

In addition to such physical alteration, the technological development that engineering makes possible has made possible social engineering, with possibilities of surveillance and ideological impression through a ubiquitous media that has the potential to dramatically alter the space of human freedom, and, if nothing more, to force a reconsideration of our relationships with others, especially those with political or economic power.

Finally, the contemplative mode of ethics, combined with the recognition that means require effort and may not be escaped, can prompt engineers to think about how this complex of material changes and ethical innovations can lead to a reconsideration of good engineering. Efficiency in engineering design often focuses on limited waste of energy or materials, but a second order analysis of efficiency could assess the ratio between the energy and materials needed to perpetuate technological elements and their systems and the energy and materials directed toward non-instrumental goals. Such reflection could reshape the practice of engineering and the design of the infrastructure that we use. Moreover, it may serve as a check to attempts to ask engineering to provide far beyond what it can, for many in modern society seem to hope its products will replace divinity and provide transcendence (Dupuy 2013, p. 2).

The considerable and continual reshaping of the world—and humans ourselves—enabled by engineering requires philosophical reconsideration of critical concepts including humanity, nature, and technology, which perhaps previously could have been unconsciously accepted. At the same time, the constructive dimension of engineering action shows the inadequacy of ethics as simply critique. Technical decisions in the construction and use of engineering and technology ultimately require a determination of what should be done, judgments that are different in nature from identifying what cannot or should not be done. In the same way, a properly critical ethics of technology must offer constructive, substantial guidance on the good life and how to act and not resign itself to delineating epistemic limitations and dispositions, acts, or consequences that are to be avoided.

References

Anders, G. (1961). Commandments in the atomic age. In C. Fatherly & G. Anders (Eds.), *Burning conscience* (New York: Monthly Review Press). Reprinted in C. Mitcham & R Mackey (Eds.), *Philosophy and technology: Readings in the philosophical problems of technology*. New York: Free Press, 1972.

Aristotle. (2002). *Nicomachean Ethics* (J. Sachs, Trans.). Newburport: Focus Publishing.
Borgmann, A. (1984). *Technology and the character of contemporary life*. Chicago: University of Chicago Press.
Bormgann, A. (2006). *Real American ethics: Taking responsibility for our country*. Chicago: University of Chicago Press.
Dupuy, J.-P. (2002). Detour and sacrifice: Ivan Illich and René Girard. In L. Hoinacki & C. Mitcham (Eds.), *Challenges of Ivan Illich: A collective reflection* (pp. 189–204). Albany: State University of New York Press.
Dupuy, J. -P. (2013). *Mark of the sacred* (M. B. Debevoise, Trans.). Stanford: Stanford University Press.
Ellul, J. (1964). *The technological society* (J. Wilkinson, Trans.). New York: Vintage Books.
Feenberg, A. (1995). *Alternative modernity: The technical turn in philosophy and social theory*. Berkeley: University of California Press.
Fuller, S. (2009). *Sociology of intellectual life: The career of the mind in and around the academy*. London: Sage.
Girard, R. (1966). *Deceit, desire, and the novel: Self and other in literary structure* (Y. Freccero, Trans.). Baltimore: Johns Hopkins Press.
Girard, R. (1986). *The scapegoat* (Y. Freccero, Trans.). Baltimore: Johns Hopkins Press.
Guthrie, W. K. C. (1962). *History of Greek philosophy: The earlier Presocratics and the Pythagoreans* (Vol. 1). Cambridge: Cambridge University Press.
Haberl, H. et al. (2006). From LTER to LTSER: Conceptualizing the socioeconomic dimension of long-term socioecological research. Ecology and Society 11(2). http://www.ecologyandsociety.org/vol11/iss2/art13/. Accessed 13 Jan 2107.
Institution of Civil Engineers. (1828). Royal charter. In *Charter, supplemental charters, by-laws, and list of members of the Institution of Civil Engineers*. London: Institution of Civil Engineers. As cited in Mitcham "Philosophical Inadequacy", p. 345.
Johnstone, J. (2012). Capabilities and technology. In P. Brey, A. Briggle, & E. Spence (Eds.), *The good life in a technological age* (pp. 77–90). New York: Routledge.
Jonas, H. (1984). *The imperative of responsibility: In search of an ethics for the technological age*. Chicago: University of Chicago Press.
Latour, B. (1987). *Science in action*. Cambridge, MA: Harvard University Press.
Maffi, L. (Ed.). (2001). *On biocultural diversity: Linking language, knowledge, and the environment*. Washington: Smithsonian Institute Press.
McDonnell, M. J., & Pickett, S. T. A. (Eds.). (1993). *Humans as components of ecosystems: The ecology of subtle human effects and populated areas*. New York: Springer.
McIntosh, R. (1985). *Background of ecology: Concept and theory*. New York: Cambridge University Press.
Mitcham, C. (2009). A philosophical inadequacy of engineering. *The Monist, 92*(3), 339–356.
Mitcham, C. (1994). *Thinking through technology: The path between engineering and philosophy*. Chicago: University of Chicago Press.
Mitcham, C., & Briggle, A. (2012). Theorizing technology. In P. Brey, A. Briggle, & E. Spence (Eds.), *The good life in a technological age* (pp. 35–51). New York: Routledge.
Soper, K. (2009). Introduction: The mainstreaming of counter-consumerist concern. In K. Soper, M. Ryle, & L. Thomas (Eds.), *The politics and pleasures of consuming differently* (pp. 1–21). Basingstoke: Palgrave Macmillan.
Spence, E. (2012). Consumption and sustainability: A neo-epicurean approach to a sustainable good life in a technological age. In P. Brey, A. Briggle, & E. Spence (Eds.), *The good life in a technological age* (pp. 168–180). New York: Routledge.
Verbeek, P.-P. (2011). *Moralizing technology: Understanding and designing the morality of things*. Chicago: University of Chicago Press.
Winner, L. (1980). Do artifacts have politics. *Daedalus, 109*, 121–136.

Chapter 17
New Trends in Engineering Ethics – A Japanese Perspective

Hidekazu Kanemitsu

Abstract Engineering ethics education is widespread in Japan and is offered by many engineering academic and professional societies. Two elements, namely, the use of an analogy between ethical and design problems and the use of specific methods for ethical decision-making, like a seven-step guide, are the characteristics of current Japanese engineering ethics education. This form of ethics education helps students view ethical problems in a familiar light and elicits realistic solutions from them. However, this method lacks the normative sources to evaluate the moral design of problems, and sometimes fails to motivate students to learn engineering or/and engineering ethics by emphasizing on the negative aspects of engineering. To compensate for this lack, scholars expressed the need for introducing "aspirational ethics" in addition to the existing "preventive ethics." Apparently, virtue ethics and positive psychology provide fruitful insights in realizing this aim. However, one needs to be wary of being too optimistic or too pessimistic. The author believes that the philosophy of technology can provide a new framework for engineering ethics. Especially, mediation theory by Peter-Paul Verbeek proposes a third way that is neither too pessimistic nor too optimistic by claiming that we need to "accompany" technological development, not aiming to merely reject or accept a new technology; the theory also provides a perspective of normative sources to evaluate moral design. Future engineering ethics would do well to incorporate this philosophy of technology further.

Keywords Engineering ethics · Aspirational ethics · Philosophy of technology · Postphenomenology · Technological mediation

H. Kanemitsu (✉)
Kanazawa Institute of Technology, Nonoichi, Ishikawa, Japan
e-mail: kane@neptune.kanazawa-it.ac.jp

A. Fritzsche, S. J. Oks (eds.), *The Future of Engineering*, Philosophy of Engineering and Technology 31, https://doi.org/10.1007/978-3-319-91029-1_17

17.1 Current Situation of Engineering Ethics in Japan

Engineering ethics education has become widespread in Japan because with the internationalization of engineering education, the Japan Accreditation Board for Engineering Education (JABEE) established in 1999 specified the criteria for education in engineering ethics as follows. The criteria were set with the aim to develop

> an ability to understand the effects and impact of professional activities on society and nature, and the professional's social responsibility. (JABEE 2015)

For example, Sato and Harada (2005) surveyed the reactions of institutes to the criteria suggested and reported that approximately 76% of the institutes opened an engineering ethics course in some form in the early 2000s.[1]

It is a fact that the need for engineering ethics arose after various accidents, for example, the Tokaimura nuclear accident (1999), the cover-up of recalls by MITSUBISHI MOTORS (2000, 2004), and so on. However, we cannot deny the influence of internationalization on engineering education, especially the establishment of JABEE and the criteria for engineering ethics as a part of it. In other words, Japanese engineering ethics emerged not only from the need within the country but from external pressure as well.

17.1.1 Ethics as Design

In the early stages, when engineering ethics was introduced in Japan, the system and the content of engineering ethics were imported from other countries, especially the United States. For example, the method proposed by the American philosopher Caroline Whitbeck is frequently used in the field of engineering ethics in Japan. For example, the Japanese Society for Engineering Education (JSEE) has set up a Research Committee on Engineering Ethics, which surveys the syllabi of engineering ethics related courses in Japan, and it defines an ability to "design" one's conduct to resolve ethical problem as crucial, following Whitbeck (Kobayashi and Fudano 2014).

Whitbeck criticizes the existing ethics and applied ethics saying that they ignored the "agent perspective." In other words, she strongly criticizes dealing with ethical problems from a "judge perspective," which she considers is excessively abstract and rationalistic (Whitbeck 1998, p. 72). For ethics education, Whitbeck emphasizes the benefits of using the analogy between ethical and design problems, and she insists on the need for an engineering ethics education based on the "agent perspective" (Whitbeck 1995, pp. 299–300).

[1]According to the survey, 47.9% of the institutions started a new course in engineering ethics, 15.5% already had a corresponding course, and 12.7% modified the existing courses and started a new course. Therefore, 76.1% of the institutions now offer a course in engineering ethics in some form or the other.

Whitbeck aims not only at acquiring moral knowledge but also at developing the moral skills needed to respond suitably to ethical problems, which are grasped through an analogy with design problems.

She identifies several features of design problems that are significant for ethical problems (Whitbeck 1995, p. 302; 1998, pp. 57–60):

1. For interesting or substantive engineering design problems, there is rarely, if ever, a uniquely correct solution or response, or indeed, any predetermined number of correct responses.
2. Even though no unique correct solution exists, some possible responses are clearly unacceptable and some solutions are better than others: there are wrong answers even if there is no unique right answer.
3. Solutions may have advantages of different sorts, such that where there are two candidate solutions, neither may be incontrovertibly better than the other.
4. A proposed solution must do all of the following:
   - Achieve the desired performance or end
   - Conform to specifications or explicit criteria for this act
   - Be reasonably secure against accidents and other miscarriages that have severe negative consequences
   - Be consistent with existing background constraints, for example, ensure that no one's human rights are violated and that infringement of other rights is minimized.

Whitbeck is able to relate some lessons from design problems to ethical problems.

First, the unknowns and uncertainties in a situation should be considered. Although ambiguities and uncertainties are often underemphasized in ethics, in cases where crucial ambiguities cannot be fully resolved early in the problem situation, there is a need to consider the ambiguity as a defining characteristic of the situation (Whitbeck 1998, p. 62).

Second, the development of possible solutions is distinct from the definition of the problem and may therefore require more information. This is one of the features that distinguishes ethical problems from formal decision problems, which have multiple-choice answers and in which the nature of the problems is fully defined. In real life, however, we need to develop possible solutions; therefore, ethical problems usually need to be dealt with using open-ended statements rather than seeking single solutions from multiple-choice answers. Furthermore, before proposing the possible solutions, agents must frequently clarify the problem. In other words, the agents need to accumulate additional information to develop possible solutions. "One of the important characteristics of a responsible or wise response to a practical problem is appropriate investigation of a problem before attempting to solve it" (Whitbeck 1998, p. 63).

Third, it is often important to begin by simultaneously pursuing several possible solutions in order to avoid feeling helpless when faced with insuperable obstacles and avoid expending too much energy on too broad span of activities. This lesson concerns acting under time constraints. In addition to keeping options open, agents should avoid investing their energy too broadly. "Pursuing several options contrasts with representing the ethical problem as a static situation with static solutions; the problem becomes simply one of selecting the right alternative and doggedly pursuing it" (Whitbeck 1998, p. 66).

Finally, the dynamic character of problem situations implies that both the problem situation and the individual's understanding of it are likely to change and develop over the course of time. "When the dynamic character of the ethical situation is neglected, people often mistake doing the wrong thing and then making the best of the bad situation with taking an action that is justified in some circumstance" (Whitbeck 1998, p. 66).

Using the analogy between ethical and design problems is useful when people aim for education that helps students to develop the ability to deal with real-life ethical problems. By using this analogy, students can think of ethical problems in a practical manner, without developing any aversion or fear toward these problems.

In fact, this way of thinking is influential in Japanese engineering ethics education.

17.1.2 Seven-Step Guide to Ethical Decision-Making

To deal with real-life ethical problems, Japanese engineering ethics education uses specific methods as well. A seven-step guide, developed by the American philosopher Michael Davis, is well known in Japan as a concrete guide to ethical decision-making (Davis 1999, pp. 166–167):

1. *State the problem*. For example, "there's something about this decision that makes me uncomfortable" or "do I have a conflict of interest?"
2. *Check the facts*. Many problems disappear upon closer examination of the situation, while others change radically.
3. *Identify relevant factors*. For example, persons involved, laws, professional code, other practical constraints (e.g. under $200).
4. *Develop* [*a*] *list of options*. Be imaginative, try to avoid "dilemma"; not "yes" or "no" but whom to go to, and what to say.
5. *Test the options*. Use such tests as the following:

   - harm test: does this option do less harm than alternatives?
   - publicity test: would I want my choice of this option published in the newspaper?
   - defensibility test: could I defend choice of option before a Congressional committee or committee of peers?
   - reversibility test: would I still think choice of this option good if I were adversely affected by it?
   - colleague test: what do my colleagues say when I describe my problem and suggest this option as my solution?

- professional test: what might my profession's governing body or ethics committee say about this option?
- organization test: what does the company's ethics officer or legal counsel say about this?

6. *Make a choice* based on steps 1 to 5.
7. *Review steps* 1 to 6. What could you do to make it less likely that you would have to take such a decision again?
   - Are there any precautions you can take as an individual (announce your policy on question, change job, etc.)?
   - Is there any way to have more support the next time?
   - Is there any way to change the organization (for example, suggest policy change at the next departmental meeting)?

As you can see, this practical guideline to ethical decision-making is very useful in education by the so-called case-method. When we deal with engineering ethics cases, this method will help analyze the problem practically and logically, without allowing one's feeling to influence decision-making. Using the case-method in engineering ethics education is a popular approach in Japan, and several institutes have adopted this case-method using the seven-step guide. For example, the Kanazawa Institute of Technology (KIT), one of the leading universities in the field of engineering ethics in Japan, has been using this method for more than 10 years, revising the original version.[2]

17.1.3 Criticism

These two elements, namely the use of analogy between ethical and design problems and the use of a specific method for ethical decision-making, are the characteristics of the current engineering ethics education in Japan.

For a practical implementation of ethics education, institutes use existing cases as examples, such as the National Society of Professional Engineers (NSPE) in the United States provides various ethics resources, including videos on engineering ethics cases, and a few Japanese universities too have added their own cases recently.[3]

[2] KIT introduced "Science and Engineering Ethics" as a compulsory course for Junior in 2006, and since then have continued using this method. Research Committee on Engineering Ethics, JSEE, also adapted the Seven-Step Guide in their model syllabus to enhance the ability of students to make ethical judgements and solve problems (Kobayashi and Fudano 2014).

[3] For example, KIT made a video *Solar Blind,* which was expected to contribute to discussion in the Engineering Ethics class as part of a support program for distinctive university education (by the Ministry of Education, Culture, Sports, Science and Technology) and the Yumekobo project (Student project at KIT) in 2009. The author will elaborate on this case in Sect. 17.4.

This method of learning engineering ethics has several advantages since students can learn to deal with ethical problems the same way they would approach design problems and propose realistic solutions. The agent perspective puts the students in the position of an engineer faced with ethical problems such that they can empathize with the engineer and understand his or her dilemma. In addition, they can provide concrete solutions by using the practical know-how of the Seven-Step Guide.

However, it has its drawbacks, too. Kohji Ishihara, a Japanese philosopher, points out the danger of overemphasis on the agent perspective (Ishihara 2002, pp. 234–235):

> To overemphasize the "complexity (of the situation)" and "agent perspective" can pose an ethical threat. If we consider "constraints" in each situation seriously, it appears possible that most unethical behaviors can be justified. Therefore, ethicists should not be too compassionate about agents in each situation; instead, they should take a bold stand. From the vantage point of the judge, ethicists should boldly judge actions that the agent had to take and should promote the establishment of the guidelines that agents can use in similar situations. (My translation)

The agent perspective recommends adopting a pragmatic approach and considers ethical problems as practical problems. However, by rejecting rational foundationalism and other justificatory approaches, it lacks the necessary normative sources to evaluate the norms and values at stake, as well as the moral design itself. Therefore, even if we agree on the significance of the view of ethical problems as design problems and on moral agency as a skill, we need to consider the normative sources to evaluate moral design.[4]

17.2 New Trends in the Field of Engineering Ethics

In order to consider the moral design, we need another perspective on engineering ethics. In fact, there are some new trends in the field of engineering ethics. Before examining the kind of perspective adequate to consider moral design, let us overview a new aspect of engineering ethics.

17.2.1 Introduction of Aspirational Ethics and Virtue Ethics

Traditional engineering ethics has mostly focused on prevention of harm to the public. Harris et al. call this type of ethics as "preventive ethics" (Harris 2013; Harris et al. 2013). Preventive ethics attempt to prevent harm resulting from professional misconduct or danger from engineering products or/and engineering processes. Famous cases used as examples in engineering ethics, such as *DC-10*, *Ford Pinto*, *Challenger*, deal with such issues.

[4] For more details about the agent-centered approach, see Kanemitsu (2008).

Table 17.1 Aspirational ethics and preventive ethics

	Aspirational ethics	Preventive ethics
Direction	Positive	Negative
Aim	To promote human well-being	To not cause harm to the public
Element	Virtues	Duties and utilities
Effect	Motivate engineers	Demoralize engineers

To realize public safety and health, it is crucial to prevent harm and danger. However, preventive ethics tend to emphasize on the negative aspects of engineering. In fact, negative cases are used to reprimand engineers, and the code of ethics mostly consists guidelines for preventing harm and danger.

In contrast, Harris et al. propose "aspirational ethics" to emphasize a more positive dimension of engineering. Aspirational ethics attempt to encourage or motivate engineers or engineering students to use their professional ability to promote human well-being. Aspirational ethics are not only *motivational* but also *discretionary*. Engineers have a considerable range of freedom in how they promote public welfare (Harris et al. 2013, p. 18) (Table 17.1).

Of course, we do find a positive moral prescription in a rule such as "hold paramount public welfare." However, it provides little direction for conduct. Therefore, Harris et al. consider that it is more appropriate to use vocabulary of professional character for talking about aspirational ethics (Harris et al. 2013, p. 19). They use expressions related to the professional character and pick up a case of "good engineer" instead of a "disaster case."

In this context, Harris believes that virtue ethics is a more appropriate vehicle for expressing some aspects of engineering professionalism. First, there are "technical virtues or technical excellences." Technical excellences are capacities and sensitivities closely related to the technical side of engineering, and they cannot be expressed well in rules. "Sensitivity to risk" is an example (Harris 2008, pp. 158–159). Second, there are "non-technical excellences," toward which good engineers should aspire, for example, traditional virtues such as honesty, professional integrity, and so on. However, Harris considers three non-technical excellences especially important in the professional virtue portrait, or professional character, of a good engineer: "Techno-social sensitivity (awareness of the social context of technology)," "respect for nature," and "commitment to the public" (Harris 2008, pp. 159–162). Certainly, these virtues cannot be adequately represented by rules, especially not ones that list the dos and don'ts.

Stovall (2011) also believes that a virtue ethics approach can enrich pedagogy of engineering ethics courses and help foster a sense of pride and responsibility among students. In addition to the virtues that Harris pointed out, Stovall emphasizes the virtue of "professional self-awareness." This is an analogue to Aristotle's "phronesis" or "practical wisdom." Stovall argues that the virtuous professional is the successful professional, just as the virtuous life is the happy life for Aristotle.

A question then arises, how do we teach these virtues? Although the university is not the whole world, the goal of education is to prepare an individual to take up the active duties of life (Stovall 2011, p. 129). Stovall points out that teaching the history of their profession, its roles at key points in the development of society, biographical sketches of critical figures of the profession, and stories of influential organizations, films, and corporations can encourage students to view their professions as instrumental to both the development and maintenance of societies, and validate their own particular social identity. In fact, identifying a subject of study with the subject learning is an important methodological driving force for the virtue ethics approach (Stovall 2011, pp. 130–131).

Harris believes that humanities and social sciences play an important role in promoting professional virtues. For example, the philosophy of technology and Science and Technology Studies (STS) can promote techno-social sensitivity. Moreover, philosophical writers and authors in general can promote the virtue of respect for nature. "[T]he humanities and social sciences are not merely peripheral to the professional education of engineers. Rather, they are the best means of training engineers in some of the professional virtues that comprise the professional virtue portrait of the good engineer" (Harris 2008, p. 163).

In fact, the "HONDA CVCC engine" case is used in Japan to encourage students.[5] It would certainly help if such cases can be collected to introduce the insights of humanities and social sciences into engineering ethics education.

17.2.2 Use of Positive Psychology

Some researchers have attempted using the insights from "positive psychology" for engineering ethics education. Han (2015) attempts to apply virtue ethics and positive psychology to engineering ethics education and to develop a new conceptual framework for more effective education: He believes that a theoretical framework of virtue ethics and positive psychology can contribute to motivating students toward self-improvement by connecting the notion of morality and *eudaimonic* happiness. He suggests concrete educational methods: moral modeling in engineering and activity-based education such as the use of Institutional Review Board (IRB) processes, a mock student research-integrity committee, and so on.

Jun Fudano, one of the leading professors in Japanese engineering ethics education, also tries to develop an engineering ethics education with a "well-being" model (Fudano 2015). His emphasis is on "good work" being an important element of engineering ethics education; he also considers the insight from happiness studies, especially from the "PERMA" model by Seligman (2011). Seligman argued that happiness comprises P-Positive Emotion, E-Engagement, R-relationship, M-Meaning, and A-Achievement (or Accomplishment). It is true that the code of ethics stipulates "hold paramount public welfare," but does not elaborate on what

[5]For details of the "HONDA CVCC engine" case, see HONDA. Introducing the CVCC. http://world.honda.com/history/challenge/1972introducingthecvcc/index.html. Accessed December 3, 2016.

exactly is welfare. To consider this, Fudano uses the PERMA model: Individual happiness comes from the belief that we lead a meaningful life. The greatest happiness, however, is when we work for and contribute to something beyond ourselves. Therefore, an engineer is able to not only bring welfare to society (social happiness) but also achieve individual happiness by doing his/her work ethically. Fudano believes that this model based on well-being in engineering ethics is more effective than any other model, such as the social contract model, traditional professional responsibility model, and so on. He then tries to measure the effect of this model with tools such as the PERMA Profiler and the SWLS (Satisfaction with Life Scale).

17.3 Need for Philosophy of Technology

It is true that in engineering ethics education, we need to encourage or motivate engineers, or engineering students, and not merely emphasize on prevention of disasters. Using positive psychology seems to an effective way to achieve that goal. However, some people consider this to be a highly optimistic solution: they feel that the model naively identifies subjective happiness with happiness itself.

Han himself admits that "although the mere pursuit of happiness emphasized by mainstream positive psychology, which does not have a firm moral philosophical foundation, can be helpful to promote subjective happiness, it cannot necessarily contribute to objective well-being and human flourishing, which needs moral justification from the objective and universal perspective" (Han 2015, p. 445). Therefore, he introduced the concept of "authentic happiness" (Seligman 2011): not only subjective happiness, which is akin to hedonic feelings, but also meaningfulness of life is necessary to achieve authentic happiness. Fudano regards this meaningfulness in the life of engineers as crucial for their happiness and tries to combine it with social goodness.

However, an overemphasis on the engineer's happiness will easily lead to less emphasis on the negative aspects of engineering toward society. For example, how can we talk about authentic happiness of engineers after the Fukushima Daiichi nuclear disaster? Although the usefulness of the well-being model of engineering ethics is acknowledged, there is also the danger of abuse that can deceive an engineer into believing that he/she is doing good things for the society and hence can feel happy.

We need to avoid being too pessimistic or too optimistic regarding engineering ethics. Insight from the philosophy of technology can play an important role here in creating a new framework that takes into consideration normative sources for evaluating the moral design. This study should take into account the notion of "technological mediation" introduced by Peter-Paul Verbeek.

Verveek adopts the postphenomenological approach that Don Ihde advocated. According to the postphenomenological point of view, the subject and object constitute each other:

> [P]ostphenomenology can be viewed as an offshoot of phenomenology that is motivated by the postmodern aversion to context-independent truths and the desire to overcome the radical separation of subject and object, but that does not result in relativism. From the postphenomenological perspective, reality cannot be entirely reduced to interpretations, language games, or contexts. (Verbeek 2005, p. 113)

The key concept of his approach is "mediation." According to Verbeek, technological artifacts take part in the human act and decision-making by mediating human perception and human praxis. These analyses from technological mediation have significant implications for engineering design: "Designers should focus not only on the functionality of technologies but also on their mediating roles. The fact that technologies always mediate human actions charges designers with the responsibility to anticipate these mediating roles" (Verbeek 2006, pp. 377–378). This implies that a designer should design a product "not only on the basis of the desired functionality but also on the basis of an informed prediction of its future mediating role and a moral assessment of this role" (Verbeek 2006, p. 372).

From the perspective of mediation theory, designing should be regarded as a form of "materializing morality" (Verbeek 2006). Accordingly, one needs to give more serious consideration to the moral dimension of designing.

Verbeek' recently work talks about the way of "moralizing technology":

> Rather than working from an external standpoint vis-à-vis technology, aiming only to either reject or accept a new technology, the ethics of technology then aims to *accompany* technological developments [...], experimenting with mediations and looking for ways to discuss and assess how these mediations could fit with the way humans live. (Verbeek 2011, p. 95)

I think this is the crucial point of future engineering ethics because this implies a third way that is neither too pessimistic nor too optimistic. Verbeek also presents the real task of the ethics of "technology accompaniment":

> Its primary task is to equip users and engineers with adequate frameworks to understand, anticipate, and assess the quality of the social and cultural impacts of technologies. (Verbeek 2011, p. 165)

To achieve this task, of course, we need insight from the philosophy of technology, especially of mediation theory. Therefore, the insight from the philosophy of technology needs to be introduced into engineering ethics to consider the moral design and to avoid being too pessimistic or too optimistic.

17.4 Practical Example

A practical example applying the above discussion will help one understand the current situation and the future direction of engineering ethics.

17.4.1 Case Example

As mentioned earlier, the case method is considered effective in engineering ethics education. Engineers must make decisions about a particular design that will affect the lives and well-being of many people in a specific case. "The study of cases helps students understand that professional ethics is not simply an irrelevant addition to professional education but, rather, is intimately related to the practice of engineering" (Harris et al. 2013, pp. 20–21).

Gilbane Gold is a famous hypothetical case developed by NSPE and commonly used in Japan. Although it is a good case, its "American" characteristics do not fit perfectly with Japanese contexts such as the social status of its engineers, its business culture, and so on, which are different from those in the United States. Therefore, some faculty and students of the Kanazawa Institute of Technology, in 2009, developed a new case, *Solar Blind*, a hypothetical one to fit the Japanese context. Using both *Gilbane Gold* and *Solar Blind*, we are now able to introduce global contexts effectively in engineering ethics education.

A brief outline of *Solar Blind* is as follows: A young Japanese engineer is facing an ethical dilemma as the leader of a development team for an innovative product, *Solar Blind*, an indoor, easy-to-use solar power generation system. His team is in charge of designing a control unit, the key component of the system. On encountering a problem concerning overheating, he is unable to find a solution to rectify it, despite all efforts, before the release date of the product into the market. He suggests postponing its release date, but the company opposed his suggestion.

17.4.2 Teaching Example

The purpose of engineering ethics education is not to help students analyze an ethical problem from the "judge perspective," but to help them find concrete solutions from the "agent perspective." Therefore, students are asked to answer what they would do in the place of the young engineer in the *Solar Blind* case.

Before studying the case, they have learned the analogy between design and ethical problems, so they are expected not to hesitate confronting ethical problems. In addition, they are expected to "design" their conduct based on the Seven-Step Guide.

Usually, the class is organized into several small groups to discuss the case according to the Seven-Step Guide. All the facts and relevant factors are examined before developing the options for conduct considering the situation of accidents, stakeholders, specifications (limits and objectives), and so on. After the discussion, the students are given the assignment to find a solution for the case.

17.4.3 The Focus of the New Trends

Although the analogy between design and ethical problems lets students consider the ethical problems from a familiar and realistic perspective, some students hesitate to confront ethical problems exactly because of its reality. They say, "I can imagine the situation" or "I am scared to be an engineer," and so on, when they discuss the case.

This is where aspirational ethics come into play. When we study the case, in fact, we can encourage or motivate students to use their professional ability to promote human well-being. For example, we can emphasize that a correct decision made by the young engineer in the *Solar blind* case will bring happiness not only to the company and society, but to his own self as well.

However, an overemphasis on the engineer's happiness will easily being blind to the negative aspects of engineering. In fact, the most important thing the young engineer in *Solar Blind* must do is prevent accidents. We need to avoid being too pessimistic or too optimistic.

The author thinks that the philosophy of technology, especially the mediation theory by Verbeek plays an important role in creating a new framework, neither too pessimistic nor optimistic. Mediation theory considers the problem from a realistic perspective like the current engineering ethics, because it aims to "accompany" the technology. From the viewpoint of mediation theory, the focus of *Solar Blind* is to consider how the new product can mediate our life: Does it produce desirable results? Does it threaten people's autonomy? To what extent does it contribute to environmental problems? In the first place, what are the current environmental problems?

In this way, mediation theory will consider the change in our lifestyle or environmental issues in the *Solar Blind* case by understanding, anticipating, and assessing the quality of the social and cultural impacts of new technologies, and not consider merely how successful engineers can be developed. It expands the current engineering ethics by providing normative sources for evaluating moral design.

17.5 Conclusion

Current engineering ethics education in Japan adopts an agent-centered approach and education by case-method using specific methods like the seven-step guide. This is certainly effective in letting students consider ethical problems from a familiar and realistic perspective. However, it lacks consideration of normative sources to evaluate moral design and sometimes fails to motivate students to learn engineering or/and engineering ethics.

The concept of aspirational ethics is a supplement needed for the existing preventive ethics. Virtue ethics and positive psychology may provide fruitful insights to realize aspirational engineering ethics. However, we should be aware of the danger of being too optimistic or too pessimistic.

The philosophy of technology presents another new framework for engineering ethics. Especially, mediation theory by Peter-Paul Verbeek shows the third way that is neither too pessimistic nor too optimistic by claiming that we need to "accompany" technological development; this philosophy of technology also provides the perspective of normative sources to evaluate moral design.

To accompany the technology, we need to describe the art of technology. A descriptive inquiry is significant to accompany the technology leading to the expansion of conventional engineering ethics that tend to deal with only individual professional responsibility. This means that we need to teach engineers or engineering students not only about conventional issues but also about issues in the philosophy of technology. Future engineering ethics should incorporate the philosophy of technology further.

Acknowledgments This work was supported by JSPS KAKENHI Grant Number JP 16 K02143 and JP16K00976.

References

Davis, M. (1999). *Ethics and the university*. London: Routledge.

Fudano, J. (2015). What is the well-being of the ethical engineer?. *Journal of IEE Japan, 135*(5), 283–286. https://doi.org/10.1541/ieejjournal.135.283.

Han, H. (2015). Virtue ethics, positive psychology, and a new model of science and engineering ethics education. *Science and Engineering Ethics, 21*, 441–460. https://doi.org/10.1007/s11948-014-959-7.

Harris, C. E., Jr. (2008). The good engineer: Giving virtue its due in engineering ethics. *Science and Engineering Ethics, 14*, 153–164. https://doi.org/10.1007/s11948-008-9068-3.

Harris, C. E., Jr. (2013). Engineering ethics: From preventive ethics to aspirational ethics. In D. P. Michelfelder, N. McCarthy, & D. E. Goldberg (Eds.), *Philosophy and engineering: Reflections on practice, principles and process* (pp. 177–187). Dordrecht: Springer.

Harris, C. E., Jr., et al. (2013). *Engineering ethics: Concepts and cases* (5th ed.). Boston: Wadsworth.

Ishihara, K. (2002). The necessity of introducing external perspective into engineering ethics. In N. Kamo (Ed.), *Research materials for social philosophy* (pp. 226–237), I.

JABEE. (2015). *JABEE Common criteria for accreditation of professional education programs applicable in the years 2015*. http://www.jabee.org/english/evaluation_accreditation/documents/. Accessed 29 Nov 2016.

Kanemitsu, H. (2008). Agent-centered approach in engineering ethics: A consideration of the methodology of applied ethics. In K. Ishihara & M. Shunzo (Eds.), *Applied ethics: Perspectives from Asia and beyond. Center for applied ethics and philosophy* (pp. 97–105). Sapporo: Hokkaido University.

Kobayashi, Y., & Fudano, J. (2014). Commentary on learning and educational objectives of engineering ethics education. *Journal of JSEE, 62*(4), 81–87. (in Japanese).

Sato, Y., & Harada, S. (2005). Survey results on JABEE. *Journal of JSEE, 53*(3), 101–112. (in Japanese).

Seligman, M. (2011). *Flourishing: A visionary understanding of happiness and well-being*. New York: Free Press.
Stovall, P. (2011). Professional virtue and professional self-awareness: A case study in engineering ethics. *Science and Engineering Ethics, 17*, 109–132. https://doi.org/10.1007/s11948-009-9182-x.
Verbeek, P. P. (2005). *What things do: Philosophical reflections on technology, agency, and design*. University Park: Pennsylvania State University Press.
Verbeek, P. P. (2006). Materializing morality: Design ethics and technological mediation. *Science Technol Human Values, 31*(3), 361–380. https://doi.org/10.1177/0162243905285847.
Verbeek, P. P. (2011). *Moralizing technology: Understanding and designing the morality of things*. Chicago: University of Chicago Press.
Whitbeck, C. (1995). Teaching ethics to scientists and engineers: Moral agents and moral problems. *Science and Engineering Ethics, 1*(3), 299–308.
Whitbeck, C. (1998). *Ethics in engineering practice and research*. Cambridge: University Press.

Chapter 18
Nietzsche, Postphenomenology, and Nihilism-Technology Relations

Nolen Gertz

Abstract In part one of this paper I describe a tendency in technological production that I call the "leisure-as-liberation" model of technological design, and outline how it has evolved from Aristotle, to Marx, to today. In part two I show, by developing an outline of what I call "human-nihilism relations," how Nietzsche would criticize the idea that we could find liberation through leisure. In part three I provide a tour of Don Ihde's postphenomenology, in order to set up parallels between Nietzsche's human-nihilism relations and Ihde's human-technology relations. In part four I develop these parallels into a new set of relations that I call "nihilism-technology relations," in order to highlight how technologies can mediate nihilism and how nihilism can mediate technologies. I conclude with a reflection on how rehabilitating the concept of responsibility can help us to move away from this nihilistic "leisure-as-liberation" model of technological design.

Keywords Nihilism · Postphenomenology · Nietzsche · Ethics of technology · Responsibility

18.1 Leisure-As-Liberation

This paper will examine a particular trend in the design and use of technologies, a trend that I will call the "leisure-as-liberation" paradigm of technological design. Examples of this paradigm are becoming more and more ubiquitous, from the popularity of the "Netflix and chill" (Fusion 2015) model of consuming algorithmically-recommended entertainment, to Amazon's introduction of Dash Buttons for replacing household items without having to even log in to Amazon, as well as to Uber's development of driverless cars that will soon deliver us to desired locations without having to have any human interaction at all. In all such cases what we find are technologies that are designed to make real what engineers and designers

N. Gertz (✉)
Department of Philosophy, University of Twente, Enschede, The Netherlands
e-mail: n.gertz@utwente.nl

A. Fritzsche, S. J. Oks (eds.), *The Future of Engineering*, Philosophy of Engineering and Technology 31, https://doi.org/10.1007/978-3-319-91029-1_18

perceive to be our fantasy of being liberated from the chores of everyday life, the chores that prevent us from living the lives we truly desire.

This fantasy is best exemplified in a recent advertisement for iRobot's popular robotic vacuum cleaner, the Roomba (Boston Business Journal 2016). In the advertisement we see a perfect family in a perfect home on a perfect day. In the center of the frame we find a mother and daughter inside of a couch cushion fort, a father attempting to attack the fort, and a dog obliviously chewing on a toy. Off in the corner, barely still in the frame, we find the black, puck-like object known as the Roomba. There is no explanation of the image provided, as the message is clear. Roomba works so you can play. Or, to come back to my aforementioned theme, the technology takes on the undesired chore—whether it be a Roomba cleaning a house, Netflix choosing a movie, or a Dash Button doing our grocery shopping—so that humans can be liberated to experience leisure.

We can trace this "leisure-as-liberation" model of design to as far back as Aristotle. Throughout his *Politics*, Aristotle advocates for having leisure as the guiding goal of the State and of its leaders. For example:

> …in a well-ordered state the citizens should have leisure and not have to provide for their daily wants… (Aristotle 1996: 1269a30)

> Since the end of individuals and of states is the same, the end of the best man and of the best constitution must also be the same; it is therefore evident that there ought to exist in both of them the excellences of leisure; for peace, as has often been repeated, is the end of war, and leisure of toil. (Aristotle 1996: 1334a10)

> …many necessaries of life have to be supplied before we can have leisure. Therefore a city must be temperate and brave, and able to endure: for truly, as the proverb says, 'There is no leisure for slaves'… (Aristotle 1996: 1334a15–20)

> …the first principle of all action is leisure…leisure is better than occupation and is its end; and therefore the question must be asked, what ought we to do when at leisure? Clearly we ought not to be playing, for then play would be the end of life. (Aristotle 1996: 1337b30–35)

While it is in the first quote that we find Aristotle's definition of leisure, it is in the third quote that we find the ultimate outcome of this focus on leisure. Citizens can have leisure, slaves cannot. If "citizens should have leisure," as Aristotle states in the first quote, then slavery is a necessity. In suggesting in the fourth quote that leisure is more than play, he is also pointing to his ultimate justification of slavery, for it is the citizens who are capable of the rational activity befitting a human being, while slaves are born instead for toil, or "occupation," for the toil that someone else must carry out so that citizens can have the leisure required for exercising their rationality, and thus, for liberating their humanity.

In Karl Marx we find an extension of this "leisure-as-liberation" mindset, though with the revolutionary expansion of who should be liberated, of who is capable of leisure, to include the slaves as well as the citizens. For example:

> What, then, constitutes the alienation of labor?...in his work, therefore, he does not affirm himself but denies himself, does not feel well but unhappy, does not freely develop his physical and mental energy but mortifies his body and ruins his mind. (Marx 1983:136)

> From the relationship of alienated labor to private property it further follows that the emancipation of society from private property etc., from servitude, expresses itself *politically* as the *emancipation of the workers*; not that *their* emancipation alone is at stake, but because the emancipation of the workers contains within it universal human emancipation. (Marx 1983:144)

> ...in communist society, where nobody has one exclusive sphere of activity but each can become accomplished in any branch he wishes, society regulates the general production and thus makes it possible for me to do one thing today and another tomorrow, to hunt in the morning, fish in the afternoon, rear cattle in the evening, criticize after dinner, just as I have a mind, without ever becoming hunter, fisherman, herdsman or critic. (Marx and Engels 1970:53)

If Aristotle was right that humans require leisure, but wrong that some are born for leisure and others are born for toil, then, as Marx puts it, the workers are becoming alienated, are becoming cut off from their humanity. However, as Aristotle made clear, if humans are to be liberated to experience leisure, then someone must do the undesirable work preventing our "emancipation." Whereas Aristotle's answer to this is slavery, Marx's answer is Communism, wherein, as he explains in the third quote, the State would regulate production so that the very nature of work is changed, enabling everyone to participate in emancipation by removing the alienating forced nature of labor from work.

While we share with Aristotle and Marx this "leisure-as-liberation" way of thinking, we have replaced the means they envisioned for achieving this end. This is not to suggest—though we do like to proclaim it—that we have found slavery to be wrong, or central planning to be impossible. Rather we have replaced human slaves with technological slaves, and replaced the central planning of bureaucrats with the central planning of algorithms. Though it is clear, as is shown by the success of Netflix, Amazon, Uber, iRobot, and many, many similar companies, that technological progress is providing more people with more leisure than ever before, the question remains whether we are at the same time finding in this leisure the liberation Aristotle and Marx promised us, the liberation to finally realize our humanity. To answer this question I will turn to Nietzsche and to Postphenomenology. By combining their respective analyses of what it means to be human, and of what it means to be a user of technology, I will argue that this model of technological progress is not helping us to become more liberated, but rather more nihilistic.

18.2 Human-Nihilism Relations

In the third essay of his *On the Genealogy of Morality*, Nietzsche enumerates five strategies that "ascetic priests" have used in the "grand *struggle against the feeling of displeasure*" (Nietzsche 1989: 131). According to Nietzsche, this "feeling of displeasure" is the ever-spreading feeling of being unable to face who and what we are, the feeling of being sick of being human. As Nietzsche writes:

> ...I am speaking, of course, of the cultural domain, of every kind of "Europe" on this earth. The *sick* are man's greatest danger; *not* the evil, *not* the "beasts of prey." Those who are failures from the start, downtrodden, crushed—it is they, the *weakest*, who must undermine life among men, who, call into question and poison most dangerously our trust in life, in man, and in ourselves. Where does one not encounter that veiled glance which burdens one with a profound sadness, that inward-turned glance of the born failure which betrays how such a man speaks to himself—that glance which is a sigh! "If only I were someone else," sighs this glance: "but there is no hope of that. I am who I am: how could I ever get free of myself? And yet—I *am sick of myself*!" (Nietzsche 1989: 122)

Though Nietzsche sees asceticism and its proponents—the "ascetic priests"—as contrary to life, he nevertheless believes that the purpose of the ascetic priest is preservative rather than destructive. The ascetic priests use various strategies that protect the healthy from the sick by helping the sick to manage, redirect, or ignore their hatred of life, of nature, of themselves, the hatred that Nietzsche famously names "nihilism" (Nietzsche 1989: 44).

Nietzsche does not carefully enumerate or systematize these strategies, but for my purposes here I will collect them together under the name of "human-nihilism relations" (a paraphrase of the name Don Ihde gives to the relations humans have with technologies, which I will describe in the next section). The first human-nihilism relation that Nietzsche describes is that of "self-hypnosis." Nietzsche writes:

> This dominating sense of displeasure is combatted, *first*, by means that reduce the feeling of life in general to its lowest point. If possible, will and desire are abolished altogether; all that produces affects and "blood" is avoided (abstinence from salt: the hygienic regimen of the fakirs); no love; no hate; indifference; no revenge; no wealth; no work; one begs; if possible, no women, or as little as possible; in spiritual matters, Pascal's principle *il faut s'abêtir* is applied. The result, expressed in moral-psychological terms, is "selflessness," "sanctification"; in physiological terms: hypnotization—the attempt to win for man an approximation to what in certain animals is *hibernation*, in many tropical plants *estivation*, the minimum metabolism at which life will still subsist without really entering consciousness. An astonishing amount of human energy has been expended to this end—has it been in vain? (Nietzsche 1989: 131)

If we are sick because we cannot face who and what we are, then the obvious treatment for this sickness is to avoid ourselves as best we can. Anything that is a reminder that we are alive, that we are embodied, that we feel pain, is destroyed or kept at a distance. In other words, as best we can, and for as long as we can, we try to put ourselves to sleep. Nietzsche has in mind here primarily the increasing

European interest in Buddhism, in meditation, and in the concept of Nirvana, or of achieving a total loss of self as the highest good.

The second human-nihilism relation that Nietzsche describes is that of "mechanical activity." Nietzsche writes:

> Much more common than this hypnotic muting of all sensitivity, of the capacity to feel pain—which presupposes rare energy and above all courage, contempt for opinion, "intellectual stoicism"—is a different *training* against states of depression which is at any rate easier: *mechanical activity.* It is beyond doubt that this regimen alleviates an existence of suffering to a not inconsiderable degree: this fact is today called, somewhat dishonestly, "the blessings of work." The alleviation consists in this, that the interest of the sufferer is directed entirely away from his suffering—that activity, and nothing but activity, enters consciousness, and there is consequently little room left in it for suffering: for the chamber of human consciousness is *small!* (Nietzsche 1989: 134)

Instead of making the effort required to actively avoid ourselves, we can simply use repetitive tasks to so occupy our time that we cannot have even a moment of self-reflection. It is here that we find a possible Nietzschean counter-argument to the "leisure-as-liberation" model. For Nietzsche it is not the case that we are too busy to find ourselves, and thus are in need of being liberated from mind-numbing chores in order to have the leisure necessary to rediscover our humanity. Rather, Nietzsche is here suggesting that we busy ourselves with chores precisely because they are mind-numbing, which perhaps might explain why Marx's predicted revolution has not taken place, for fear that we would then indeed have no one else but ourselves to blame for our not finding ourselves, for our not rediscovering our humanity.

The third human-nihilism relation that Nietzsche describes is that of the "petty pleasure." Nietzsche writes:

> An even more highly valued means of combating depression is the prescribing of a *petty pleasure* that is easily attainable and can be made into a regular event; this medication is often employed in association with the previous one. The most common form in which pleasure is thus prescribed as a curative is that of the pleasure of *giving* pleasure (doing good, giving, relieving, helping, encouraging, consoling. praising, rewarding); by prescribing "love of the neighbor," the ascetic priest prescribes fundamentally an excitement of the strongest, most life-affirming drive, even if in the most cautious doses—namely, of the *will to power.* The happiness of "slight superiority," involved in all doing good, being useful, helping, and rewarding, is the most effective means of consolation for the physiologically inhibited, and widely employed by them when they are well advised: otherwise they hurt one another, obedient, of course, to the same basic instinct. (Nietzsche 1989: 135)

Whereas the first two relations had in common the desire to avoid confronting who and what we are, the third relation is instead concerned more specifically with avoiding a confrontation with our powerlessness. In helping others we get to experience, not only the pleasure of being helpful, but the pleasure of being more powerful than the helped. In other words, we can evade our inferiority in the face of nature, of suffering, of death, by immersing ourselves in the superiority we can experience in the face of those in need. Hence, in keeping with the logic of the "will to power," the more we help others, the more we experience power over them, for which reason—as Kant feared—it is nearly impossible to tell if in helping others we are not instead helping ourselves.

The fourth human-nihilism relation that Nietzsche describes is that of "herd formation." Nietzsche writes:

> The "will to mutual aid," to the formation of a herd, to "community," to "congregation," called up in this way is bound to lead to fresh and far more fundamental outbursts of that will to power which it has, even if only to a small extent, aroused: the *formation of a herd* is a significant victory and advance in the struggle against depression. With the growth of the community, a new interest grows for the individual, too, and often lifts him above the most personal element in his discontent, his aversion to *himself* (Geulincx's *"despectio sui")*. All the sick and sickly instinctively strive after a herd organization as a means of shaking off their dull displeasure and feeling of weakness: the ascetic priest divines this instinct and furthers it; wherever there are herds, it is the instinct of weakness that has willed the herd and the prudence of the priest that has organized it. (Nietzsche 1989: 135–136)

This relation, the most famous of Nietzsche's human-nihilism relations, shares with the previous relation the attempt to make the sick feel powerful rather than weak. However, rather than help others, here we join with others, for it is in joining with others, in merging our interests and actions with those of people around us, that we can overcome our individual weaknesses and replace them with the newfound strength of the whole. Again following the logic of the "will to power," the more we derive strength from those around us, the more we need others, and the less able are we to leave the others. Consequently we identify ourselves more and more by our associations with the group and less and less by our differences with the group, with such phenomena becoming the basis of what has since Nietzsche come to be known as group psychology.

The fifth and final human-nihilism relation that Nietzsche describes is that of "orgies of feelings," which Nietzsche describes as, unlike the previous four, a "guilty" form of treating our nihilistic sickness. Nietzsche writes:

> To wrench the human soul from its moorings, to immerse it in terrors, ice, flames, and raptures to such an extent that it is liberated from all petty displeasure, gloom, and depression as by a flash of lightning... Fundamentally, every great affect has this power, provided it explodes suddenly: anger, fear, voluptuousness, revenge, hope, triumph, despair, cruelty; and the ascetic priest has indeed pressed into his service indiscriminately the *whole* pack of savage hounds in man and let loose now this one and now that, always with the same end in view: to awaken men from their slow melancholy, to hunt away, if only for a time, their dull pain and lingering misery, and always under cover of a religious interpretation and "justification." Every such orgy of feeling has to be *paid* for afterward, that goes without saying—it makes the sick sicker; and that is why this kind of cure for pain is, by modern standards, "guilty." (Nietzsche 1989: 139–140)

We can lose ourselves not only in mindless activities or in the presence of others, but even in our own emotions. Since at least the time of the Ancient Greeks and their notion of the Furies we have had the idea that we can be overcome with emotion, so overcome that we have legally recognized this aspect of ourselves through the category of the "crime of passion." Blinded by love or by rage, what is important is that we are blinded, momentarily unable to experience any reality other than that of the pleasure of letting ourselves go—both in the sense of unleashing an emotional outburst, and of evading any sense of the self behind the outburst.

18.3 Human-Technology Relations

Having now taken a tour through Nietzsche's diagnoses of the ways in which we try to treat rather than cure our nihilism, we can compare these human-nihilism relations with the human-technology relations found in postphenomenology. In so doing we will be able to take Nietzsche's late-nineteenth century analyses and bring them into the present in order to determine whether, and to what extent, we are using technologies today in ways that could be described as satisfying the criteria Nietzsche has provided us for recognizing when we are becoming not more liberated, but more sick. Before we can do that however we must take one more tour, this time through the work of Don Ihde.

In his *Technology and the Lifeworld*, Ihde attempts to both adopt Heidegger's phenomenological analyses of technology, and to overcome the conclusions Heidegger reaches based on those analyses. To do this Ihde opposes the descriptions of technological interactions that Heidegger provides in *Being and Time*, with the descriptions Heidegger provides in his later essay, "The Question Concerning Technology." According to Ihde, whereas in the latter work Heidegger takes a totalizing perspective on technology, attempting to find one "essence" of technology, in the earlier work Heidegger takes a more nuanced approach, focusing instead on specific technologies and specific interactions. Ihde is in this way able to develop a new line of research into what he will call "human-technology relations," thus enabling investigations into how users interact with technologies that can avoid the pessimistic pronouncements concerning "Technology" that Heidegger's later work had initiated.

Ihde was able to carry out this reversal of post-Heideggerian philosophy of technology by arguing that the essence of technology was that it has no essence. There is not one "technology," but technologies, each of which has no necessary, essential way that it must be adopted, used, or understood. Not unlike Wittgenstein's pronouncement that the meaning of a word is its use in a sentence, Ihde argues that the meaning of a technological device is its use in a project. Ihde encapsulates this idea through the concept of "multistability" (Ihde 1990: 144). Technologies have multiple possible ways of being constituted—of achieving a stability—but we cannot predict in advance which one of those ways will indeed be adopted. Designers and engineers may build their intentions into a product, but that in no way guarantees that users will share those intentions.

The multistability of technologies does not mean however that human-technology relations cannot be classified into types. Ihde identifies four primary ways in which users relate to technologies: embodiment relations, hermeneutics relations, alterity relations, and background relations. What all of these relations have in common is what Ihde takes primarily from Heidegger's *Being and Time*, the idea that in using technologies to carry out our projects there is a dynamic of revealing and withdrawing. In Heidegger's example of using a hammer, he argues that in hammering we take no notice of the hammer, but only of the project. It is when the hammer breaks that the project withdraws from view, revealing the hammer before us as an object, for which reason Heidegger argues that subject/object dualism is not our primary mode of being, but is instead how we come to see the world when being-in-the-world breaks down.

Hammering would, for Ihde, be an example of embodiment relations, of those relations where technologies expand our bodily abilities. In embodiment relations, what the technology allows us to do is what consumes our attention, while the technology itself fades from view. Ihde formalizes this relation as:

$$(\text{I} \quad Technology) \to World$$

The formalism is meant to indicate how, such as with a pair of glasses, the user and the technology become fused, relating to the world as one, for which reason I say, "I see the world," rather than "I, through my glasses, see the world."

Hermeneutic relations occur when technologies help us to understand the world. In hermeneutic relations technologies do not expand our abilities, but rather provide us with new abilities. For example, telescopes today do not merely expand the vision of the naked eye, but instead reveal to us how the universe looks to eyes capable of seeing wavelengths and spectra that human eyes are incapable of seeing. What is revealed in hermeneutic relations is what is conveyed to us through the technology, while what is obscured is the faith we must put in the technology, for it is always possible that the world the technology is revealing to us does not indeed exist, but is instead the result of a malfunction. Ihde formalizes these relations as:

$$\text{I} \to (Technology \quad World)$$

The formalism suggests that we are on the outside looking in, which is precisely why faith is necessary, even if not noticed, in such relations.

Alterity relations occur when technologies become the focus of our attention. In alterity relations we see not the world but the technology, as what had been a means to an end in the previous two relations now becomes the end itself. Alterity relations occur not only when, as Heidegger described with the hammer, technologies cease to function properly, but also when they function as toys, as obstacles, as competitors, when they are designed to occupy our attention. Ihde formalizes these relations as:

$$\text{I} \to Technology(\quad World)$$

The formalism indicates that in such relations our love/hate relationships with technologies are revealed, while the world outside our relationship disappears.

Background relations occur when technologies function completely beyond our attention. Background relations are thus the opposite of alterity relations, for here technologies, such as air conditioners, ceiling lights, and refrigerators, are meant to work without our even having to be aware that they exist. Ihde formalizes these relations as:

$$I \rightarrow World(\ Technology)$$

The formalism highlights that these technologies operate best by making it appear as if the world did not need any technological support, and thus they tend to mimic and expand not human capacities as in embodiment relations, but rather the capacities of nature.

18.4 Nihilism-Technology Relations

Having now toured both the human-nihilism relations of Nietzsche and the human-technology relations of Ihde, I propose combining these two descriptions of the world into what I will call *nihilism-technology relations*. As we have seen, both Nietzsche and Ihde are pursuing the various ways in which we experience the world, how our experience can be mediated (whether by nihilistic or by technological means), and how that mediation operates through a dynamic of revealing and withdrawing, or, as Nietzsche might put it, of occupying and evading. By combining their analyses we can see, on the one hand, how technologies can mediate our nihilism, and, on the other hand, how nihilism can mediate our technologies.

The first nihilism-technology relation is *techno-hypnosis*. While in Nietzsche's day Buddhism was the primary means for dulling our senses and putting ourselves to sleep, today we can instead find virtual reality devices sold in every electronics store. These devices offer us the fantasy of experiencing new realities, an experience which of course requires that we cease to experience the realities of our bodies and of the natural world. Similarly we find Google, Tesla, Uber, and other companies pursuing driverless cars, cars that would allow us to actually sleep while our cars do the driving for us. What these techno-hypnotic pursuits have in common is the goal of mind/body dualism, the dualism of either being able to let the mind wander while the body is ignored (VR), or being able to let the body wander while the mind is put to sleep (automation). The extreme version of this dualistic techno-hypnosis can be found in the recent developments—or, as Nietzsche might say, religions—of Posthumanism (Bostrom 2005) and Transhumanism (Fuller 2015). In both of these new -isms, –isms the proponents of which often unite with their Buddhism, we find the belief that either technology can create a new species beyond the human or can elevate humanity to a superhuman level. In either case, we find the idea of leaving our frail bodies behind and entering a post-pain, even post-death utopia.

The second nihilism-technology relation is *data-driven activity*. Whereas in the nineteenth century we were in need of other humans to order us around and give us chores to occupy us, today, as Apple might say, "there's an app for that." Netflix can tell us what to watch, Amazon can tell us what to buy, and eHarmony can even tell us who to love. Though the algorithms can seem invasive to some, as they attempt to develop profiles of us by keeping track of all of our online activities, technological progress seems to be measured more and more not by the protection of our privacy, but by the accuracy of the algorithms' predictions (Gertz 2016). Similarly the development of wearable health trackers, such as the Fitbit, are used not only to monitor the steps we are taking and the calories we're burning, but are increasingly becoming wearable gym trainers, pushing us to take more steps, to burn more calories, with some apps even uploading your data for others to see in order to shame you into increased activity. What we find in all such cases of data-driven activity, much as with mechanical activity, is activity without an actor, as either someone or some app tells us what to do, thereby freeing us from the oppressiveness of decision-making, or, as Sartre (1993) would call it, our having been condemned to be free.

The third nihilism-technology relation is *pleasure economics*. Whereas in Nietzsche's time we could only help those near to us, today we can go online and help people all over the world. On "crowdfunding" sites such as Kickstarter, GoFundMe, and Indiegogo, billions of dollars are being sent all over the world to help artists, musicians, filmmakers, and designers in exchange for small gifts of gratitude. Similarly sites such as Airbnb are allowing people to rent out their homes, sites such as TaskRabbit are allowing people to rent out their bodies, and sites such as Wikipedia are allowing people to rent out their minds. In all such cases we find the ability to, as Nietzsche would put it, expand the range of our will to power to undreamt of heights, allowing us to experience the superiority of helping those in need through not only charitable donations, but also through letting strangers have the gift of spending time in our homes, of doing our chores, and of receiving our wisdom.

The fourth nihilism-technology relation is *herd networking*. Previously we could only band together with those physically near us, but today we can connect to and form groups with anyone who has an internet connection. Social networking sites such as Facebook, Twitter, Instagram, and Reddit have millions of users, all of whom are people who—according to the Nietzschean language already inherent to social networking—are defined as "followers." Social networks operate by exploiting group psychology, leading users to "follow" the most popular accounts, and to in turn try to become more popular by following not just people, but also "trends," or the topics of the day that the most people online are also discussing. Social networking may seem to privilege uniqueness rather than conformity, as it is those who stand out who get the most noticed, but once one has been noticed, once one has achieved an online identity, then that identity must be maintained, as deviations can appear not as quirky individuality, but as a sign of phoniness. Social network users therefore become not only followers in search of followers, but brands in search of consistent content.

The fifth nihilism-technology relation is *orgies of activism*. While Nietzsche may have been describing the riots, uprisings, and revolutionary atmosphere of nineteenth century Europe, today we have flash mobs, viral memes, and the ability to take down leaders and celebrities with little more than the click of a button. With not only social networking, but with the proliferation of smart devices and cloud computing, more and more people are not only spending every minute of the day online, but they are putting more of themselves online. So-called "hacktivists" like Wikileaks and Anonymous have not only released the secret documents and e-mails of politicians and corporations, but have helped to create an atmosphere of both indignation over the nearly endless supply of revelatory hacks, but also of paranoia that we will become the target of such hacks. We are thus led to simultaneously participate in and complain about each hour's latest trending outrage. Furthermore, by not having to actually participate in a riot, by not having to even rise up out of chairs, we can unleash our orgiastic passions while remaining anonymous, enabling us to become hashtag activists who can experience the fervor of revolution without the fears that come with actually revolting.

18.5 Conclusion

To return to Ihde, I argue that we can now add these nihilism-technology relations to the list of human-technology relations, but under the more general banner of *nihilism relations*. While these nihilism relations have elements of Ihde's embodiment relations, hermeneutic relations, alterity relations, and background relations, I believe it is more useful to keep these relations separate. Their uniqueness can best be seen if we formalize nihilism relations as follows:

$$Technology \rightarrow World(\ \mathrm{I})$$

What this formalism is meant to indicate is not only our increasing detachment from the world, but our increasing attachment to letting technologies mediate, moderate, and even take up our concerns for us.

Critics may argue that we have in fact never been more engaged, more involved, more active in world affairs. Yet it must be acknowledged that, first, this increased engagement is taking place primarily (if not, in many cases, entirely) online. Second, it must be acknowledged that this increased engagement is taking place not only online, but, to return to Ihde, through hermeneutic relations, relations which allow us increased engagement in exchange for increased faith that what we are engaged with is reality and not malfunction or manipulation. That we are more and more today seeing suggestions that we are in a "post-truth" political world is a sign of our increasing participation, not in the political sphere, but in what I have called herd networking and orgies of activism. In the perpetual contest for the most followers, the most retweets, the most likes, we are in a race to share "news" rather than to read the news, a development which has led to the reduction of journalism to "clickbait" headlines.

My argument should not be taken to be one of defeatism however. What I have been describing has been meant to be seen as tendencies, not as deterministic proclamations, which I believe is also how we should read Nietzsche. Indeed if we return to Nietzsche we can find a challenge to nihilism, a challenge in the form of responsibility. Nietzsche warns that we have reduced responsibility to merely accountability, for which reason being "moral" requires not that we *take responsibility* but only that we are *held responsible*. Yet even accountability is achieved not by judging one's actions, but rather by determining our actions in advance, that is, by being able to *count on* the actions of others because we have been made to act in ways that are "calculable, regular, necessary," and, consequently, "predictable" (Nietzsche 1989: 58).

Such a morality-as-predictability is precisely what we find in the two primary philosophical approaches to technology today: Science and Technology Studies (STS) and Postphenomenology. Bruno Latour (1992), taking up Madeleine Akrich's (1992) concept of "scripts," argues that we need not worry if people are becoming less moral, as designers and engineers can *inscribe* morality into products, products which *prescribe* how users must act, such as for example occurs with seat belts, speed bumps, and electronic ticket gates. Tromp et al. (2011), using Ihde's concept of multistability, argues that while users need not necessarily use products as designed and can instead *re-inscribe* products in unforeseen ways, designers can nevertheless still create products that can be *seductive*, *persuasive*, *coercive*, or even *decisive* in making us act in "socially responsible" ways.

Nietzsche would see such technologically "prescribed" or "nudged" morality as yet further evidence of our nihilistic evasion of who and what we are. Hence counter to morality-as-predictability, Nietzsche challenges us to take responsibility for our actions, to be autonomous in the truest sense of the word, and to only make judgments and promises "like sovereigns" (Nietzsche 1989: 60). If, as I have argued elsewhere (Gertz 2014), responsibility is not best thought of morally, but existentially, that responsibility is not something we *do*, but something we *are*, then the more we outsource responsibility to technology the more nihilistic, sick, inhuman, we become.

From a Nietzschean perspective, we can now see how technologies that make us safer can make us weaker, technologies that make us predictable can make us thoughtless, and technologies that make us engaged can make us detached. The mistake that must be avoided here is to think that, because technologies can make us weaker, thoughtless, and detached, then if we remove our technologies we thereby remove our nihilism. What Nietzsche and Ihde reveal to us is that if technologies are used nihilistically, it is because we, the users, are nihilistic. The first step to becoming responsible therefore is to recognize that technologies are not the problem confronting us, we are.

It may appear that if the answer to the problem of our nihilistic use of technologies is not to remove technologies, then the answer must be to remove nihilism. However such a view is also mistaken. As Nietzsche writes:

> My formula for greatness in a human being is *amor fati*: that one wants nothing to be different, not forward, not backward, not in all eternity. Not merely bear what is necessary, still less conceal it—all idealism is mendaciousness in the face of what is necessary—but *love* it. (Nietzsche 1989: 258)

To love one's fate means to not fight how things are, to not wish things were different, to not curse our frailties and our finitude. In other words, to love one's fate is to be responsible, to not only accept the nature of reality and of humanity, but to embrace it.

And if we are nihilistic, then we must love our nihilism. We must confront rather than evade our nihilism, be responsible for our nihilism rather than blame technology. There is likely nothing more nihilistic that we do than pretend that we are not nihilistic, pretending our actions are meaningful when they are meaningless (e.g., orgies of activism) and meaningless when they are meaningful (e.g., techno-hypnosis). For Ihde, technologies mediate our relations to the world. For Nietzsche, nihilism mediates our relations to the world. Bringing these two together, I have tried to show how nihilism mediates our relations to technologies. By recognizing and confronting our nihilism we can stop waiting for technologies to liberate us, stop blaming technologies for enslaving us, and instead begin to become responsible for who and what we are. Technologies have the power to liberate, to help more people experience the leisure that Aristotle and Marx envisioned, but only if we stop preventing technologies from doing so by either worshipping them or scapegoating them.

References

Akrich, M. (1992). The description of technological objects. In W. E. Bijker & J. Law (Eds.), *Shaping technology/building society* (pp. 205–224). Cambridge: MIT Press.

Aristotle. (1996). *The politics and the constitution of Athens* (S. Everson, Ed.). Cambridge: Cambridge University Press.

Boston Business Journal. (2016). http://www.bizjournals.com/boston/blog/techflash/2016/08/irobots-newest-roomba-is-cheaper-but-still-smart.html#g1. Accessed 29 Oct 2016.

Bostrom, N. (2005). In defense of Posthuman dignity. *Bioethics, 19*(3), 202–214.

Fuller, S. (2015). *We may look crazy to them, but they look like Zombies to us: Transhumanism as a political challenge*. Institute for Ethics and Emerging Technologies. http://ieet.org/index.php/IEET/more/fuller20150909

Fusion. (2015). *'Netflix and chill': The complete history of a viral sex catchphrase*. http://fusion.net/story/190020/netflix-and-chill/. Accessed 8 Dec 2016.

Gertz, N. (2014). *The philosophy of war and exile*. Basingstoke: Palgrave-Macmillan.

Gertz, N. (2016). Autonomy online: Jacques Ellul and the Facebook emotional manipulation study. *Research Ethics, 12*(1), 55–61.

Ihde, D. (1990). *Technology and the lifeworld*. Bloomington: Indiana University Press.

Latour, B. (1992). Where are the missing masses? The sociology of a few Mundane Artifacts. In W. E. Bijker & J. Law. (Eds.) *Shaping technology/building society* (pp. 225–258). Cambridge: MIT Press.

Marx, K. (1983). *The portable Karl Marx* (E. Kamenka, Ed.). Middlesex: Penguin Books.

Marx, K., & Engels, F. (1970). *The German ideology* (C. J. Arthur, Ed.). International Publishers.

Nietzsche, F. (1989). *On the genealogy of morals and ecce homo* (W. Kaufmann, Trans.). Vintage.

Sartre, J. -P. (1993). *Essays in existentialism* (W. Baskin, Trans.). Citadel Press.

Tromp, N., Hekkert, P., & Verbeek, P. -P. (2011). Design for socially responsible behavior: A classification of influence based on intended user experience. *Design Issues, 27*(3), 3–19.

Chapter 19
Assisting Ourselves to Death – A Philosophical Reflection on Lifting a Finger with Advanced Assistive Systems

Bruno Gransche

Abstract Artificial assistance is about to become a major intermediary between humans and the world. Siri, Alexa, Cortana etc. are positioned to become tomorrow's everyday life djinns that will grant all sorts of wishes at a magical wake word. They are positioned to unburden us even of the slightest effort of lifting a finger. At the brink of a possible future *assistive society,* the relation between human actors and the world's resistance has been transformed by interposed artificial assistants. The comprehensive saturation of all spheres of life with such comfort oriented assistants calls again for a philosophical reflection on our relation to labour, work, and action. This article focuses on the nexus of work, competence, and comfort, of automation, assistance, and autonomy. It analyses the emerging human-assistant-world relation in the light of a master-slave-thing relation leaning mainly on Hannah Arendt. The following key questions are addressed: If we proceed on the path to an assistive society, how would this development alter the behaviour of individuals within that society, what would be the consequences for the relation between humans and the world, and how would this rearrange the distribution of competencies among human actors and artificial agents? In an assistive society, the loss of the things' resistance is not regretted as an incapacitating development; on the contrary, it is celebrated as a gain in comfort. Yet, this means to submit ourselves to a regime of pseudo-magical effortlessness of a simple-and-easy society no longer capable of lifting a finger and no longer knowing what for.

Keywords Advanced assistance systems · Comfort · Competence · Action theory · Philosophy of technology

---

B. Gransche (✉)
Institute of Advanced Studies FoKoS, University of Siegen, Siegen, Germany
e-mail: Bruno.Gransche@uni-siegen.de

A. Fritzsche, S. J. Oks (eds.), *The Future of Engineering*, Philosophy of Engineering and Technology 31, https://doi.org/10.1007/978-3-319-91029-1_19

19.1 Introduction

> In the end, they will lay their freedom at our feet, and say to us, 'make us your slaves, but feed us.' (Dostoyevsky 1993, p. 26)

> If it should turn out to be true that knowledge (in the modern sense of know-how) and thought have parted company for good, then we would indeed become the helpless slaves, not so much of our machines as of our know-how, thoughtless creatures at the mercy of every gadget which is technically possible, no matter how murderous it is. (Arendt 1998, p. 3)

Artificial assistance[1] is about to become a major intermediate between humans and the world. All sorts of intelligent artificial systems preselect, pre-configure, and shape what we perceive of the world and what we identify as possible spheres of influence. Along with advanced automation, increased technical capabilities, and autonomy, more tasks than ever can be delegated to artificial assistance. Delegating repetitive and dull tasks to assistive systems frees our schedules and cognitive capacities. Instead of navigating by paper map, a GPS navigation system takes over. This offers the possibility to chat on the phone while driving. Doing so repeatedly changes the capabilities to navigate without GPS assistance because competence formation and task actualisation are interlinked. This article analyses the relation among human actors, the world, and advanced artificial assistants; it is about how delegating tasks is decisively changing the delegator and how artificially assisted action is transforming the relation between the human actor and the world s/he acts in.

The focus of this reflection is on a '*futurible*', a *possible future* (Jouvenel 1964), that can be considered as quite plausible in light of current developments: In line with many time diagnostics like *risk society* (Beck 1986), *knowledge society* (Bindé 2006; Hubig 2000), *labour society* (Arendt 1998) etc. this futurible could be labelled '*assistive society*'. Talking about a futurible is a hypothetic endeavour – the future is principally open – yet it can be well founded if based on recent research and innovation advancements. The futurible of an *assistive society* starts from several current phenomena and extends them for heuristic purposes to a further, not yet reached, but probably imminent stage. We witness ever more autonomous, less task-specific, general personal artificial assistants that are not just one robot or device but a whole assemblage of hard- and software, resources, sensors, information, (cloud) services etc.[2] This assistive society will be comprehensively saturated with intelligent, interconnected, and personalised artificial assistive systems. These systems will be omnipresent in every aspect of life such as work, leisure, health, disability care, sports, intimacy, relationship, education etc. For brevity reasons and in order not to confound with special medical-ethical aspects the schematic focus of this reflection

---

[1] In this article, if not otherwise specified, 'assistance' and 'assistant' will be used exclusively for intelligent artificial assistive systems, assemblages of components, interfaces and services that work together to provide unspecific everyday life support.

[2] "Alexa—the brain behind Echo—is built in the cloud, so it is always getting smarter. The more you use Echo, the more it adapts to your speech patterns, vocabulary, and personal preferences." Amazon (2016b).

shall be *voluntary, non-dependent, and comfort oriented everyday life assistance.* Current examples of everyday life assistance systems are the so-called 'Intelligent Personal Assistants (IPA)' *Siri* (also 'Knowledge Navigator'*)* from Apple, *Cortana* from Microsoft, *Echo* with *Alexa* from Amazon, *Now* from Google, the robotic "day-to-day companion" *Pepper*[3] from Aldebaran, the 'social robot' *Jibo,*[4] the "companion robot" *Buddy*[5] from Blue Frog Robotics etc. The futurible of the assistive society is about the then truly omnipresent successors of Alexa and their kind that might actually live up to today's promises of their developers.

If we proceed on the path to the assistive society, how would this development alter the behaviour of individuals within that society, what would be the consequences for the relation between humans and the world, and how would this rearrange the distribution of competencies among human actors and artificial agents?

In order to answer these questions a set of preliminary questions will be addressed: What kind of relation is an assistive relation? How does human assistance differ from artificial assistance? What is the guiding principle of assistance: comfort or enabling? Why do human 'masters' who delegate work to artificial 'slaves' tend to be converted into slaves of another kind? How comes that excessive assistance might incapacitate instead of enable the assisted? How does unasked imposed assistance actually victimise the human operators and at times enrage them, as if they suspected that tolerating unneeded help is a first step to need it all too soon? This article discusses this possible conversion of voluntary comfort-oriented assistance into assistance dependency.

19.2 Vogli Sempre Poter: Will to Competence or Will to Comfort

19.2.1 Assisted Action

Simplified, terms such as *work,*[6] *task accomplishment, problem-solving* – generally speaking, the *overcoming of resistance* – signifies a twofold relation between human actors and the world. The actors use their competencies, capabilities, and means to

[3] "Pepper is a human-shaped robot. He is kindly, endearing and surprising. We have designed Pepper to be a genuine day-to-day companion, whose number one quality is his ability to perceive emotions. Pepper is the first humanoid robot capable of recognising the principal human emotions and adapting his behaviour to the mood of his interlocutor." Aldebaran Robots (2016).

[4] "Jibo isn't an appliance, it's a companion, one that can interact and react with its human owners in ways that delight. (Mashable)"Jibo (2016).

[5] "BUDDY is the revolutionary companion robot that improves your everyday life. Open source and easy to use, BUDDY connects, protects, and interacts with each member of your family." Blue Frog Robotics (2016).

[6] In this article the words *work* and *labour* are used interchangeably unless otherwise specified and except in the context of Hannah Arendt's special concepts of *Labour* (Arbeit), *Work* (Herstellen) and *Action* (Handeln), see Arendt (1998, chapter 5). I do not necessarily argue that artificial systems can actually *act* in Arendt's sense but that task delegation to those systems influences the users' capabilities to act.

transform the world according to their goals. The world differs from these goals and resists the transforming efforts of the actor. The balance of the means and the world's resistance determines the output. If the means are adequate, the action will be successful and the problem is solved. If the resistance is too high or the goal too ambitious, and if the available means are inadequate, then the action fails. Leonardo da Vinci, a true *Renaissance man* who lived up to the ideal of the *homo universalis* like no other and who possessed extraordinary competencies to pursue his goals, postulated the consequences: "Chi non può quel che vuol, quel che può voglia; [...] Che da quel che non può suo voler toglia." (Carrer 1836, p. 75)[7] For a Renaissance genius wanting less just because the competencies are inadequate to achieve the initial goals, cannot be a satisfactory solution. A way to avoid 'eating humble pie' – and a way the Renaissance up to Humboldt venerated – is to work on the means, to train, learn and to educate oneself. This is a tiresome endeavour, *a fortiori* for the many who are not omnitalented like Leonardo. Yet for him, the mission of constant self-improvement is clear and so he concludes his untitled poem: "Vogli sempre poter quel che tu debbi."[8] Wanting less is not an option.

If lesser goals are unacceptable and competence improvement is limited or too tiresome, then a third option has to be found: assistance. If you organise help, you have neither to want less nor to arduously train yourself. As Hannah Arendt stressed, the *human condition* is plurality.[9] Humans are always already acting with others. They take not only their own single, mostly unsatisfactory means into account but also those of their co-actors. This solution of inter-human cooperation either comes with the obligations of aligning the respective goals. This then again urges to mutually adjust the purposes but allows for wanting different instead of just lower goals. Alternatively, this option comes with the obligation of compelling possible co-actors into submission and then benefit by their competencies via servitude. Enslaving others, setting the desired goals, leaving the problem solving to the slaves (be it by their multitude, training or both), and just enjoying the accomplished effects exchanges the work of overcoming the world's resistance with the work of overcoming other humans' resistance. Assistance is a serving relation; it is about delegation of labour. The delegation of work against the world's resistance comes at a price that might corrupt the promises of current assistance technology.

[7] "Who is not able to achieve what he or she wants, wants what he or she can achieve. [...] So he or she withdraws his or her wanting from what he or she cannot achieve." (my translation).

[8] „You always have to want to be able to do what you have to." – or – "Always desire to possess the necessary competencies." (my translation).

[9] "Action, the only activity that goes on directly between men without the intermediary of things or matter, corresponds to the human condition of plurality, to the fact that men, not Man, live on the earth and inhabit the world. While all aspects of the human condition are somehow related to politics, this plurality is specifically the condition—not only the *conditio sine qua non*, but the *conditio per quam*—of all political life." Arendt (1998, p. 7). Interestingly the Romans even linguistically identified 'living' with 'to be among men' *(inter homines esse)*.

So far, three ways to align a mismatch of goals and means have been identified: a) wanting less, b) improving one's own means, or c) expanding the hypothetically twofold actor-world relation to a multifold co-actors-world relation – by either agreeing consensually on goals or by exchanging cooperation with submission.

19.2.2 Advanced Assistive Systems

Advanced assistive systems join this human-world relation as new players. They are special in many ways and differ a lot from tools or mechanical machines, which human actors always used as parts of their means. Hammer, combustion-engine, washing machine, phone etc. are manifestations of human inventive skills. They extend the portfolio of means, allow for different and higher goals and create new (producing, handling, operating and maintaining) skills. The technology of the tool and device type massively altered and extended the scope of possible goals, the impact, and force of human means. Yet, they did not decisively alter the twofold actor-world or the multifold inter-human co-actors-world relation.

Today, we are no longer exclusively operating machines or using tools, we also act with and within intelligent systems. These systems derive from developments in the fields of intelligent systems, interconnected devices, ubiquitous computing, internet of things, cyber-physical systems, learning algorithms, deep learning etc. These systems are characterised by a vast amount of autonomy (not in the Kantian but in a technical sense): they choose and adjust their means, settings, programmes, timing etc. following sensor-based and self-compiled profiles of the user, situation, and the 'world'.

The actor still chooses the goal he wants to actualise but the intelligent systems autonomously select how to actualise them. When interacting with advanced assistance systems, the users theoretically maintain the autonomy over the goal setting but delegate their strategic and operative autonomy to the system. In other words, humans set the goal; the assistants choose the means and execute the action. In artificially assisted actions, the schematic human-world relations turn into human-technology-world relations.[10] This type of relation is often called the human-technology or – because of the essential role of computer technology – the human-computer interaction.[11] The assistive relations between humans and intelligent assistive systems, which are at the core of the assistive society futurible, are already today widely distributed cases of these human-technology relations. In this

[10] With the 'right' technology concept – if only chosen broadly enough – there is no human-world relation without any technical aspects. For instance, the concept of technology as a medium grasps technology as an all-pervading medium like water surrounding the fishes. Or if – as another example – being human means being a *homo technologicus,* then human-technology-world relation would be a pleonasm. However, such broad technology concepts then tend to lose their suitability for most analysis. See Gransche (2015, p. 146–176).

[11] These relations are widely called *interaction* even though the relation type of 'interaction' is just one of the possible relations, besides co-action for instance. See Gransche et al. (2014).

human-assistant-world actions, especially when the assistants autonomously 'decide' what the assisted actually wants, the role of the actor shifts from the human to the artificial system. This shift of agency is accompanied by a shift of the roles of the involved parties converting the human *actor* into a *user*.

19.2.3 Open Sesame – Submission or Magic

Voluntarily teaming up to achieve a goal turns individuals into *co-actors*. The acceptance of help converts an actor into a *help recipient*. Using assistive systems accordingly turns an *actor* into a *user* and an *assisted person*. In the human-assistant relations, the artificial assistants provide means and skills, which are a) not in the possession of the user or b) in the possession of the user, but s/he struggles to actualise them (because their actualisation is too tiresome, risky, expensive etc.), and that make the chosen goals accomplishable at all or with more comfort. Artificial assistance is less like the co-action of a team effort but more like the delegated or imposed action of a master-slave relation. The master-user sets the goals and the slave-assistant executes the work, while the master-user enjoys the actualisation of the goals without effort. Machines like a hammer-drill technically augment action by adding their means to the actors' means such as the rotation and the beat to the handling, strategy, and goal of the craftsman. Intelligent systems like personal assistants take over the entire action for a given goal. *Acting with* (technology) becomes an *action delegation* (to technology). Similarly, *acting with other humans* converts into *letting slaves act instead* in the master-slave mode. This is the case because the guiding principle of unspecific all-task assistance is not *enabling support* of the human action, but a *comforting substitute* for it. The prevalent character of this independent, comfort-oriented assistance is not to improve human competencies to reach higher goals and it is not "vogli sempre poter". Instead, it makes everyday life *simple and easy:* "Like a personal assistant, Jibo proactively helps you, to make everyday tasks simpler and easier." (Jibo 2016) It is fulfilling the wish of becoming a master without – at least at first glance – having to subject humans as slaves. The "advent of automation" emptied vast parts of today's factories, as Hannah Arendt predicted in 1958:

> Closer at hand and perhaps equally decisive is another no less threatening event. This is the advent of automation, which in a few decades probably will empty the factories and liberate mankind from its oldest and most natural burden, the burden of laboring and the bondage to necessity. (Arendt 1998, p. 4)

If automation is the delegation of specific and repetitive tasks to machines, then today's emerging learning systems are about to take over the tasks of everyday life. The intelligent systems of the assistive society aspire to 'liberate mankind from the burden of acting[12] for themselves' – from the burden to even *lift a finger*:

[12] Acting is used here in the general sense of the word, not in Arendt's rather specific definition. See Arendt (1998, chapter 5).

> Use Echo to switch on the lamp before getting out of bed, turn on the fan or space heater while reading in your favorite chair, or dim the lights from the couch to watch a movie—all without lifting a finger. (Amazon 2016b)

Imposing hard labour on others is historically well known from kingdoms, empires, slavery, colonialism, or capitalism.[13] Enjoying the fruits of labour "without lifting a finger" was hitherto only available actually by submission or imaginably by magic. The advertisers of today's assistive systems are strategically connecting them to the latter by evoking connotations of the land of milk and honey (Schlaraffenland). Similarly, Clarke's third law postulates that "Any sufficiently advanced technology is indistinguishable from magic." (Clarke 1973, p. 21) *Open Sesame* is seamlessly replaced with *Ok, Google!*. Amazon advertises: "When you want to use Echo, just say the wake word 'Alexa' and Echo responds instantly." (Amazon 2016b) The new assistants are positioned to become like actual Djinns: "All you have to do is ask." (Amazon 2016b).

This resemblance with magical fairy tales for now concerns only the promises of the system developers. Alexa was just released for German Beta testers by the time these lines were written. It is currently indeed capable of finding your next train connection, but not to buy the ticket for you. Yet, the Echo assistant is not a machine or device with its set and programmed features. It is an access point, a mere interface to a vast variety of hard- and software components, to cloud-based services and information that are continuously changing. This assistant, just as *Siri* and others, already is *open tasked* and *open skilled* and this unspecific character will surely intensify in their successors of futurible assistive societies. The number of times you need to lift your finger shrinks daily with every garage gate, smart fridge, logistic robot, sensor floor, streaming speaker, and with every skill or app that is uploaded to the cloud.[14] Advanced assistive systems are set to become a highly personalised entry point to the internet of information, services, and things. Would that not mean living the dream, the long-cherished wish of freedom from labour?

> [T]he wish to be liberated from labor's 'toil and trouble,' is not modern but as old as recorded history. Freedom from labor itself is not new; it once belonged among the most firmly established privileges of the few. In this instance, it seems as though scientific progress and technical developments had been only taken advantage of to achieve something about which all former ages dreamed but which none had been able to realize. However, this is so only in appearance. (Arendt 1998, pp. 4–5)

---

[13] "[E]ven though capitalism's ideology is indeed antithetical to slavery, in practice capital nonetheless not only subsumed and reinforced existing slave production systems throughout the world but also *created new systems of slavery* on an unprecedented scale, particularly in the Americas." Hardt and Negri (2000, p. 122).

[14] "Skills add even more capabilities like ordering a pizza from Domino's, requesting a ride from Uber, opening your garage with Garageio, and more. Enabling skills lets your Echo do even more—simply discover and enable the skills you want to use in the Alexa App. New skills are being added all the time."Amazon (2016b).

19.3 Magic Backfires

Arendt points to a myopia that only sees what is "in appearance." It was necessary for her to debunk the promise of freedom from labour through automation. The situation at the brink of an assistive society is similar. Being helped, supported, and served, enjoying comfort – what could possibly go wrong? The plan to pervade everyday life with advanced assistive technologies deserves an equally critical examination as the phenomenon of automation. Hannah Arendt considered automation a development of freeing the "laboring society" of their defining characteristic within "a few decades" (from 1958). According to her, "a fundamental aspect of the human condition is at stake" (Arendt 1998, p. 4) and nothing less is the case for everyday task delegation in an assistive society within 'a few decades' from now.

'Be careful what you wish for.' This proverbial advice is based on the insight that desirable goals come at a price even if they seem free. Magic wish fulfilment tends to backfire. "The fulfilment of the wish, therefore, like the fulfilment of wishes in fairy tales, comes at a moment when it can only be self-defeating." (Arendt 1998, pp. 4–5) Another *Renaissance man,* only a few centuries after Leonardo, Johann Wolfgang von Goethe, exemplified this insight in his poem *The Sorcerer's Apprentice.* The apprentice frees himself from the labour of carrying water and cleaning, delegating the work to magically animated helpers. Yet, he realises soon that it was much easier to start such autonomous processes than to control or even stop them. Together with the prosthesis winged Icarus, the Sorcerer's Apprentice is *the* iconic admonition of technicists' hubris, imprudence, and prematurity. In times of increasingly self-controlled technology, of autonomous robots, learning algorithms, biofacts[15] etc., this warning is more crucial than ever. In today's sociotechnical interconnectedness, every snowball is always already a potential avalanche. Assistive assemblages that create new possibilities create by their complexity new potential harms ipso facto. Since in complex sociotechnical systems, the potential harm is just not yet actualised harm, and accidents are *Normal Accidents* (Perrow 1984). In principle, the momentums of the Sorcerer's Apprentice and Icarus are inherent in complex assemblages. The close entanglement with all spheres of everyday life of assistive systems calls for Daedalic precaution.

19.3.1 Brave New Comfort

Arendt problematizes automation and labour freedom as a "threatening event" (Arendt 1998, p. 4) and addresses much more profound problems than user prudence, Daedalic precaution, or safety considerations.

> The modern age has carried with it a theoretical glorification of labor and has resulted in a factual transformation of the whole of society into a laboring society. [...] It is a society of

[15] Biofacts are *biological artifacts* with mutation and growth capabilities, see: Karafyllis (2003).

laborers which is about to be liberated from the fetters of labor, and this society does no longer know of those other higher and more meaningful activities for the sake of which this freedom would deserve to be won. (Arendt 1998, pp. 4–5)

Arendt's time diagnostic of the labouring society from 1958 evokes a duo of earlier fictitious societal ideas: Aldous Huxley's *Brave New World* from 1932 and George Orwell's 1984 from 1948:

Orwell warns that we will bc overcome by an externally imposed oppression. But in Huxley's vision, no Big Brother is required to deprive people of their autonomy, maturity and history. As he saw it, people will come to love their oppression, to adore the technologies that undo their capacities to think. In short, Orwell feared that what we hate will ruin us. Huxley feared that what we love will ruin us. (Postman 1985, Foreword)

This concise contraposition by Neil Postman – whose book *Amusing Ourselves to Death* delivered the inspiration of this article's title – contrasts these iconic utopias. Postman's media critique is about the possibility that Huxley was right. Arendt seems to be on the same side fearing similar to Huxley that loving freedom of labour and approaching it by automation might ruin us. Not because being unemployed impoverishes us, but because labour has become a basic part of the human condition in the modern age. Striving to avoid labour leads to inventions – for technology is with Ortega y Gasset "the effort to save effort"[16] – but succeeding to do so would lead to a crisis: "What we are confronted with is the prospect of a society of laborers without labor, that is, without the only activity left to them. Surely, nothing could be worse." (Arendt 1998, p. 5). A widely assistive society would have delegated not only *labour* (here Arendt's *Arbeit*) and *work* (here Arendt's *Herstellen*) to automated machines or capitalistic slavery but potentially most parts of *action* (here Arendt's *Handeln*). If *labour* (again Arendt's *Arbeit*) is decisive for the society of labourers and freedom from labour would inflict the worst crisis, then what would freedom from 'lifting a finger' do to a society of actives? Unlike an Orwellian suppressor hindering the people to pursue self-chosen meaningful tasks, individuals then – like in a Huxleyan soma-state – would lack any motivation and conception of meaningful activities. It would be the worst case, having delegated not only operative control (choice of means) and strategic control (choice of procedure) to assistive systems. The autonomy of purpose (choice of goals, wishing) would be withered. Ortega y Gasset elaborated on the relation of technology and wishes (Técnica y deseos). For him the technician/man with technological capacities (el técnico) is like a *nouveau riche* (el nuevo rico) when it comes to goals and wishes:

He (man the technician) holds in his hands the means to achieve his wishes, but he does not know how to wish. At heart he notices that he wishes nothing, that he on his own is incapable to orient his desire and to decide among the countless things that the environment

[16] "Tenemos, pues, que la técnica es, por lo pronto, el esfuerzo para ahorrar el esfuerzo o, dicho en otra forma, es lo que hacemos para evitar por completo, o en parte, los quehaceres que la circunstancia primariamente nos impone." Ortega y Gasset (1964, p. 333). "Let us record the fact that technology is, for the time being, the effort to safe effort or, in other words, it is what we do to avoid entirely or in parts the tasks that the circumstances primarily impose on us." (my translation).

> offers to him. Therefore, he looks for an intermediary that orients him and finds it in the predominant wishes of others. Here is the reason why the nouveau riche first buys a car, a player piano, and a phonograph. He puts in charge the others who wish for him. (my translation of: Ortega y Gasset 1964: S. 343–344)[17]

Assuming that a futurible omni-assisted user classifies as an Ortegan *técnico* this point can be specified as follows: The assisted nouveaux riches hold the means to realize their wishes (– the land of milk and honey where no one even lifts a finger). But – and this is the crux – they are not able to have wishes. They know deep insight that they do not wish for anything, unable to direct their desire and to decide among the countless things that the assistive systems offer. This assisted nouveaux riches have delegated even their normative autonomy to others (to assistive assemblages) that wish for them – or – that 'decide' on their own what the user actually wishes for. All assistance, support, help, all improvement of means is vain if there is no idea about a worthy goal to pursue. A mismatch of means and goals is no longer the problem then, but a fundamental lack of goals and a basic incapacity of goal setting.[18]

The extensively pampering services of the assistive systems do not just assist the wish realization but the wish generation as well. Unsurprisingly, current artificial assistants or services come with sophisticated recommendation skills to obligingly compensate for that directive. They offer a choice of millions and billions of songs, podcasts, audiobooks, videos, pictures, and all sorts of services – just one magical "wake word" away. In the past, we might have only needed operative or strategic assistance such as a mp3-library simplifying the search of a CD in the cupboard or a mechanical Jukebox relieving the activation of a vinyl on the turntable.[19] In the assistive society, we would also need help for the choice of goal and purpose itself. If even the goal choices are artificially assisted, it becomes easier to predict the results of these 'choices': One aspect of personalised assistance, which is currently a hype, is indeed the prediction of needs and desires of users. Intelligent systems track personal preferences and habits in order to predict when (the digital represen-

---

[17] "Tiene en la mano la posibilidad de obtener el logro de sus deseos, pero se encuentra con que no sabe tener deseos. En su secreto fondo advierte que no desea nada, que por sí mismo es incapaz de orientar su apetito y decidirlo entre las innumerables cosas que el contorno le ofrece. Por eso busca un intermediario que le oriente, y lo halla en los deseos predominantes de los demás. He aquí la razón por la cual lo primero que el nuevo rico se compra es un automóvil, una pianola y un fonógrafo. Ha encargado a los demás que deseen por él." (Ortega y Gasset 1964: S. 343–344)

[18] Arendt's student Richard Sennett analyzed "the special human condition of being *engaged*" and emphasizes "the desire to do something well, concretely, for its own sake." Sennett (2008, p. 20, 144–145). He shows how rewarding labor for its own sake is and positions competences and mastery – not comfort – as a way of life, thus reminding on the Renaissance concept of *sprezzatura* (see note 30). "History has drawn fault lines dividing practice and theory, technique and expression, craftsman and artist, maker and user; modern society suffers from this historical inheritance. But the past life of craft and craftsmen also suggests ways of using tools, organizing bodily movements, thinking about materials that remain alternative, viable proposals about how to conduct life with skill." Sennett (2008, p. 11).

[19] Sometimes the offered assistance is more than questionable: What is an 'eject' button on a CD player's remote control good for?

tation of) the user might desire a certain assistive intervention. These services then offer a recommendation before the user has actually formed his desire. Therefore, the systems disrupt the goal setting process of the user early on. Is this not capitalism's great dream: the provision of satisfaction before the actual desire has emerged? The assistant shapes the formation of desire according to the offered satisfaction. The user's personalised profile restricts the offered options and his current desire formation is interrupted based on the past user history. The assistants' artificially learned recommendations are always deduced on past wishes, past desires, past situations.[20] The company owned assistants do not completely reflect learned preferences, but also feed into their recommendations what they are programmed to sell. The interconnected, sensor-loaded, learning assistants are indeed *Servants of Two Masters* (Goldoni), they are both assistants to their users as well as agents of their creators. Freeing time would not be such a severe problem if the unburdened preserved or gained the skills and ambitions to choose their purposes, to direct all their technical might towards desirable differences, towards higher goals. Even *Vogli sempre poter* is insufficient if the "society does no longer know of those other higher and more meaningful activities for the sake of which this freedom would deserve to be won." (Arendt 1998, pp. 4–5) Therefore, the goal setting competence, the purpose autonomy has to be an educational focus in an assistive society. Otherwise, all we will want is either Huxley's soma of not wanting anything at all or just what the assistants guess or are told to propose us. Compared to the latter, Postman's fear of television *amusing us to death* seems comparably harmless because television as a one-to-many medium does not know the individuals' moving profile, sleep pattern, eating or sexual preferences, consumption profiles etc.

[20] The name *Echo* involuntarily reminds on this orientation towards the past. The mystical Echo (Ovid, Metamorphoseon, Lib. III.) was an extraordinarily chatty nymph ("vocalis nymphe", 357). She was transformed into a stone and condemned to merely repeat the last syllables of other voices, condemned to not being able to refuse an answer ("quae nec reticere loquenti", 357) nor to speak first ("nec prior ipsa loqui didicit", 358) and to resonate everything ("resonabilis Echo", 358). In short: The output of an *echo* is strictly restricted by the input. It is a self-amplifying system with the deceitful illusion of a dialogue ("alternae deceptus imagine vocis", 385). What you say is what you get ("quot dixit, verba recepit", 384) – or – you only here what you ask for. How could one ever learn something this way? To bring in a concept of inquiry by Martin Heidegger (who called Hannah Arendt in letters *his coquettish dryad/nymph,* "neckische Waldnymphe"): The *enquired answer* (das Erfragte) coincides thus with the *asked question* (das Gefragte), which makes gaining knowledge impossible because in order to learn something new (different Erfragte), it would not suffice to ask differently (change the Gefragte) but you would have to be another *inquirer* (Fragender). See Heidegger (1977, §2), Gransche (2015, pp. 361–363). We shall not forget that Echo was in love with Narcissus and that the illusion of a counterpart that always confirms one-sided inquiries (see Pariser (2011)) and affirms preexistent beliefs reinforces narcissism. In other words, conversing with echo might metamorphose you into a narcissist.

19.3.2 Assistants as Artificial Slaves

Considering what was said above, the master-slave figure seems to describe the human-technology relations in an assistive society more adequately than the collaborative mode. Schematically, the master autonomously sets the goals and enjoys the yield; the slave has to work to actualise these goals. Thinking about the human-assistant-world relation as a master-slave-thing relation sounds a clarion call against wishing for an effortless enjoyment all too prematurely. An assistive society has to keep on educating its goal setting, purpose, or normative competencies while delegating the operative and strategic choices to others. With Hegel in mind, the advice can be added to carefully reflect on those other two delegations as well. Work, the "activity of giving shape and form" (Hegel 1807, p. 195) is closely related to the respective skills to rework the world, to actively relate with the things. This working lies especially in the spheres of operative and strategic choices, of choosing and using plans, procedures, methods, machines, tools, and techniques. The German verb *bilden* combines the meaning of *to shape/to form* and to *educate*, so it means forming things as well as competencies. The English translation "…labour shapes and fashions the thing" (Hegel 1807, p. 195) is – as always – just one possible translative interpretation. The original reads: "Die Arbeit hingegen ist gehemmte Begierde, aufgehaltenes Verschwinden, oder sie bildet." (Hegel 2006, 135 [115]) "Sie bildet" can be translated in an educational sense as '…labour shapes and fashions the (skills of the) labourer.' Work – in the form of single-handedly executed tasks, specifically actualised competencies – *shapes and forms* not only the things but consequentially the executing instance as well. If this instance is a human person, then this shaping is far reaching, by learning and training, which ultimately can result in mastery thus offering "viable proposals about how to conduct life with skill." Sennett (2008, p. 11) If the instance is a tool or machine, the shaping is much less, only by attrition or abrasion. In the case of intelligent learning systems – i.e. artificial assistants – the series of task executions shapes the 'skills' of these systems, by personalisation or adaptation.

> The more you use Echo, the more it adapts. (Amazon 2016b)
>
> Artificial Intelligence algorithms learn your preferences to adapt and fit into your life. (Blue Frog Robotics 2016)
>
> Your robot evolves with you. Pepper gradually memorises your personality traits, your preferences, and adapts himself to your tastes and habits. (Aldebaran Robots 2016)

The impression of an independent assistant-slave is hardened not only by the fact that the performance improves while serving (unlike knives which usually pejorate), but also by the fact that the entire class of assistive systems improves through any single task execution or servile act of one of their members. Advanced assistants are not individual machines but an assemblage of interconnected components. Users do not possess this vast array of distributed skills and services; they only interact with one of many interfaces. One buys an Echo but is served by Alexa and – ultimately – by Amazon. Mistaking the single interface in the living room or pocket with the entire assistive system might be tempting because that would be the expectation of a human assistant, but it would be extremely naïve.

The research team of the *Human To Robots Laboratory*[21] attempts to teach Baxter robots how to manipulate things in *The Million Object Challenge*. A multitude of robots experimentally learns how to grab objects and stores the learned skills in a database. By the end of the experiment, every robot connected to this pool will have learned the skills to manipulate a million objects, even if one component has only learned a few or none of these skills by experimentation. According to this approach, every single assistant user 'educates' the assistant systems' skills, which then serves every user again. This is a major difference between the notion of human slaves and artificial assistants. Technically interconnected servants learn collectively, every skill one develops improves the entire class of instances.

Another rather obvious difference between human and artificial co-actors (serving or not) is that the latter are not fully recognised as social partners, and maybe never will, because social acceptance is based on mutual recognition. In other words, only acknowledged acknowledgement or recognised recognition is of any value to social beings. Artificial systems are hardly recognised to be worthy of acknowledging their users, with two exceptions: First, with the advancements in technology, people are tempted to project human characteristics onto technology up to a point where parasocial[22] and technophile or even techno-intimate[23] relations become possible. Humans are apparently very good in taking over both sides of a bilateral relation, compensating the complete lack of reaction from the counterpart: no matter whether this relation is about loving the yellow press presentation of Angelina Jolie or loving God. In this context, it seems possible that some humans might actually feel respected and acknowledged even by artificial systems.

A second exception concerns the close interweavement of assistive systems with actual humans as part of the assemblage. It has become almost impossible for the user to detect whether a signal, reaction, recommendation etc. comes from an algorithm or an actual human person. „Digital information is really just people in disguise."(Lanier 2013, p. 15) Amazon, for instance, created a service infrastructure called *Mechanical Turk – Artificial Artificial Intelligence*[24] that delegates micro-tasks – so called HITs – *Human Intelligence Tasks* – to real humans giving the impression of almost magical task achievement.[25] The *Artificial Artificial Intelligence*

[21] Brown University (2016b); It is not really surprising that the *Humans To Robots Laboratory* team uses Amazons Echo to interact with Baxter robots as well: Brown University (2016a).

[22] Parasocial relationships mean an affective relationship with a media character, like actually loving Dr. House, James Bond or Lara Croft (see Rubin (2015). These relationships obviously work without reciprocity, why potential parasocial relationships with artificial agents or robots would not suffer from their inability to return any affection.

[23] Levy (2008).

[24] Amazon (2016a).

[25] "However, Amazon is also exploring how to get non-elite service jobs out of the way of the Siren Servers of the future. The company offers a Web-based tool called Mechanical Turk. The name is a reference to a deceptive eighteenth century automaton that seemed to be a robotic Turk that could play chess, while in fact a real person was hidden inside. The Amazon version is a way to easily

is really just human intelligence in disguise of an artificial service. Where the humanness of an activity is effectively hidden, the humanness of the working conditions is especially precarious. Using artificial slaves and thus live a master's life without the bad conscious of subjecting human beings into slavery does not add up. Today's magically free services and cheap products produce their own kind of human slavery or highly precarious servitude hidden by an internet browser until at times a factory in Bangladesh burns down and lifts the veil of human entanglement and reveals how "empty the factories" after the "advent of automation" really are.

The last major difference between human and artificial servitude: the roles in human master-slave, magnate-butler, chef-assistant etc. relations are clearly assigned. The roles of the user-assistant relation, on the other hand, are quite uncertain. *User* is a revealing name here because it points at the behavioural dependency of a *user* (non-substance addict). Apart from the maintenance 'care' that artificial systems demand, the human-technology relation calls for using, operating and interaction skills on the human side; the German word for this human competence is *bedienen*. The phrase "Den Assistenten bedienen." translates either as "Operating/controlling/handling the assistant." or as "Serving the assistant." That means the assisted has to serve the assistant. What assistive systems developers withhold is that unlike human slaves, the artificial assistants not only execute but also cause labour. These systems, including the infrastructure they need to unfold their potential, need to be purchased, installed, maintained, troubleshot, supported and served with resources like energy, information, connection etc.

Immersing oneself in assistive assemblages with most intimate information of all spheres of life is not exactly an independent, enabled, empowered relation to the world. It is much closer to submitting oneself to a regime of pseudo-magical effortlessness, to a reign of comfort. Induced by Huxley's *Brave New World Revisited* (Huxley 1958) Dostoyevsky comes to mind: "In the end, they will lay their freedom at our feet, and say to us, 'make us your slaves, but feed us.'" (Dostoyevsky 1993, p. 26)

19.4 Comfort and Competencies in an Assistive Society

Two schematic types of assistance can be distinguished so far. The first type is *enabling assistance*, which can be currently found predominantly in medical or therapeutic contexts but not in voluntary day-to-day assistance. Technical systems like pacemakers, respiration apparatuses, exoskeletons, brain-computer interfaces and others enable, enhance, restore, cure, or substitute lost, weakened or never

outsource – to real humans – those cloud-based tasks that algorithms still can't do, but in a framework that allows you to think of the people as software components. The interface doesn't hide the existence of the people, but it still does try to create a sense of magic, as if you can just pluck results out of the cloud at an incredibly low cost." Lanier (2013, pp. 169–170).

developed abilities.[26] Other examples are rehabilitation assistive systems supporting movements with the aim of not being needed after a certain training time. This assistance focusses on enabling capabilities, on improving a patient's means and skills to autonomously reach goals (again) in the future. This assistance is involuntary in the sense that the assistance is used just as long and just as much as necessary to survive or life relatively self-determined without pain. No one would implant a pacemaker only because it is more comfortable to technically control one's heartbeat while getting nervous.

A second type is *comforting assistance*. Even if the first type aims at comfort as well (more or less and often secondary to very tiresome enabling processes), the second type of assistance is geared towards ease and relief, to unburden the users of all sorts of effort. It aims to simplify everyday life, to enhance efficiency, to take over the hard, boring, dangerous, exhausting parts of actions, and – generally speaking – to reduce the resistance of the world (e.g. technically overcoming gravity with an escalator instead of exhausting muscle power). Comfort-oriented assistance is voluntary in as far as the choice using an escalator is motivated by laziness and not because disability makes stair climbing impossible. The latter type is assumed prevalent in a futurible assistive society. *Pepper, Buddy, Echo, Jibo, Cortana, Siri, Alexa;* all fall into the category of the comfort-oriented type.[27]

In the here assumed futurible assistive society a plethora of advanced assistive systems for everyday life ease will be part of our lives. Comfort is *the* central theme and goal of everyday life artificial assistance (type two), in the Jibo-sense of "making everyday tasks simpler and easier." There are two basic normative positions to it: a) that will not be a problem or b) that will be a problem. The position depends on the respective concepts of *competence* and *comfort.*

Competence can be understood as a *routinely available, habitual skill, ability,* in the form of a dispositive (Hubig 2007, p. 214). The dispositive has two conditions that have to be fulfilled to competently solve a problem: a *structural* condition and an *actualisation* condition. The competence *car driving,* for instance, acquires some cognitive structural conditions as steering knowledge and understanding other drivers and some physical structural conditions such as strength, visual and reactive abilities. In order to actualise these structural conditions and successfully solve the task of *car driving*, actualisation conditions need to be met such as the presence of an actual (and not a toy) car, an adequate road and fuel in the tank, no traffic jam or flooding. The interrelation of these two conditions is crucial because the structural conditions are influenced by the repeated single actualisations of the task; just as a sequence of attempted marathon actualisations (that is physical training) builds the

[26] Assisted suicide, assisted conception etc. are complicated and medico-ethical special problems that cannot be addressed here, despite the – for some misleading – title *Assisting Ourselves to Death* that is a reminiscence to Postman's *Amusing Ourselves to Death* and not to assisted suicide.

[27] The two types are schematic and simplified for heuristic reasons. As said competence-oriented assistance aims as well at ease and relief and comfort-oriented assistants can just as well be used in a medically enabling context. But Alexa etc. are not designed nor advertised for e.g. paraplegic people.

competence of running a marathon where the structural conditions for that endeavour (muscles, stamina, running techniques etc.) are hitherto missing. This double character of the competence dispositive reveals that skills are built in the respective task accomplishment. Competence arises from the actual problem-solving attempt, by single-handedly overcoming the resistance of the world. A delegation of task execution to other instances is an avoidance of contact possibilities with the world's resistance. The effort delegation to artificial assistants results in a loss of resistance, a loss of the conditions for competence development. This loss is much worse than just losing a specific competence because it is a loss of the possibilities to develop competencies in general. This delegation to assistants deprives the assisted of its actualisation instances that are essential to competence formation. It is not possible to learn car driving with a toy car nor to learn horse riding on a unicorn. The actualisation conditions are just not met.

Overcoming resistances, working against counteraction (like gravity or inertia) is how we build competencies in the first place. Resistance, therefore, is a prerequisite for competence. Less resistance, however, is at the heart of comfort. The major aim of comfort-oriented assistance is resistance-reduction to spare the effort of even having to lift a finger. In an assistive society, the loss of resistance is not regretted as an incapacitating development. On the contrary, it is celebrated as a gain in comfort. This relation between competence, delegation, and comfort reveals that pseudo-magical effortlessness of assistive systems provides a disabling *easy and simple comfort* and not an effortlessness due to superior means. The Latin root of the word *conforto* initially meant *to profoundly strengthen, to encourage a lot.* In that sense comforting means *making strong and able, enhance competencies* (both physical and mental). Concerning a potential imbalance of means and goals it means making tasks simpler and easier *by enabling and empowering the actor* and not by avoidance, delegation, letting others do the task instead. A virtuous musician, for instance, who invested a lot of effort in forming his skills, comfortably plays sophisticated parts. This effortlessness does not need delegation to a third instance because it rebalanced means and goals in a way that allows for an independent comfortable twofold actor-world relation. The performance of mastered competencies of a virtuoso then itself gives the impression of complete effortlessness and thus a sense of magic to observers. The *Renaissance men* – this time Baldassare Castiglione – idealised this performed effortlessness in even the most demanding actions with the untranslatable concept of *sprezzatura.*[28]

[28]Approximately *sprezzatura* can be grasped as effortlessness in most demanding tasks, especially in art, literature, poetry, fencing, dancing etc. It is a sort of careless grace, of nonchalance and an indispensable attribute of the perfect Renaissance gentleman (*Corteggiano*).

19.5 Conclusion

Many current developments make a future assistive society possible that is extensively pervaded with comfort-oriented assistive systems that take over tiresome tasks and make everyday life *simple and easy*. The artificial assistants are not performing up to the developers' promises yet, but paths heading for them are on the horizon. This article reflected on consequences for the human-world relation and on whether this relation seems likely to be continuously transformed to a human-assistant-world relation by interposing artificial assistive systems between human actors and the things.

A far-reaching delegation of tasks – that is of choices and executions on the operative and strategic level of actions – has severe consequences for the delegators. The delegation has not only profound effects on the possibilities of competence formation but it also affects the highest part of our autonomy, the normative choices of goal and purpose. Depending on the design principles of the artificial assistants, the comfort-orientation of this threefold action constellation can have hardly desirable outcomes for an assistive society. Current developments in voluntary everyday life assistance indicate a predominant orientation on incapacitating comfort by task-avoidance instead of enabling comfort by competence formation. Assistance can foster or hinder skill formation depending on whether it is positioned to grant enabling support for human action or comforting substitution of it. If the actor-assistant-world relation is understood in correspondence to a master-slave-thing relation, it becomes apparent how work delegation and servant interposing between oneself and the resistance of things damages the own competencies and inverts an initial or alleged independence into an actual dependence. An initial or alleged voluntary assistance thus converts into an actual assistance dependency. If means improvement, goal-setting competence, and the necessary skills needed to overcome the world's resisting forces are not carefully fostered, if we no longer skilfully engage, then masters turn into slaves, delegators into users, and the assisted have to serve the assistants.

Due to constraints in time and energy, we cannot master every competence we might need to pursue any goal we might want. The Renaissance man is an impossible figure in the twenty-first century: a lifetime is not even enough to master the state of the art of only one discipline let alone gaining *sprezzatura* as a *homo universalis*. Certain realisation instances fade away and with them our possibilities to form the respective competencies. Today's technical gas sensors have made the profession of the gas smeller obsolete because they stopped the need for humans to walk the streets with a smelling pipe to detect gas leaks. This is surely not a problem. We gain and lose continuously various skills according to our *vita activa*, depending on what we ourselves do and do not. If we maintain a well-considered balance between losing and gaining competencies, then there is no problem.

However, after the advent of automation and at the brink of all-task, unspecific, omnipresent assistive assemblages we risk losing this balance. We risk losing the possibility of resistance and thus the very conditions of competence formation itself. An assistive society has to reflect on crucial competencies, whose losses are impossibly acceptable. Professional gas smelling surely is not part of this, but literacy might be. One of the biggest challenges on the way to an assistive society will be to identify, agree on, and protect those indispensable competencies and never delegate them to assistants or whatsoever slaves or servants. The challenge is to prevent losing the resistance experiences and the sense "of those other higher and more meaningful activities." (Arendt 1998, pp. 4–5) The challenge is to gain an enabled comfortable relation with the world but to avoid all too easy and simple comfort at crucial tasks.

Maybe the focus on the assistance role of intelligent support systems is already problematic. Artificial systems are designed and developed in a series of roles other than assistant. Perhaps roles like an artificial team member, coach, trainer, or teacher help steering towards an enabling function of the support rather than the comfort orientation of assistant, servant, butler, or slave. Ultimately, is has to be reflected whether competence-oriented assistants should be endowed with a higher-level directive to protect the skills of the assisted, if necessary against their own comfort addicted laziness. Would that not be egregious, an artificial assistant that at times denies comforting services in order to preserve or build the users' competencies? Would we engage with occasionally unruly, sassy assistants that re-delegate demanding but enabling action parts back to the assisted?

How are we able and willing to influence the direction of the possible future developments in the assisted society? Following the reflections of this essay, we will have to choose from two schematic options of either a simple-and-easy assistive society – that would be simple minded and easily overwhelmed – of incompetent dependent masters assisted to a death-like soma state, or of a reflective assistive society of up-to-date skilled people. It all starts with lifting a finger.

References

Aldebaran Robots. (2016). *Who is Pepper?* https://www.aldebaran.com/en/cool-robots/pepper. Accessed 10 Mar 2016.

Amazon. (2016a). Mechanical Turk. *Artificial Artificial Intelligence*. https://www.mturk.com/mturk/welcome. Accessed 1 Nov 2016.

Amazon. (2016b). *Amazon Echo*. https://www.amazon.com/Amazon-Echo-Bluetooth-Speaker-with-WiFi-Alexa/dp/B00X4WHP5E. Accessed 30 Oct 2016.

Arendt, H. (1998). *The human condition* (2nd ed.). Chicago: University of Chicago Press.

Beck, U. (1986). *Risikogesellschaft. Auf dem Weg in eine andere Moderne*. 5. Aufl. Frankfurt am Main: Suhrkamp.

Bindé, J. (Ed.). (2006). *Towards knowledge societies*. Paris: United nations educational scientific and cultural organization (UNESCO world report).

Blue Frog Robotics. (2016). *BUDDY*. http://www.bluefrogrobotics.com/en/buddy/. Accessed 10 Mar 2016.

Brown University. (2016a). *Amazon Echo + ROS + Baxter*. http://h2r.cs.brown.edu/amazon-echo-ros-baxter/. Accessed 1 Nov 2016.

Brown University. (2016b). *Million object challenge*. http://h2r.cs.brown.edu/million-object-challenge/. Accessed 1 Nov 2016.

Carrer, L. (Ed.). (1836). *Lirici italiani del secolo decimosesto,*. con annotazioni di Luigi Carrer. Venezia: L. Plet.

Clarke, A. C. (1973). Hazards of Prophecy: The failure of imagination. In *Profiles of the future: An inquiry into the limits of the possible* (2nd ed.). New York: Harper & Row.

Dostoyevsky, F. (1993). *The grand inquisitor: With related chapters from the Brothers Karamazov*. Indianapolis: Hackett Pub..

Gransche, B. (2015). *Vorausschauendes Denken: Philosophie und Zukunftsforschung jenseits von Statistik und Kalkül* (1st ed., Edition panta rei). Bielefeld: Transcript.

Gransche, B., Shala, E., Hubig, C., et al. (2014). *Wandel von Autonomie und Kontrolle durch neue Mensch-Technik-Interaktionen: Grundsatzfragen autonomieorientierter Mensch-Technik-Verhältnisse*. Stuttgart: Fraunhofer Verlag.

Hardt, M., & Negri, A. (2000). *Empire*. Cambridge, MA: Harvard University Press.

Hegel, G. W. F. (1807). Phenomenology of mind/spirit (J. B. Baillie, Trans.). http://www.gwf-hegel.org/PhenText/compare.html. Accessed 1 Nov 2016.

Hegel, G. W. F. (2006). *Phänomenologie des Geistes* (Philosophische Bibliothek, Vol. 414). Hamburg: Meiner.

Heidegger, M. (1977). *Sein und Zeit (1926): I. Abteilung: Veröffentlichte Schriften 1914–1970* (Gesamtausgabe, Vol. 2). Frankfurt am Main: Klostermann.

Hubig, C. (Ed.). (2000). *Unterwegs zur Wissensgesellschaft. Grundlagen, Trends, Probleme*. Berlin: Ed. Sigma.

Hubig, C. (2007). *Die Kunst des Möglichen II: Grundlinien einer dialektischen Philosophie der Technik; Ethik der Technik als provisorische Moral* (Vol. 2). Bielefeld: Transcript.

Huxley, A. (1958). *Brave New World Revisited*. https://www.huxley.net/bnw-revisited/

Jibo. (2016). *Jibo, The World's First Family Robot*. https://www.jibo.com/. Accessed 10 Mar 2016.

Jouvenel, B. D. (1964). *L'Art de la conjecture*. Monaco: Éditions du Rocher.

Karafyllis, N. C. (Ed.). (2003). *Biofakte: Versuch über den Menschen zwischen Artefakt und Lebewesen*. Paderborn: Mentis.

Lanier, J. (2013). *Who owns the future?* London: Allen Lane.

Levy, D. (2008). *Love and sex with robots: The evolution of human-robot relations*. New York: Harper Perennial.

Ortega y Gasset, J. (1964). Meditaciòn de la técnica (1939). In: *Obras Completas. Tomo V (1933–1941)*. Madrid.

Pariser, E. (2011). *The filter bubble: What the internet is hiding from you*. New York: Penguin Press.

Perrow, C. (1984). *Normal accidents: Living with high-risk technologies*. New York: Basic Books.

Postman, N. (1985). *Amusing ourselves to death: Public discourse in the age of show business*. New York: Viking.

Rubin, A. M. (2015). Parasocial relationships. In C. R. Berger, M. E. Roloff, S. R. Wilson, J. P. Dillard, J. Caughlin, & D. Solomon (Eds.), *The international encyclopedia of interpersonal communication* (pp. 1–5). Hoboken: Wiley.

Sennett, R. (2008). *The craftsman*. London: Penguin.

Chapter 20
Engineering Privacy on the Scaffolds: An Existentialist Examination of Privacy by Design

Kristen Psaty

Abstract What is digital privacy and who in society should be responsible for protecting it? This paper considers the challenges in defining digital privacy as well as the newest movement in privacy law and regulation, the concept of "privacy by design," which calls upon engineers to integrate privacy into technology throughout the construction process. This paper seeks to advance two objectives. First, it turns the focus of inquiry to the experience and decision-making process of the engineer. Through a combination of contemporary technological thought, an existential philosophical lens, and an informed legal perspective, this paper seeks to understand the relationship between the computer engineer, the user, and privacy. Second, this paper hopes to add to the philosophical debate surrounding the contours of privacy as it relates to the engineer's role in forging it. This analysis seeks to reveal, in part, the nature of privacy and the necessary conditions for its implementation in technology.

Keywords Engineering for people · Privacy by design problems

20.1 The Debate Over Modern Digital Privacy Rights

Some of the most contentious debates in the field of technology law and ethics erupted in disputes over digital privacy rights. One of the most well-known instances includes the 2016 Federal Bureau of Investigations (FBI)-Apple encryption controversy, which centered on whether United States Courts could compel technology hardware manufactures to assist in unlocking cellphones protected by encryption.

Foreword
All ideas and points of view belong to me, and are no way representative of any of my employers.

In dedication to the future.

K. Psaty (✉)
Privacy and Public Policy, Facebook, Menlo Park, CA, USA

A. Fritzsche, S. J. Oks (eds.), *The Future of Engineering*, Philosophy of Engineering and Technology 31, https://doi.org/10.1007/978-3-319-91029-1_20

The issue gained international attention after a December 2015 terrorist attack occurred in San Bernardino, California; killing 14 and injuring 21 others (Perez-Pena and Schmidt 2015). Shooter Syed Farook's iPhone iOs 9 was recovered, though it was protected with an access code and enabled with a feature that would erase all phone data after 10 unsuccessful password attempts (Rubin et al. 2017). This hardware feature rendered the device inaccessible by law enforcement. A U.S. magistrate ordered Apple to assist the government in unlocking the iPhone. However, Apple CEO Tim Cook publicly contested the order saying, "The implications of the government's demands are chilling. … We believe it would be in the best interest of everyone to step back and consider the implications."(Cook 2016). The incident sparked some of the most intense debates on technology ethics in history, and illuminated the engineer's increasingly powerful role as arbiter of personal information and privacy rights. This paper contemplates the framework by which regulators perceive privacy harm and the challenges with engineering privacy into technology design.

20.1.1 Problems Regulating Digital Privacy

There is great confusion regarding what digital privacy is and whose responsibility it is to protect it. The unsettled privacy law landscape does little to clarify norms and standards.

So far, tensions in existing legal frameworks have proven unstable, as is the case with American and European privacy regimes, which reached a breaking point with the collapse of the Safe Harbor Agreement (Clark 2015). In October of 2015, the European Court of Justice ruling in *Schrems v. Data Protection Authority* struck down an international agreement between the U.S. and the EU that allowed companies to transfer the personal information of users between countries (Scott 2015). The dissolution of the Safe Harbor Agreement illuminated the need to harmonize international standards as well as the underlying differences related to digital privacy. Subsequently, the EU-U.S. and Swiss-U.S. Privacy Shield Frameworks were put into place in order to adequately enable international data transfers under EU law (The International Trade Administration, U.S. Department of Commerce 2016). However, with the change in U.S. Executive Administration, President Trump has already threatened U.S. suitability to process foreign data (Lomas (2017)). Section 14 "Privacy Act" of the President's Executive Order "Enhancing Public Safety in the Interior of the United States" mandates that applicable agency privacy policies, "exclude persons who are not United States citizens or lawful permanent residents from the protections of the Privacy Act," contrary to the spirit of guaranteed protections negotiated under the Privacy Shield (The White House Office of the Press Secretary 2017). Further cross-border anxiety has continued within the privacy law community with the May 25, 2018. General Data Protection Regulation (GDPR) compliance deadline and associated enforcement uncertainty.

It is then to be expected that the engineering community constructing global technological systems have openly expressed confusion over lack of clear and uniform guidance. For instance, The Internet Architecture Board (I.A.B.),[1] "Privacy Considerations for Internet Protocols," concedes, "Different people have radically different conceptions of what privacy means, both in general and as it related to them personally." (IAB 2016; Cooper et al. 2013). Furthermore, the IAB observes, "privacy as a legal concept is understood differently in different jurisdictions." (Cooper et al. 2013). Compounding this lack of clarity are the epistemic challenges derived from the rapidly shifting technological landscape. The high acceleration rate of modern innovation has meant engineers are constantly being posed with novel ethical applications, often corresponding to a loss of prescriptive guidance. Without a clear normative process, engineers and their creations must emerge to answer the question of what future privacy institutions should look like.

20.2 Engineering and Privacy-Choices

Without explicit and uniform guidance on privacy, engineers must create and define their own standards and precedents in order to protect privacy. Just as Tim Cook urged his company, the government and law enforcement during the San Bernadino debate, "to step back and consider the implications," (Cook 2016) regarding unlocking criminal's iPhones, engineers must make choices that dictate the extent to which technology impacts personal privacy.

There is a current policy shift toward engineer-driven standards called 'Privacy by Design' (PbD), an "approach to protecting privacy [by] embedding it into the design specifications of technologies," including into physical infrastructures themselves.[2] (Hustinx 2010). An international group of privacy regulators formally adopted PbD in 2010 at the 32nd International Conference of Data Protection and Privacy Commissioners (Cavoukian 2010). The resolution declares that enlisting engineers as stakeholders in privacy protection is necessary stating, "With the increasing complexity and interconnectedness of information technologies, nothing short of building privacy directly into the system design and process can suffice." (Cavoukian 2010)

PbD is now championed by regulators and policy advocates as a major solution to digital privacy concerns.[3] (Klitou 2014). Engineers in a variety of contexts are

[1] The IAB provides long-range technical direction for Internet development, a direct evolution of the Interne Configuration Control Board (ICCB), created by the 1979 program manager of DARPA during early development of the Internet.

[2] The concept of privacy by design has its roots in the publication of a report on "Privacy Enhancing Technologies" in a joint effort by the Information and Privacy Commissioner of Ontario, Canada and the Dutch Data Protection Authority in 1995.

Privacy by design is built on 7 Foundational Principles: 1. Proactive not Reactive; Preventative not Remedial; 2. Privacy as the Default Setting; 3. Privacy Embedded into Design; 4. Full Functionality: Positive-Sum, not Zero-Sum; 5. End-to-End Security – Full Lifecycle Protection; 6. Visibility and Transparency – Keep it Open; 7. Respect for User Privacy – Keep it User-Centric.

[3] The operationalization of privacy still rests on the premise that privacy can be engineered. As some critics have asserted, not every privacy threat posed by disruptive technology is compatible with being designed or engineered away.

actively experimenting with operationalizing the principles of PbD.[4] (Carnegie Mellon University 2017). Still, the mandate to cast privacy into digital creations is problematic. PbD is foiled when engineers design for users across international boundaries, where integrating "privacy" means different things to different users. The result is an increased need for thoughtful engineering decision-making.

20.2.1 Difficulties Defining an Applicable Privacy Model

In order to design privacy into a user-facing system, some of its characteristics must be understood. Though it does not lend itself to easy reduction, one of the universal and immutable qualities of a pragmatic view on privacy is that it relies on a situational contact and cannot exist in a vacuum. A necessary condition for privacy is that it involves the interaction of at least two parties. The two-fold relationship between an "observer" and an "observed" is the most paradigmatic example of a privacy relationship. Digital privacy and digital privacy harms are frequently more caveated and can include non-natural or hypothetical parties as defined by modern privacy laws. For example, a business entity can perpetrate a privacy violation against an individual.[5] (Federal Trade Commission 2016). Further, an unauthorized disclosure, where personal information was only made available to a hypothetical actor and not necessarily seen by a human being, can also suffice as a recognized privacy incident.[6] (U.S. Department of Health and Human Services 2016). Still, it seems it seems impossible for a single person or entity to violate their own

---

[4]Carnegie Mellon University has implemented a Master of Science in Information Technology (MSIT) in Privacy engineering. The program is the first of its kind and aims to integrate privacy perspectives spanning product design, software development, cyber security, human computer interaction, as well as business and legal considerations.

[5]Examples of corporations implicated in privacy actions abound. One example can be found in the 2016 Federal Trade Commission settlement with Vulcun under privacy related charges that it unfairly replaced a popular web browser game with a program that installed applications on consumers' mobile devices without their permission.

[6]For example, in the American personal health information sector the Health Insurance Portability and Accountability Act (HIPAA) Breach Notification Rule, 45 CFR §§ 164.400–414, requires HIPAA covered entities and their business associates to provide notification following a breach of unsecured protected health information. A breach under the HIPAA Rules is defined as, "…the acquisition, access, use, or disclosure of PHI in a manner not permitted under the [HIPAA Privacy Rule] which compromises the security or privacy of the PHI." (See 45 C.F.R. 164.402).

"Access" to the personal information is broadly construed to encompass situations where electronic protected health information (ePHI) could have only have *potentially* been viewed. This example is described in regulatory guidance as a breach in incidents where ePHI is encrypted as the result of a ransomware attack, unless the covered entity can overcome the presumption by demonstrating that there is a "…low probability that the PHI has been compromised." The entity must then comply with the applicable breach notification provisions, including notification to affected individuals without unreasonable delay, to the Secretary of HHS, and to the media (for breaches affecting over 500 individuals) in accordance with HIPAA breach notification requirements. See 45 C.F.R. 164.400–414.

privacy. To create privacy, or privacy protective products, then it seems necessary that an engineer could not craft or "design" privacy without considering another person either. An awareness of the user is necessary to posit privacy into a digital system.

20.3 Existentialism and Engineering – *Engineering for the Digital Other*

If legal institutions and policies are earnestly entreating engineers to breathe privacy into their products, or as in the Apple encryption debate, calling upon them to mediate the privacy values of society, it is worth examining what it is we are asking of them at the individual level. Using an existentialist lens to examine the engineering process and the choices engineers make as individual actors who are increasingly divorced from normative and regulatory advice, we may reveal in part, *what privacy is*, as well as what is necessary for its implementation.

In *Being and Nothingness*, seminal Existentialist thought-leader Jean Paul Sartre, sets up the unique problems and insights that arise from the existence of others. A modern application of Sartre's framework exposes the relationship of an engineer with the hypothetical user for whom a technological program or interface is designed. For Sartre, the presence of the Other is seismic to our mode-of-being. He explains, "The appearance of the Other in the world corresponds therefore to a fixed sliding of the whole universe," as well as, "a decentralization of the world which undermines the centralization which I am simultaneously effecting." (Sartre 1992a, p. 343). In other words, the awareness of another person alters the way in which we perceive and interact with the world.

For Sartre, the existence of an articulable Other is critical to the formation of reflexive consciousness. This awareness precipitates such things as self-recognition, an understanding of one's own power, and even new modes of being. (Sartre 1992b, p. 350–357). Sartre explains this shift in consciousness most aptly with the example of shame. He explains:

> In fact, no matter what results one can obtain in solitude by the religious practice of shame, it is in its primary structure shame before somebody. I have just made an awkward or vulgar gesture. This gesture clings to me; I neither judge it nor blame it. I simply live it. I realize it in the mode of for-itself. (Sartre 1992e, p. 302)

An awareness of the Other, however, alters the individual's consciousness and mode-of-being. Sartre continues, "But now suddenly I raise my head. Somebody was there and has seen me. Suddenly I realize the vulgarity of my gesture, and I am ashamed." (Sartre 1992e, p.302).

Sartre's recount of shame is relevant to the problem inherent in designing privacy for several reasons. The first is in understanding the implications of seeing the Other while engineering, as well as impediments involved with seeing the Other in a digital space. Second because the structure of shame, which cannot exist in solitude, is similar to that of privacy, the limitations and consequences associated with the absence of the Other may also be instructive for engineers who are trying to craft privacy protective systems or make privacy-related decisions.

20.3.1 Discerning the Other in a User

In Sartre's analysis, he discusses apprehension of the Other as the sight of a physically embodied person. What he could not have contemplated would be apprehension of the Other as manifested in the digital form of today.

To understand the digital apprehension of the Other, it is worthwhile to examine the engineer's experience of shame in a digital space. Instead of making an awkward or vulgar gesture as Sartre describes, one can imagine the engineer has hacked a server, sent a ransomware demand, or purposefully built a toxic, privacy invasive system. How does she feel shame and apprehend the Other? Manifested digitally, a user is often represented in the negative, through empty form fields, such as a space to type a name, credit card or other personal information. These are the cracks and spaces an engineer leaves in anticipation of user interaction. Frequently, this may be the only meaningful interaction an engineer has with the Other while creating or constructing. It is worthy to explore to what extent this interaction is sufficient in triggering meaningful, reflexive consciousness within an engineer.

20.3.2 The Hypothetical User and the Problems Posed by a Digital Other

The hypothetical nature of the user limits the recognition of the Other in a digital space. This is especially problematic for Sartre who asserts that, "if the Other is accessible to us only by means of the knowledge which we have of him, and if this knowledge is only conjectural, then the existence of the Other is only conjectural." (Sartre 1992c, p. 340). Apprehending a hypothetical person will not suffice. As Sartre explains:

> In short, if the Other is to be a probable object and not a dream of an object, then his objectness must of necessity refer not to an original solitude beyond my reach, but to a fundamental connection in which the Other is manifested in some way other than through the knowledge which I have of him. (Sartre 1992c, p. 340)

Without a type of fundamental connection, the paradigm shift presented by the Other is not possible.

Other factors preventing the recognition of the Other in a digital space might have also been novel for Sartre. First, engineers must overcome a problem of scale, where the hypothetical user is nearly impossible to predict or distinguish in a scenario such as a high volume website or service. Further, there is a temporal problem. In the digital space, the moments of construction and use, or of mutual engagement between user and builder, are far removed temporally. Additionally, the engineer's interaction with the Other is spatially disconnected. Rarely does a programmer create in the presence of his intended audience. Even if she intends to engage with the system herself, and considers her own expectation of privacy, she will lack the reflexive consciousness produced by the existence of the Other, not to mention her expectations will likely be highly divergent from a typical user.

Moreover, even if the engineer is able to ascertain the Other in the process of creation, she may be resistant to doing so. Sartre elucidates the Other as, "the hidden death of my possibilities" with a metaphor about the inability of an individual to abscond himself in the presence of the Other. Sartre explains, "Both the obscurity of the dark corner and my possibility of hiding there are surpassed by the Other." (Sartre 1992d). Sartre goes on to say:

> When, before I have been able to make a move to take refuge there, he throws the light in the corner. This in the shock which seizes me when I apprehend the Other's look, this happens – that suddenly I experience a subtle alienation of all my possibilities, which are now associated with objects of the world, far from me in the midst of the world. (Sartre 1992d, p. 354)

The engineer may be, in part, resistant to seeing the Other, or so imagining the Other as seeing her, in order to avoid the subtle alienation Sartre describes. However, it may not be completely impossible for the engineer to see the user as the Other since the technological systems they are constructing are built in part as a result of user needs and include constraints based on considerations of user interface and design.

20.4 Significance of the Other: Indispensible for Shame and for Privacy

Applying the example of shame and the Other to an analysis of PbD also offers subtle clues about the contours of privacy. In particular, the structure of shame as a two-party process is similar to that of privacy. Sartre explains, "Shame is by nature recognition. I recognize that I am as the Other sees me." (Sartre 1992e, p. 302). Privacy similarly involves a dualism. In privacy, we recognize that we are being seen, or are capable of being seen. This involves a shift in consciousness whereby we are forced to see ourselves as observable. We are compelled to recognize ourselves, or our personal information, as the Other sees it.[7] Sartre presses, "Thus shame is shame of oneself before the Other; these two structures are inseparable." (Sartre 1992f, p. 303). Similarly privacy is privacy before another and the two are similarly inseparable.

Building on these similarities may provide some clues as to its inherent difficulty in operationalizing and implementing privacy. Applying Sartre's observations on

[7]Perhaps this is part of the contour of what is private, that which triggers this shift in self-observation.

the existence of the Other to engineers and users, or digital representations of people, illuminates the underpinnings of privacy as well as some of the inherent impossibilities related to mandated PbD.

Shame and privacy share similar properties in structure, situational operation, and the accompanying reflexivity. If the concrete Other is necessary for inducing the requisite reflexive consciousness for shame, then understanding the factors that enable shame may shed light on what enables or prevents the experience of privacy. This problem may in turn lead to access to the mode-of-being necessary to transfer privacy into technology. If the Other is necessary for privacy in the same way it is necessary for shame, ascertaining the Other in the digital ecosystem may be necessary if we ask engineers to bake privacy into their creations. Otherwise, it may be categorically impossible to ask engineers to meaningfully consider and design privacy.

In his analysis of the Other, Sartre also described the inability to perceive the Other while simultaneously formulating new ideas or concepts not readily apparent to the senses. As he explained, we cannot, "perceive and imagine simultaneously; it must be either one or the other." (Sartre 1992g, p. 347). Innovation similarly rests on the formulation of new ideas or concepts not readily apparent to the senses. If building privacy requires perceiving the Other, then an innovating engineer or designer will be categorically unable consider PbD during the inception of new technology. If this is the case, the problems associated with requesting engineers design privacy into technology may extend to all aspects of engineering decision making from the construction up to the very conception and innovation driving technology.

Clearly, reinforcing the needs of users and systems is imperative. This may mean engineering leaders have an onus to help their engineering teams learn the human condition presented by the Other as users. In addition, it would seem critical to understand implications in edge cases in order to conceive of the impact and potential for precedent, as in the Apple encryption debate over privacy standards and policy.

20.5 Engineers of the Future: Engineer Driven Standards and New Institutions

The challenges related to regulating technology are not new. In 1816, former American President Thomas Jefferson observed, "Laws and institutions must go hand in hand with the progress of the human mind." He explains:

> As [progress of the human mind] becomes more developed, more enlightened, as new discoveries are made, new truths disclosed, and manners and opinions change with the change of circumstances, institutions must keep pace with the times. We might as well require a man to wear still the coat which fitted him when a boy as civilized society to remain ever under the regimen of their barbarous ancestors." (Jefferson and Tompkinson 1816).

As technology continues to progress, it is likely the accompanying institutions will need to change as predicted. Increasingly engineers must not only define privacy for us, but also cast it into their creations. Because of the continued confusion in privacy law and regulation, the constant flux of the technological landscape and the policy push toward PbD, engineers must not look to be told the appropriate privacy standards when building, but instead forge and develop those standards. Engineers who build in the solitude of their own choices must be equipped with their own ethical decision making tools and consistently seek to consider the Other when engineering.

Future research may need to carefully explore the possibilities for engineers to see the Other while programming software applications or otherwise constructing technology in a way that is sufficient to be moralizing. As engineers continue to scaffold and build above the surface level of the policy advocate and lawmaker's waters, it is imperative that the engineer's role in safeguarding and developing digital privacy rights emerge as well.

Acknowledgments Thank you to the gamut of engineers interviewed in the process of writing this piece. Special thanks to George Jakobsche for sparking my love of technology law, to Greg Pemberton for stoking it, to Santa Clara University School of Law High Technology Law Institute for fostering it, and to my readers Karl Bozicevic, Kyle Psaty, Celine Purcell, Evan Selinger and Andrew Watts. Deep and humble thanks to Jill Gordon, Lydia Moland, Valiere Dionne, and the Colby College Philosophy Department.

References

Carnegie Mellon University. (2017) *Master of science in information technology in privacy engineering*. http://privacy.cs.cmu.edu/. Accessed 1 Feb 2017.

Cavoukian, A. (2010). *Privacy by design resolution*. Resource document. 32nd international conference of data protection and privacy comissioners. https://www.ipc.on.ca/site_documents/pbd-resolution.pdf. Accessed 15 Oct 2016.

Clark, K. (2015). *The EU safe harbor agreement is dead, here's what to do about it*. Resource document. Forbes. http://www.forbes.com/sites/riskmap/2015/10/27/the-eu-safe-harbor-agreement-is-dead-heres-what-to-do-about-it/#5401f2971719. Accessed 15 Oct 2016.

Cook, T. (2016). *A message to our customers*. http://www.apple.com/customer-letter/. Accessed 15 Oct 2016.

Cooper, A., Tschofenig, H., Aboba, B., Peterson, J., Morris, J., Hansen, M., & Smith, R. (2013). *Privacy considerations for internet protocols: RFC 6973*. Resource document. Datatracker. https://datatracker.ietf.org/doc/rfc6973/?include_text=1. Accessed 15 Oct 2016.

Federal Trade Commission. (2016). *Tech company settles FTC charges it unfairly installed apps on android mobile devices without users' permission*. https://www.ftc.gov/news-events/press-releases/2016/02/tech-company-settles-ftc-charges-it-unfairly-installed-apps. Accessed 1 Feb 2017.

Hustinx, P. (2010). *Privacy by design: Delivering the promises*. Resource document. Identity in the information society. https://www.ipc.on.ca/english/privacy/introduction-to-pbd/. Accessed 15 Oct 2016.

IAB. (2016). *Internet architecture board*. https://www.iab.org/. Accessed 15 Oct 2016.

Jefferson, T., & Tompkinson, H. (1816). *Quotations on the Jefferson memorial*. Resource document. Library of Congress. https://www.monticello.org/site/jefferson/quotations-jefferson-memorial. Accessed 15 Oct 2016.

Klitou, D. (2014). *A solution, but not a panacea for defending privacy: The challenges, criticism and limitations of privacy by design*. Resource document. Lecture notes in computer science. http://download.springer.com/static/pdf/612/chp%253A10.1007%252F978-3-642-54069-1_6.pdf?originUrl=http%3A%2F%2Flink.springer.com%2Fchapter%2F10.1007%2F978-3-642-54069-1_6&token2=exp=1463269172~acl=%2Fstatic%2Fpdf%2F612%2Fchp%25253A10.1007%25252F978-3-642-54069-1_6.pdf%3ForiginUrl%3Dhttp%253A%252F%252Flink.springer.com%252Fchapter%252F10.1007%252F978-3-642-54069-1_6*~hmac=569e0d4e52675a460416e634350c55b969018b5d631fcfd520dd5c5eee358c1f. Accessed 15 Oct 2016.

Lomas, N. (2017). *Trump order strips privacy rights from non-U.S. Citizens, could Nix EU-US data flows*. Tech Crunch. https://techcrunch.com/2017/01/26/trump-order-strips-privacy-rights-from-non-u-s-citizens-could-nix-eu-us-data-flows/. Accessed 1 Feb 2017.

Perez-Pena, R., & Schmidt, M. S. (2015). F.B.I. treating San Bernadino attack as terrorism case. Resource document. *New York Times*. http://www.nytimes.com/2015/12/05/us/tashfeen-malik-islamic-state.html?_r=0. Accessed 15 Oct 2016.

Rubin, J., Queally, J., & Paresh, D. (2017). FBI unlocks San Bernardino shooter's iPhone and ends legal battle with Apple, for now. *Los Angeles Times*. http://www.latimes.com/local/lanow/la-me-ln-fbi-drops-fight-to-force-apple-to-unlock-san-bernardino-terrorist-iphone-20160328-story.html. Accessed 1 Feb 2017.

Sartre, J. P. (1992a). *Being and nothingness: A phenomenological essay on ontology* (Barnes, H. E., Trans.). New York: Washington Square Press, p. 343.

Sartre, J. P. (1992b). *Being and nothingness: A phenomenological essay on ontology* (Barnes, H. E., Trans.). New York: Washington Square Press, p. 350–357.

Sartre, J. P. (1992c). *Being and nothingness: A phenomenological essay on ontology* (Barnes, H. E., Trans.). New York: Washington Square Press, p. 340.

Sartre, J. P. (1992d). *Being and nothingness: A phenomenological essay on ontology* (Barnes, H. E., Trans.). New York: Washington Square Press, p. 354.

Sartre, J. P. (1992e). *Being and nothingness: A phenomenological essay on ontology* (Barnes, H. E., Trans.). New York: Washington Square Press, p. 302.

Sartre, J. P. (1992f). *Being and nothingness: A phenomenological essay on ontology* (Barnes, H. E., Trans.). New York: Washington Square Press, p. 303.

Sartre, J. P. (1992g). *Being and nothingness: A phenomenological essay on ontology* (Barnes, H. E., Trans.). New York: Washington Square Press, p. 347.

Scott, M. (2015). *Data transfer pact between U.S. and Europe is ruled invalid*. Resource document. *New York Times*. http://www.nytimes.com/2015/10/07/technology/european-union-us-data-collection.html?_r=0. Accessed 15 Oct 2016.

The International Trade Administration, U.S. Department of Commerce. (2016). *Privacy shield framework, "Privacy shield overview"*. https://www.privacyshield.gov/Program-Overview. Accessed 1 Feb 2017.

The White House Office of the Press Secretary. (2017). *Executive order: Enhancing public safety in the interior of the United States*. https://www.whitehouse.gov/the-press-office/2017/01/25/presidential-executive-order-enhancing-public-safety-interior-united. Accessed 1 Feb 2017.

U.S. Department of Health & Human Services. (2016). *Fact sheet: Ransomware and HIPPA*. https://www.hhs.gov/sites/default/files/RansomwareFactSheet.pdf. Accessed 1 Feb 2017.

The manufacturer's authorised representative in the EU is Springer Nature Customer Service Centre GmbH, Europaplatz 3, 69115 Heidelberg, Germany. If you have any concerns regarding our products, please contact ProductSafety@springernature.com

Printed and bound by CPI Group (UK) Ltd, Croydon, CR0 4YY
30/09/2026
02225601-0012